GLOBAL PRODUCTION AND TRADE IN EAST ASIA

GLOBAL PRODUCTION AND TRADE IN EAST ASIA

edited by

Leonard K. Cheng
Hong Kong University of Science and Technology

and

Henryk Kierzkowski
Graduate Institute of International Studies, Switzerland

KLUWER ACADEMIC PUBLISHERS
Boston / Dordrecht / London

Distributors for North, Central and South America:
Kluwer Academic Publishers
101 Philip Drive
Assinippi Park
Norwell, Massachusetts 02061 USA
Telephone (781) 871-6600
Fax (781) 871-6528
E-Mail <kluwer@wkap.com>

Distributors for all other countries:
Kluwer Academic Publishers Group
Distribution Centre
Post Office Box 322
3300 AH Dordrecht, THE NETHERLANDS
Telephone 31 78 6392 392
Fax 31 78 6546 474
E-Mail <orderdept@wkap.nl>

Electronic Services <http://www.wkap.nl>

Library of Congress Cataloging-in-Publication Data
Global production and trade in East Asia / edited by Leonard K.Cheng and Henryk Kierzkowski.
p.cm.
Based largely on papers presented at an international conference held in Hong Kong on Oct. 25-27, 1999.
Includes bibliographical references and index.
ISBN 0-7923-7330-8
1. Free trade--East Asia. 2. Industries--Technological innovations--Economic aspects--East Asia. 3. International division of labor. 4. Globalization--Economic aspects--East Asia. 5. International business enterprises--East Asia. I. Cheng, Leonard K. (Leonard Kwok-Hon) II. Kierzkowski, Henryk.

HF2360.5.G55 2001
382'.095--dc21

2001020380

Printed on acid-free paper.

Printed in the United States of America

TABLE OF CONTENTS

ACKNOWLEDGMENTS

This book is based largely though not exclusively on papers presented at an international conference held in the Hong Kong Special Administrative Region (SAR) on October 25–27, 1999. The conference, titled "Global Production: Specialization and Trade," was co-sponsored by three academic units, a semi-official organization, and a government department, namely,

(a) The Center for Asian Pacific Studies at the Lingnan University of Hong Kong,
(b) The Center for Economic Development at the Hong Kong University of Science and Technology,
(c) The Graduate Institute of International Studies in Geneva,
(d) The Hong Kong Trade Development Council, and
(e) The Trade and Industry Branch of the government of the Hong Kong SAR.

Dr. William K. Fung of Li & Fung, Ltd., chairman of the Hong Kong Committee for Pacific Economic Cooperation, supported the conference with generous financial contributions and enthusiastic encouragement. The conference was also graced by the presence of Mr. Donald Y. K. Tsang, financial secretary of the Hong Kong SAR, and by three keynote speakers who spoke at the conference's public forum. These speakers were Mr. Roy C. P. Chung of TechTronic Industries Co., Ltd.; Ms. Cherry K. C. Ling of the Trade Department of the Hong Kong SAR; and Professor Peter A. Petri of Brandeis University.

Several scholars who did not participate in the conference kindly contributed papers and case studies to this book.

We are grateful to the conference's organizing committee, which was chaired by Professor Edward K. Y. Chen, chairman of the Hong Kong Trade Policy Committee, and to the above-mentioned sponsors and supporters for making the conference a reality. We are also indebted to the authors for their fine contributions to this volume, and for their cooperation throughout the editorial process. Finally, we would like to thank our English editor Ruth Barzel for improving each chapter's exposition and style and for enhancing the book's overall consistency. She has made the book easier and more pleasing to read.

Leonard K. Cheng and Henryk Kierzkowski

LIST OF CONTRIBUTORS

Baldwin, Robert E.
Hilldale Professor
Frank W. Taussig Professor
Department of Economics
University of Wisconsin
Room 6462, Social Services Building
Madison, WI 53706, U.S.A.
Email: rebaldwi@facstaff.wisc.edu

Cheng, Leonard K.
Professor
Department of Economics
Hong Kong University of Science and Technology
Clear Water Bay, Kowloon
Hong Kong, China
Tel : (852) 2358 7621
Fax : (852) 2358 2084
Email: leonard@ust.hk

Chyn, Regina
Research Associate
Council for Economic Planning and Development
3 Pao Ching Road, Suite 5176
Taipei 100, Taiwan
Email: reginaychyn@sun.cepd.gov.tw

Chu, Wan-wen
Research Fellow and Deputy Director
Sun Yat-Sen Institute for Social Sciences and Philosophy (ISSP)
Academia Sinica
Nankang, Taipei
Taiwan 11529
Tel : 886-02-2789-8127
Fax : 886-02-2785-4160
Email: wwchu@gate.sinica.edu.tw

Deardorff, Alan
John W. Sweetland Professor of International Economics
The University of Michigan
Department of Economics and School of Public Policy
458 Lorch Hall
Ann Arbor MI 48109-1220, U.S.A.
Tel : 1-734-764-6817
Fax : 1-734-763-9181
Email: alandear@umich.edu

Fung, William K.
Group Managing Director
Li & Fung Limited
11th Floor, LiFung Centre
888 Cheung Sha Wan Road
Kowloon, Hong Kong
Tel : (852) 2300-2338 (direct line)
(852) 2300-2300 (general)
Fax : (852) 2300-2000
Email: williamkfung@lifung.com.hk

Greenaway, David
Professor of Economics
University of Nottingham
Nottingham, England

Ho, Kong Weng
Assistant Professor
Department of Economics
Faculty of Arts & Social Sciences
The National University of Singapore
10 Kent Ridge Crescent
Singapore 119260
Tel : (65) 874 6231
Fax : (65) 775 2646
Email: ecshokw@nus.edu.sg

Hoon, Hian Teck
Associate Professor
Department of Economics
Faculty of Arts & Social Sciences
National University of Singapore
10 Kent Ridge Crescent
Singapore 119260
Email: ecshht@nus.edu.sg

Jones, Ronald W.
Professor of Economics
University of Rochester
1301 Highland Ave.
Rochester, NY 14620
U.S.A.
Tel: 1-716-275-2688
Email: jonr@troi.cc.rochester.edu

Kierzkowski, Henryk
Professor of Economics
Graduate Institute of International Studies
132, rue de Lausanne
CH-1211 Geneva
Switzerland
Tel: (41-22) 734-89-50
Fax: (41-22) 733-30-49
Email: kierzkow@hei.unige.ch

Kimura, Fukunari
Associate Professor
Faculty of Economics
Keio University
2-15-45 Mita, Minato-ku
Tokyo 108-0073, Japan
Tel : (O) 81-3-3453-4511 (x 3215)
(H) 81-424-65-7540
Fax : (O) 81-3-3798-7480
(H) 81-424-67-6782
Email: fkimura@econ.keio.ac.jp

Kohli, Ulrich
Professor of Economics
University of Geneva
102 boulevard Carl-Vogt
CH-1211 Geneva 4 Switzerland
Tel: (41-22) 705-8285
Fax: (41-22) 705-8293
Email: ulrich.kohli@ecopo.unige.ch

Lloyd, Peter J.
Professor
Asian Economics Centre
Department of Economics
Faculty of Economics and Commerce
The University of Melbourne
Parkville, Victoria 3052
Australia
Tel: 61-3-9344-5291
Fax: 61-3-9344-6899
Email:
p.lloyd@ecomfac.unimelb.edu.au

Milner, Chris
Professor of Economics
University of Nottingham
Nottingham, England

Ng, Francis
Trade Team, Development Research Group
The World Bank
Washington, D.C. 20433 U.S.A.

Schive, Chi
President
Taiwan Academy of Banking and Finance
39 Chi-Nan Road, Section 2,
5th Floor Taipei, Taiwan, 104
Tel: 886-2-2397-3219
Fax: 886-2-2327-8882
Email: chischive@mail.tabf.org.tw

Sung, Yun-Wing
Professor
Economics Department
Chinese University of Hong Kong
Shatin, N.T., Kowloon
Hong Kong, China
Email: ywsung@cuhk.edu.hk or
b015719@mailserv.cuhk.edu.hk

Tung, An-Chi
Associate Research Fellow
Institute of Economics
Academia Sinica
No. 130, Sec. 2 Yan-Jiou-Yuan Road
Taipei, Taiwan 11529, R.O.C.
Email:
actung@ieas.econ.sinica.edu.tw

Wan, Henry Jr.
Professor
Department of Economics
Cornell University
Ithaca, NY 14853
U.S.A.

Yeats, Alexander
Trade Team, Development Research Group
The World Bank
Washington, D.C. 20433
U.S.A.
Fax: 1-202-522-1159

Chapter 1

INTRODUCTION

Leonard K. Cheng and Henryk Kierzkowski

The word "globalization" is often used in reference to the 21st-century world economy. This new term describes a multitude of processes that transcend international borders, draw national economies closer together, and result in rapid trade expansion. Powerful market forces and stunning technological innovations drive the process of globalization and erode the traditional power of governments. Fear and hope keep pace with each other when it comes to the economic, social, and political consequences of the ongoing changes. We need to document the process of globalization, understand its mechanics, and identify the policy implications that arise as a result of its profound impact on the world economy and on the welfare of billions of people.

Should a national economy embrace or resist globalization? The stakes of policy choice are high, and an objective analysis of the causes and effects of globalization rather than ideological warfare is called for. While we need powerful analytical tools to understand globalization, the analysis cannot remain abstract but must be embedded in real experience. East Asia is an ideal region in which to apply economic models to the study of globalization, since globalization has had such a marked and profound effect on this part of the world. The region's economy is closely integrated with the world economy, and the growth and trade performance of the region's economic leaders continues to astound observers. What lessons do the experiences of East Asia's victors hold for less developed economies, including the region's economic followers? Do these followers stand a better chance of catching up with the leaders by embracing or resisting globalization? The financial crisis that erupted so unexpectedly in Asia in 1997 has shown that even the most successful economies are not immune to severe shocks. It is reasonable to ask whether globalization may give rise to even bigger risks for countries that are only now opening up to trade and foreign investment.

It goes without saying that no single work can address all the aspects of globalization. We have thus chosen to focus on the most important issues. It is our opinion that with the dramatic improvements that have taken place in telecommunications, banking, transportation, and other services, the organization of the production process is changing as well. We therefore concentrate on an increasingly important phenomenon known as "international production fragmentation," with a special emphasis on empirical studies of the phenomenon in East Asia. International production fragmentation refers to the fragmentation of the production process across national borders, or the dispersion of vertically related processes in two or more countries or regions that are required to produce a final output. It is also known to some researchers as "international production sharing" (see, e.g., Ng and Yeats in Chapter 5) and to the practitioners of global trading as "distributed manufacturing" or "dispersed manufacturing." The phenomenon is mainly the result of advances in transportation and communication technologies and of the liberalization of trade policies that make the vertical division of labor practical. However, as demonstrated by Cheng and Fung in this book, in some cases it may be a consequence of the application of rules of origin to implement discriminatory trade policies and export quotas.

Building upon the book *Fragmentation: New Production and Trade Patterns in the World Economy* (edited by Sven Arndt and Henryk Kierzkowski, 2001) that summarizes earlier studies on the subject, this book focuses on three related themes: (1) further theoretical explorations of aspects of fragmentation, (2) policy implications of fragmentation, and (3) empirical aspects of fragmentation in East Asian economies.

Three papers deal with the first theme. Jones and Kierzkowski, continuing their earlier work on the role of service links in the fragmentation process, address the horizontal aspects of international fragmentation. They examine how changes in the costs of service links may affect vertical fragmentation in a horizontal array of industries, how the existence of this array of industries may affect technical progress in the service-link industry, and how fragmentation may affect the timing of producers' adoption of technology. They illustrate their theoretical insights with several real-world examples.

Given the phenomenon of fragmentation, Wan examines the efficacy of different network structures (e.g., open or closed) between "subcontractors" and "assemblers" and illustrates his analysis with examples from East Asia. An interesting question is whether open systems work better than closed systems, or vice versa.

Using the standard trade models, Deardorff has shown in earlier work the efficiency gains from fragmentation. Motivated by the Asian financial crisis of 1997–98, however, in this book he demonstrates the potential risks associated with fragmentation. If fragmentation requires the bearing of greater risk, then countries engaging in fragmentation to increase efficiency are also more vulnerable to shocks to the global economy. This seems to be consistent with the experience of many East Asian economies during the Asian financial crisis, but Taiwan was not much affected by the crisis because its export market in the United States was not affected by it at all. Even the badly damaged East Asian economies, such as South Korea, have recovered remarkably quickly. Thus, the question of whether participating in international fragmentation is too risky needs to be explored from different angles.

Deardorff's focus is on macroeconomic risks, but risks associated with fragmentation are ever present at the microeconomic level. In fact, firms engaging in such activities must be able to satisfactorily contain risks; otherwise the activities will not take place at all. In order for the "disintegration" of the vertically related physical processes (either within the same country or across countries) to occur, there must be effective coordination to achieve "virtual integration" of the entire system.

Ideas about physical disintegration of the parts and virtual integration of the whole, despite the modern terminology that is used to express them, are not entirely new. They can be found in the extensions of classical trade theories to uncertainty. Naturally, the extent of division of labor depends on the reliability of international markets, even though the literature on trade and uncertainty considers horizontal division of labor almost exclusively. Nevertheless, as emphasized by some of the empirical papers and case studies in this book, arms-length transactions in the spot market are not necessarily the best way to ensure reliability (quality and delivery time). To do so effectively may require alternative transaction modes. Long-term relationships not only can contribute to reliability but also may lead to better deals than the spot market offers. Sometimes direct control under the roof of a single firm is the best way to go, but other times arms-length transactions with periodic inspection by one of the independent parties (say, the final buyer or a trading company that is ultimately responsible to the buyer) are preferable.

The standard trade models have assumed that reliability is assured in order to focus on the cost-price relationship and the resulting patterns of efficient division of labor. Since most basic trade models are static, no distinction can be made between spot market transactions and repeated market transactions. As such, traditional trade theories can use only broad brush strokes to paint pictures of international fragmentation based on costs of production; they cannot be expected to provide predictions about the contractual arrangements, firm structures, and networks between the different parties that perform vertically related tasks. To understand actual firm behavior, which has an impact on the observed patterns of fragmentation, one must pay particular attention to the costs of different modes of transactions between the parties. The case studies in this book have helped to uncover many interesting empirical observations about contractual arrangements, organization structure, and incentive systems. In the future, researchers may wish to consider taking on the task of explaining these and other empirical observations.

Fragmentation has many implications for trade policies, the second theme of this book. Baldwin examines the implications for the traditional trade-distorting measures (tariff escalation and rules of origin) and rules of administered protection (safeguards, antidumping duties, and countervailing duties). He also explores the implications for new policy issues that may receive more attention in the future (competition policy, investment policy, and labor and environmental standards).

Complementing Baldwin's wide coverage of trade policies, Lloyd focuses on the implications of increasing fragmentation for the implementation of rules of origin. He shows the production distortion these rules cause and offers a reform proposal based on the concept of value added. This proposal has potentially important implications for shaping the future agenda of multilateral trade negotiation.

The other articles and all the case studies in this book are devoted to our third theme. They address empirical aspects of fragmentation, with a particular focus on East Asia. In the last 20 years or so, this region's comparative advantages have evolved toward those of the North American Free Trade Area(NAFTA) and the European Union (EU) in technology-intensive and human-capital-intensive products, while the region continues to enjoy a substantial comparative advantage in products that are unskilled labor intensive.[1] By analyzing data on trade in components and parts, Ng and Yeats reveal much about fragmentation in East Asia. For instance, East Asia's trade in components and parts has been growing very rapidly. By 1990 its share of such trade exceeded that of Europe and North America. Moreover, intra–East Asia trade in components and parts is extremely intensive. These examples indicate that international production fragmentation is more pronounced in East Asia than in other parts of the world. Why is this? Is it because wages in the region vary substantially (unlike in the EU, for example), making international fragmentation especially profitable?

The empirical studies on individual economies and the case studies cover Japan, South Korea, the People's Republic of China (PRC), Taiwan, Hong Kong, and Singapore, each with a different focus. Using firm-level data, Kimura studies the relationship between fragmentation, internalization, and inter-firm linkages in Japan's manufacturing sector. Schive and Chyn study similar issues faced by Taiwan's high-tech industries and pay particular attention to the importance of global logistics service for these industries. Tung's case study of Taiwan's integrated circuit industry sheds more light on the transaction-costs considerations of fragmentation. Specifically, production networks and human networks in Taiwan make vertical disintegration (i.e., fragmentation) possible and profitable. Taiwan's bicycle industry, which is analyzed in Chu's case study, is not a high-tech industry, but the importance of a well-developed local parts supplier network to bicycle assemblers is similar to its importance for high-tech manufacturers. In both cases, as well as in Schive and Chyn's study of Taiwanese high-tech firms and in Kimura's study of Japanese manufacturing firms, local fragmentation within national borders may be as important as (if not more important than) international fragmentation.

Kimura's case study of Fujitsu, Inc.'s production of hard-disk-drive (HDD) production illustrates an example of intra-firm international fragmentation. Most of the services required to link vertically related production processes are provided within the firm. Kimura discovered that, paradoxically, HDD plants that are very capital intensive are located in the Philippines and Vietnam.

Even though the coexistence of domestic fragmentation and international fragmentation can be explained in terms of different transportation costs, it seems that information barriers and the time required to establish trust may better explain this phenomenon. As a firm extends its dealings beyond its own country, it faces macroeconomic risks such as exchange-rate fluctuations, foreign-exchange control, and other unpredictable effects of foreign governments' actions. In addition, it takes time to determine what foreign producers are capable of and whether they are trustworthy. Cheng's case study of Li & Fung, Ltd., a major trading company based in Hong Kong, reports that the trading company sends inspectors to its manufacturing subcontractors to ensure that product quality and delivery time agreements are being met, but that it has no desire to employ and manage directly the many workers subcontractors employ.

In their case study of two electronics component firms in Singapore (a local firm and a subsidiary of a multinational enterprise) Hoon and Ho attempt to determine whether international fragmentation should take the form of setting up subsidiaries or buying from a local contract manufacturer. They identify the minimization of governance costs on the one hand and the effective control of proprietary technology on the other as two conflicting considerations. Moreover, a firm trying to resolve the conflict is contemplating hiring subcontract labor to work in its own plant. That is similar to the strategy adopted by Li & Fung, except that Li & Fung does not own the plants, either. The difference could have arisen because proprietary technology is not an issue for the trading company's products. In any event, these examples indicate that the transaction-costs perspective is crucial to understanding not only the selected modes of transactions but also their evolution over time.

Can theories of fragmentation and trade be formally tested? In the "restricted profit function" framework, Kohli tests the validity of an early model of fragmentation and trade (i.e., that of Sanyal and Jones 1982) using data from South Korea. This may open a new area of research—systematic testing of the theories of fragmentation and trade.

Hoon and Ho study the changing nature of fragmentation in Singapore along with its inward and outward foreign direct investment (FDI). They focus specifically on the implication of fragmentation on the relative wages of skilled and unskilled workers in Singapore. The issue of whether trade has affected relative wages has received much attention in trade policy debates in North America.

Whenever service links are used to help move a product from producers to consumers, some value will accrue to the service providers. Sung estimates the division of value added generated by export-processing production in the PRC between the PRC and Hong Kong, where export processing originates and trade services are provided.

Cheng and Fung's analysis of Hong Kong's textile and clothing industries illustrates the distortions caused by the rules of origin used to implement the Multi Fiber Agreement (MFA). It also provides specific examples to show how these rules have resulted in artificial trade in the form of multiple tripping of clothing components between Hong Kong and the PRC.

In today's global economy a firm can supply its foreign markets with either exports or direct production in the markets, or both. Thus, any measure of the extent of competition in a country posed by firms based in another country will have to take into account both the firms' exports and direct production in the former country. Production fragmentation means that this is true not only for final commodities but also for intermediate inputs. Greenaway, Lloyd, and Milner develop new measures of global competition by focusing on both intra-industry FDI and trade flows. This is an admirable effort to make empirical measures reflect the reality of the global economy.

It is useful to understand the evolution of global production and international fragmentation from the point of view of a firm that thrives on these activities. TechTronics Industries is a global firm based in Hong Kong that specializes in power tools and related appliances. It explores niche markets all over the world, beginning with those it is familiar with. At the same time, it searches for production capabilities and business partners. This case study, written by Kierzkowski, describes the firm's gradual shift in competitive strategy. The firm began as an original equipment manufacturer, became an original design manufacturer, and finally re-created itself as an original brand manufacturer. The evolution reflects its initial start-up in Hong Kong in 1980, which by that time was no longer competitive in traditional labor-intensive products but was not yet ready to begin manufacturing very sophisticated products. It also reflects the firm's conscious effort to move up the technology ladder, and while doing that to establish its own brands. Naturally, this firm acquires expertise, technology, and parts from wherever these commodities are available at the best prices.

This volume cannot fail to convince readers of the richness of East Asia's experience as it finds new ways to integrate with the global economy. A region that is so dependent on trade must constantly adjust its practices in response to a shifting external environment as well as to internal conditions. By and large, East Asian economies have not avoided a deeper integration with the world economy. In fact, they have actively looked for ways to take this process to a new level.

Integration from the East Asian perspective stands in sharp contrast to the EU model, particularly when it comes to governments' role in the process. The integration of the EU is largely a result of the European governments' collective policy initiatives, which are given a great deal of political weight. Similar policy initiatives are absent in East Asia (with the exception of the Association of East Asian Nations [ASEAN]). Indeed, the PRC, Japan, South Korea, Taiwan, and Hong Kong are not parties to any regional trade agreements. Nevertheless, the East Asian economies have attained a high degree of integration (recall the findings of Ng and Yeats) based on market forces. Apparently, governments in the region have shown willingness and flexibility in accommodating the market's needs.

East Asian governments' willingness to recognize the importance of the market has taken on a new dimension with the collapse of the planning system in a number of East Asian countries. It is conceivable that North Korea has already entered or will soon enter the beginning-of-the-end stage of its political and economic development. As political rapprochement between North and South Korea begins, an interesting question arises as to how these two economies will be integrated. Will the two Koreas take the approach adopted by East and West Germany, which has resulted in a rather heavy burden for the unified Germany? Or, given the very substantial existing factor price differences between the two Koreas, will they take a market-based approach (such as the one taken by Hong Kong and the PRC)? Only time will tell.

Despite recent signs of resistance, globalization is here to stay, and international production fragmentation will thus continue to grow in importance. We hope that this book, by examining many key facets of international production fragmentation, will succeed in stimulating further research on related theoretical, empirical, and policy issues.

Endnote

[1]In "Exports and Comparative Advantages of the World's Three Major Regions: 1980–1995," Cheng and Leung (2000) calculate the "revealed comparative advantage index" based on exports of three regions. They are NAFTA, EU, and the East Asian Economies (EAE) i.e., the PRC, Hong Kong, Japan, South Korea, and Taiwan. The remaining economies are treated as individual economies. It was found that the revealed comparative advantage index of "technology intensive products" for NAFTA declined from 1.82 in 1980 to 1.37 by 1995, and that the revealed comparative advantage for EU declined from 1.74 to 1.12 during the same period. In contrast, the index for EAE stayed within the narrow range of 1.13 to 1.26 and stood at 1.23 in 1995. The revealed comparative advantage index of human-capital-intensive products for EU declined from 1.75 in 1980 to 1.32 in 1995; the index for NAFTA was 0.75 in 1995, only slightly lower than the 0.77 level in 1980, even though it declined in the intervening years. Like that of EU, the index for EAE also declined over the period, but it stood at 1.46 in 1995. The revealed comparative-advantage index of unskilled-labor-intensive products for EAE was equal to 2.86 in 1980 but gradually declined to 1.66 in 1995. The same index for EU moved from 1.27 in 1980 to 0.89 in 1995, whereas the index for NAFTA moved from 0.65 in 1980 to 0.47 in 1995. That is to say, whereas EU moved from comparative advantage to disadvantage and NAFTA from comparative disadvantage to greater disadvantage, EAE still maintained a comfortable comparative advantage in unskilled-labor-intensive products by 1995. Similar results hold if the ASEAN countries, Australia, and New Zealand are added to EAE to make up what the authors call the Asian Pacific Rim Economies (APRE). Relative to EAE, APRE's comparative advantage in unskilled-labor-intensive products is slightly stronger, but the comparative advantages of both human-capital-intensive products and technology-intensive products are slightly weaker.

Chapter 2

FINANCIAL CRISIS, TRADE, AND FRAGMENTATION

Alan V. Deardorff

1. INTRODUCTION

Trade economists like myself routinely ignore events in exchange markets, financial markets, and the macroeconomy, but recent events have made doing that difficult. The Asian financial crisis that began in 1997 caused large contractions in the affected countries' trade. As documented in Table 1, both exports and imports fell between 1997 and 1998 in most of the countries that were hit by the crisis.[1] In the majority of the cases, these changes represent not just the effects of exchange rates on trade balances through prices of domestic vs. foreign goods, which might have been expected to reduce the quantity of imports but to increase the quantity of exports. Instead, the quantity of both exports and imports fell. Thus these countries experienced declines in their overall trade, as well as in their terms of trade, and this suggests that they lost much of the gains from trade as well. Of course, the trauma to their economies included other effects as well, many of them macroeconomic, but these effects on trade suggest that trade theory ought to have something to say about both the causes and the effects of what occurred.

My purpose in this paper is therefore to say what I can about financial crisis in the context of a simple Ricardian model of international trade. To do this I need to include some sort of financial assets in the Ricardian model, and my choice for these is simply national currencies. A financial crisis will be modeled as a change in expectations about the exchange rate between these currencies, and I will explore the effects of such a change on international trade. I will also briefly examine a possible cause of such crises, which arises naturally from an attempt to peg an exchange rate at an unsustainable level with insufficient reserves.

All of this is pretty primitive stuff. But remember, I'm a trade theorist, not (any more) a scholar of international macro or finance. My approach to the problem is therefore to make it as simple as I can manage and to hope that I can learn something.

The paper also includes a final section on fragmentation—the splitting of production processes into parts that can take place in different locations. This section is included primarily because fragmentation is something I have been working on recently in other papers,[2] and partly because it was the topic of the conference for which this paper was written. The inclusion of fragmentation adds an interesting dimension to my model here, but it is not central to its message.

Table 1: *Effects of the Asian Crisis on Trade Exports and Imports of Goods f.o.b., Millions of U.S. $*

		1997	1998	Percent Change
Indonesia	Exports	56,298	50,371	−10.5%
	Imports	46,223	31,942	−30.9%
Philippines	Exports	25,228	29,496	+16.9%
	Imports	36,355	29,524	−18.8%
Singapore	Exports	125,746	110,379	−12.2%
	Imports	124,628	95,702	−23.2%
South Korea	Exports	138,619	132,122	−4.7%
	Imports	141,798	90,495	−36.2%
Thailand	Exports	56,652	55,000	−2.9%
	Imports	52,747	36,513	−30.8%

Source: *IMF, International Financial Statistics*

2. A RICARDIAN MODEL WITH MONEY

Like any Ricardian trade model, mine has goods produced with labor alone and constant unit labor requirements that differ across countries. First, however, let me lay out the closed-economy version of the model, since that will clarify the role of money.

Autarky

The country is endowed with labor, L, which is the only factor of production, and it produces two goods, 1 and 2, the outputs of which are Q_1 and Q_2. Their production requires a_1 and a_2 units, respectively, of labor per unit of output. These units are demanded by consumers, the owners of the labor, who have identical homothetic preferences over the goods only (their labor supply is fixed), described by their common utility function $U(C_1, C_2)$. This can equivalently be written in its indirect form, $V(p, E)$, where $p = (p_1, p_2)$ is the vector of prices and E the level of expenditure, measured in domestic currency. Demand is therefore the function $D(p, E) = V_p(p, E)$.

All transactions in this economy require money, a domestic currency that has been printed by a government that for the moment plays no active role. Instead, a quantity of money M exists in the economy and must be used for transactions. It will pass back and forth between firms and workers/consumers over time as firms pay workers to produce, then sell the product to consumers. I will assume that these transactions occur in an orderly fashion over a period of time, with workers first being hired and paid, a product then being produced, and the product finally being sold to consumers. Firms must therefore have money on hand to pay workers before they can employ them, and consumers must have money before they can consume. Neither can exchange goods directly for labor, or vice versa. Nor can a unit of currency be used for multiple transactions at the same time.

We could think of starting this process with the firms holding the entire quantity of money M and using it to hire workers. However, I will add another step because it will be useful later. Consumers/workers own the existing money stock, and it is they who start with it. Firms must borrow money from them at the start of a period in order to have money to pay their workers. At the end of the period, after production has been completed, firms sell the goods to workers and get money back; they use it, finally, to repay their loans. The advantage of this arrangement is that we are assured of the existence of those who own the money, since they are the people who populate the economy, participating as both workers and consumers. Firms, on the other hand, can pass into and out of existence in response to profit opportunities, borrowing the money needed to start up, and repaying it out of revenues. In any case, the quantity of money plays a crucial role in this model, since it imposes a strict limit on the total value of, first, the labor that is hired and, at the end of the period, the goods that can be purchased.

The money wage is therefore determined immediately as $w = M/L$. These workers, employed by firms (which could be workers themselves, in which case they would be self-employed) produce either or both of the goods in quantities that they expect will sell. The firms then sell these goods to consumers for whatever prices p will clear the market, in exchange for the money, M, that the consumers earlier were paid as workers. Then the process begins again in the next period. I gather that this is in the spirit of the "cash-in-advance" economy that was introduced by Lucas (1982) and that is now commonplace in macroeconomic theory, although I am familiar with little more than the name of the concept.

It is customary to assume perfect competition in a Ricardian model, and I will do so here. But I will not necessarily assume zero profits—that equilibrium prices equal average costs—for it is possible that such prices will not clear the market. Market-clearing prices will differ from average costs if firms incorrectly anticipate demand, or if, as it will shortly, the government injects more money into the system. Perfect competition here therefore means that firms are atomistic price takers maximizing profit as best they can, and that they all expect to earn the same profit on each unit of currency spent on labor, regardless of what it produces. It will be left to the market equilibrium to determine whether either anticipated or actual profit is equal to zero.

Equilibrium in this closed economy therefore consists of a wage w, a vector of prices p, and a vector of outputs $Q = (Q_1, Q_2)$ such that the following equations hold:

$$wL = M \tag{1}$$

$$p_j = (1+\mu)wa_j \quad j = 1, 2 \tag{2}$$

$$\sum_j a_j Q_j = L \tag{3}$$

$$D_j(p,M) = Q_j \quad j = 1, 2 \tag{4}$$

The markup rate, μ, is also endogenous, so that this is a system of six equations and six unknowns, although the budget constraint underlying the demand function ensures that in equilibrium the markup is zero. That is, using (1), (3), (2), and (4) followed by the budget constraint, we get

$$M = wL = \sum_j wa_j Q_j = \sum_j \frac{1}{1+\mu} p_j Q_j = \frac{1}{1+\mu} \sum_j p_j D_j(p,M) = \frac{1}{1+\mu} M, \tag{5}$$

which implies $\mu = 0$. Thus the equilibrium is the usual Ricardian one with the variables determined recursively:

$$w = \frac{M}{L} \tag{6}$$

$$p_j = \frac{Ma_j}{L} \quad j = 1, 2 \tag{7}$$

$$Q_j = D_j\left(a_1, a_2, L\right) \quad j = 1, 2 \tag{8}$$

where M has dropped out of (equation 8) because of the zero-degree homogeneity of $D(\cdot)$.

Money does not really matter in this model, so far, as it should not, except for nominal prices and the nominal wage. However, it may seem to matter if we now allow the money supply to change. Whether it matters or not then depends on how that change is accomplished. Suppose, for example, that the government injects additional money into the system by printing it and giving it as a transfer payment to worker/consumers sometime during the period, after they are hired as workers but before they spend their wages on consumption. At their first opportunity they will spend all of the transfer, since there is no benefit here from saving, and this will bid up prices above the costs that firms have already incurred. The above solution will still hold, except that

$$p_j = \frac{(M + \Delta M)a_j}{L} \quad j = 1, 2 \tag{9}$$

for any period within which additional money, ΔM, has been transferred to consumers. If however the money had been transferred to consumers at the beginning of the period instead, before they lent to firms, then the original solution would be unchanged, except that M in each period would be the quantity of money including the transfer. More interesting, perhaps, is a monetary expansion that is spent by the government on real goods, since that may be a more plausible motive for the expansion. Suppose that government decides to purchase quantities $G = (G_1, G_2)$ of the two goods at the end of the current period, paying for them with newly printed money. If firms do not anticipate this purchase, they will produce based on other expectations—perhaps of the demand levels that they would have obtained without the government intervention. In that case, the same quantities $Q = (Q_1, Q_2)$ defined in (equation 8) will be produced. However, the market will clear at whatever prices induce consumers to spend their labor income M on the smaller quantities $Q - G$. These prices will be higher, in nominal terms, since the same money M is spread over fewer goods, and they will be tilted toward a higher relative price for whichever good the government purchases in higher proportion than consumers.

Of course it is not really money that matters here for real variables, but only the fact that the money is being injected by means of real government purchases that differ proportionally from private purchases. Nonetheless, it serves to remind us that while money itself may be neutral, the reasons for its expansion may not be.

With prices rising here over time, consumers would perhaps prefer to move their consumption to an earlier date. But they cannot do this, under the assumptions of the model, since they are constrained, both in the aggregate and individually, to consume no more than the fixed money supply will purchase. However, suppose that prices were falling rather than rising. Then consumers might want to postpone consumption. I will consider this possibility briefly here.

To handle intertemporal substitution, I must first make some assumption about time preference. I make the simplest one possible, that consumers discount consumption in future periods at a constant rate δ per period. This is a dangerous assumption, of course, since it seems to invite consumers to postpone consumption indefinitely if the circumstances permit it. However, most of the circumstances I consider will not.[3]

Suppose then that the government reduces the money supply at a constant rate $-d$ per period by what we have seen to be the simplest means available: taxing consumers $d\%$ of their money holdings at the start of each period, before they lend it to firms who use it to hire labor. And suppose that the government destroys the tax proceeds. The above solution then holds, with the money supply M falling at the rate d per period, and both prices and the wage also falling over time at that rate.

Or at least this holds so long as $d < \delta$, for in this case it will continue to be true that the extra consumption that a consumer could get by waiting—the percentage d permitted by the lower price—is too small to compensate the impatient consumer for consuming less than his wage.

But suppose that $d > \delta$, meaning that the government deflates money and prices at a rate faster than the rate of time preference. Then it would seem that we are in exactly the situation alluded to above, where consumers will want to postpone consumption indefinitely. Indeed we are in such a situation, except for one thing: if consumers don't buy the goods that have just been produced, firms will be unable to hire labor for production during the next period, and a shortage will emerge, raising prices after all.

That is, if prices were expected to be lower next period than they are this period by a fraction greater than the rate of time preference, then consumers would choose to buy nothing in the current period, and the goods market would not clear. Therefore, if it is known in this period that the government will tax away the fraction d of the money supply next period, then prices this period must already fall below the levels that otherwise would have been appropriate for the current money supply. How much below? Enough so that consumers become indifferent between buying today and waiting. At these lower prices they cannot spend all of M on current goods, since that would more than exhaust the supply of goods. But that may be manageable, since they can simply hold on to part of M to spend next period.

However, this means that firms no longer get all of M as revenues with which to pay back their loans from consumers. And consumers therefore have less to re-lend to firms at the start of next period for them to use to hire labor next period. It follows that next period's wage must fall even more than it might have as a result of the tax alone. This in turn causes next period's prices to be lower as well, complicating matters further. To allow for all of this, we would need to look further ahead and find a time path for both prices and wages that is consistent with the path of the money supply, taking into account the constraint on the rate of deflation and its effects on demands for goods.

To solve for such a path would take me too far afield (and out of my field, as well). But I conjecture that the following results would hold: an attempt to reduce the money supply at a constant rate $d > \delta$ indefinitely will, under my assumptions, cause the economy to implode, as consumers will indeed attempt to postpone consumption indefinitely. In contrast, an attempt to reduce the money supply at such a rate only temporarily, with it then settling at a new lower level from then on, may be workable. However, the price level in that case will not simply follow the same path as the money supply. Instead it will drop much closer to its long-run level as soon as information about the new path is first obtained. Consumers will hoard a fraction of the money supply in anticipation of the further decline in prices that they know will occur. And finally, I suspect that if such a path to a lower money supply is traversed too quickly, the bottom may fall out of prices anyway. This will happen in spite of the fact that an ultimately stationary money supply is in the works. For hoarding holds back money from firms, lowering wages and therefore prices, and causing a sort of speculative deflationary bubble.

Trade

I turn now to an economy that trades. The model that I have in mind in principle is a two-country model, with each country modeled like the one above, but with different relative productivities and therefore comparative advantage, and also with different currencies. Like any Ricardian model, this one would admit of two kinds of equilibrium, one in which both countries specialize and another in which one does not. In addition, finding which solution obtains for particular parameters and even specifying the equilibrium would get a bit messy, because of the need to distinguish notationally both two countries and two goods. I therefore opt to avoid many of these complications here, using instead the familiar trade theorist's assumption of a small open economy. That is, in the two-country model in the background, the foreign country is so large that it necessarily must produce both goods, and its prices, both nominal and relative, are only imperceptibly sensitive to any changes in its trade with the small country that we study explicitly.

Thus I continue to examine a country with labor force L, unit labor requirements a_1 and a_2, consumer demand function $D(p, E)$, and money supply M. But this money is now its own currency, and while it (and only it) is required for all domestic transactions, it is not acceptable abroad. The larger world, therefore, has its own prices $p^* = (p_1^*, p_2^*)$ expressed in the different "world" currency, and we take p^* as given and impervious to the home country's trade or other transactions.

Trade could now be introduced by simply allowing domestic firms to sell abroad as well as, and at the same time as, at home. They would use their foreign-currency revenues to buy foreign goods and then bring them home to sell to domestic consumers. I will not do that, however, because my objective is to model financial crisis, and I think that requires someone to take an open position across currencies. Therefore, to allow for this in a simple way, I will assume that it takes time, one period, to complete an export or an import. This time-cost of trade could arise most obviously from transportation, but one could think of it equally as arising from other requirements of trade, such as negotiating the deal with the foreign customer, or dealing with bureaucratic red tape. The point is that, while I will not allow for any explicit resource cost of getting goods from one country's market to the other, I will assume that it takes one period of time. This will be true for trade in both directions.

For exports, then, the story is as follows. Firms borrow domestic money from consumers and hire labor as before to produce either or both of the two goods. Once produced, they can sell these goods in the current period to domestic consumers. Or alternatively, they can ship them abroad at zero resource cost and sell them there in the next period for next period's foreign price in foreign currency. At that time they can instantly exchange that foreign currency with others who wish to convert in the other direction on a global and costless foreign exchange market. This leaves exporters with domestic currency with which to pay back their loan. The result of all of this is that, by exporting, firms gain access to the foreign market, but this delays by one period their receipt of the domestic currency they need in order to pay back their original loan. At the same time, exporting also subjects them to the potential uncertainty of both next period's prices and next period's exchange rate.

The delay itself is costly because of the rate of time preference of consumers and perhaps for reasons of risk as well. Consumers will not be willing to postpone consumption to next period unless they can get at least that fraction, δ, as a real rate of interest on the loan to firms. In addition, the uncertainty of trade may add a risk premium to what lenders require. To allow for this I simply assume a domestic real interest rate, r per period, that is at least as great as δ. Firms therefore must pay back in real terms $(1 + r)$ times what they borrow to produce for export. They will therefore not choose to export at all unless they expect to get a (real) price that is at least that much higher than they could get by selling their product at home in the current period. On the other hand, if they could get even more than this, then they would decline to sell at all in the domestic market. Thus for exporting to happen at all, along with serving the domestic market, the expected foreign price of an exported good, converted at the expected exchange rate, must be exactly $r\%$ higher than its domestic price next period. If domestic prices are rising at a rate of inflation p per period, then the expected foreign price next period must be higher than the current domestic price by the nominal interest rate, $i = r + p$.

The story for imports is a little different. I could assume that foreign producers behave in exactly the same way as domestic exporters—borrowing from foreign consumers, producing and shipping their product to the home country where they sell it a period later, and exchanging the revenues for their own currency at that time to repay their financiers. However, since I wish to model a small country that may be even more unfamiliar to the world than the world is to it, I will assume instead that the perceived risk on such a transaction is prohibitively high. Instead, for our country to import, it is domestic firms that must undertake the transaction, and to do so they must borrow foreign currency from lenders abroad. The real interest rate for doing so, which I will call r^*, is not simply the foreign real rate of conventional interest, but instead includes a perhaps substantial risk premium that is also motivated by the unfamiliarity of foreign lenders with the credit-worthiness of domestic borrowers.

Thus, like r for exporters, r^* for importers is an extra real cost that they must expect to cover in order to import. They will import only if the expected domestic price next period of the imported good, converted to foreign currency at the expected exchange rate next period, is $(1 + r^*)$ times next period's foreign currency price of the imported good abroad. Of course, if the expected price were even higher than this, then competition among potential importers would bring it down. If there is foreign inflation at rate π^* per period, so that the foreign nominal interest rate is $i^* = r^* + p^*$, then imports will require the expected foreign-currency-converted domestic price to equal $(1 + i^*)$ times the current foreign price.

Together, then, the needs of both exporters and importers therefore require that relative costs at home differ from exogenous foreign relative prices by $(1 + r)(1 + r^*) = (1 + \varphi)$ in order for trade to take place. If they do not, then the small home country will not trade. We can think of φ as the total financing costs of trade. To get an equilibrium in which trade occurs, I therefore assume that

$$\frac{a_1}{a_2}(1+\varphi) < \frac{p_1^*}{p_2^*} \tag{10}$$

Thus the small economy has a comparative advantage in good 1, of sufficient size for it to overcome these financing costs of trade, and indeed to specialize completely in good 1.[4]

With this assumption fully determining the pattern of specialization and trade, the rest of a trading equilibrium can be specified fairly simply, although it is now necessary to take careful account of time. In general, let $x(t)$ be the value of any variable x in period t, including those already introduced, and let any variable with a star be that of the world. Specifically, let X_j be the quantity of good j exported by the home country, and X_j^* the quantity exported by the foreign country and thus imported at home. Finally, let e be the exchange rate, defined as the home country's domestic currency price of foreign currency.

A trading equilibrium then consists of a domestic wage w and prices $p = (p_1, p_2)$, in domestic currency, levels of output Q_1, exports X_1, and imports X_2^*, and the exchange rate e, all of these indexed by time t, such that the following equations (11–18) all hold:

$$w(t)L(t) = M(t) \tag{11}$$

$$a_1 Q_1(t) = L(t) \tag{12}$$

$$p_1(t) = (1+\mu(t))w(t)a_1 \tag{13}$$

$$[1+i(t)]p_1(t) = e(t+1)p_1^*(t+1) \tag{14}$$

$$p_2(t+1) = [1+i^*(t)]e(t+1)p_2^*(t) \tag{15}$$

$$e(t)p_1^*(t)X_1(t-1) = p_2(t)X_2^*(t-1) \tag{16}$$

$$D_1(p(t), M(t)) = Q_1(t) - X_1(t) \tag{17}$$

$$D_2(p(t), M(t)) = X_2^*(t-1) \tag{18}$$

The first three of these equations represent the cash constraint on wages, the full employment condition, and the markup of price over cost, exactly as in the closed economy except that here they are simplified by specialization.

Equation (14) reflects the requirement discussed above, that exports yield an $i\%$ higher price next period than domestic sales this period, expressed in the same currency exchanged next period. Equation (15) does the same for imports, using the foreign nominal interest rate on domestic borrowing from abroad, i^*. Equation (16) is the exchange market equilibrium condition, or balance of payments constraint. The left side is this period's demand for domestic currency by last period's exporters, while the right side is this period's supply of domestic currency by last period's importers. Finally, equations (17) and (18) equate domestic demand to supply, the latter originating last period as foreign exports in the case of good 2, which the home country does not produce.

This system of equations is not particularly simple, largely because current values depend both on last period's decisions to trade and on expectations about next period's prices and exchange rates. I will therefore not attempt to characterize the solution for arbitrary time paths of exogenous variables. Instead I will look primarily at steady-state solutions and will then use the structure of the model to discuss, more loosely I'm afraid, the subject of financial crisis.

Consider first the stationary solution. That is, suppose that M, L, p^*, i, and i^* are all constant over time. Then other variables will be constant as well, and there will be no inflation, so that $r = i$ and $r^* = i^*$. A stationary solution will satisfy all of the above equations with their time indices removed, as follows:

$$wL = M \tag{11'}$$
$$a_1Q_1 = L \tag{12'}$$
$$p_1 = (1+\mu)wa_1 \tag{13'}$$
$$(1+r)p_1 = ep_1^* \tag{14'}$$
$$p_2 = (1+r^*)ep_2^* \tag{15'}$$
$$ep_1^*X_1 = p_2X_2^* \tag{16'}$$
$$D_1(p,M) = Q_1 - X_1 \tag{17'}$$
$$D_2(p,M) = X_2^* \tag{18'}$$

The solution to this problem is relatively straightforward. First, domestic relative prices are fully determined by equations (14') and (15'):

$$\rho \equiv \frac{p_1}{p_2} = \frac{1}{(1+\varphi)}\frac{p_1^*}{p_2^*} \tag{19}$$

With homothetic preferences, this relative price determines the ratio of the goods consumed domestically, C_1/C_2,

$$\frac{C_1}{C_2} = \frac{D_1(p,M)}{D_2(p,M)} = f\left(\frac{p_1}{p_2}\right) = f(\rho) \tag{20}$$

as well as the consumption shares, β_j,

$$\beta_j = \frac{p_j C_j}{M} = \frac{p_j D_j(p,M)}{M} = \beta_j(\rho) \tag{21}$$

Then, again using the budget constraint that underlies the demand functions, we note that

$$\begin{aligned} M &= wL = wa_1 Q_1 = \frac{p_1}{1+\mu} Q_1 = \frac{1}{1+\mu} p_1 (C_1 + X_1) \\ &= \frac{1}{1+\mu}\left[p_1 C_1 + \frac{1}{1+r} e p_1^* X_1 \right] = \frac{1}{1+\mu}\left[p_1 C_1 + \frac{1}{1+r} p_2 X_2^* \right] \\ &= \frac{1}{1+\mu}\left[p_1 C_1 + \frac{1}{1+r} p_2 C_2 \right] = \frac{1}{1+\mu}\left[\beta_1 M + \frac{1}{1+r}(1-\beta_1) M \right] \\ &= \frac{1+\beta_1 r}{(1+\mu)(1+r)} M \end{aligned} \tag{22}$$

Eliminating M from equation (22) and solving for the markup, μ, we find that it is no longer zero:

$$\mu = -\frac{\beta_2(\rho) i}{(1+r)} = \mu(\rho) \tag{23}$$

This is negative, meaning that producers get a price for their good in the current period that is below current cost. This is as it should be, since they get a higher price for exports to compensate them for the delay and its interest cost, and their total revenue must equal total cost, because of consumers' budget constraint.

This markup (or markdown, as it turns out) would be fully determined by parameters if preferences were Cobb-Douglas, but otherwise it depends on the relative prices determined in equation (19) as shown in equation (23). With the markup known, it is straightforward to characterize the rest of the solution. For example

$$w = \frac{M}{L} \tag{24}$$

$$p_1 = (1+\mu(\rho)) \frac{a_1}{L} M \tag{25}$$

$$e = \frac{(1+r) p_1}{p_1^*} = \frac{(1+r)(1+\mu(\rho)) a_1 M}{p_1^* L} = \frac{(1+\beta_1(\rho) r) a_1 M}{p_1^* L} \tag{26}$$

Equation (24) comes from (11'). Equation (25) comes from equations (13') and (24). Equation (26) comes from equations (14'), (25), (23) and the property that $\beta_1 + \beta_2 = 1$. As for levels of output and consumption, these are determined essentially as in the standard real Ricardian model. With specialization in good 1, the home country produces $Q_1 = L/a_1$. It then trades, in order to consume the ratio of the goods demanded at domestic relative price p_1. The only slight twist is that trade is balanced neither at foreign prices p^* nor at domestic prices p, but rather at the mixture indicated in the foreign exchange market clearing condition (equation 16'), which mixes the foreign price of the export good with the domestic price of the import good. This is illustrated in Figure 1, where consumption occurs at point C, a tangency of an indifference curve with a price line whose slope is the ratio of domestic prices, p_1/p_2, since these are what consumers face in the domestic market. However, trade must be balanced at the domestic-currency prices received by exporters, ep_1^*, and by importers, p_2, and thus at the higher ratio $ep_1^*/p_2 = (1 + r)(p_1/p_2)$. Therefore consumption must lie on a price line with slope given by this ratio, connecting it to the production point at Q_1. This is possible because consumers have extra income at domestic prices beyond what would be needed to purchase Q_1. This extra income takes the form of interest on their loans to exporters.

One could put this model through the same sorts of paces as the standard Ricardian model, changing the parameters of labor endowment, technology, foreign prices, and tastes to find out what would happen, but I doubt that

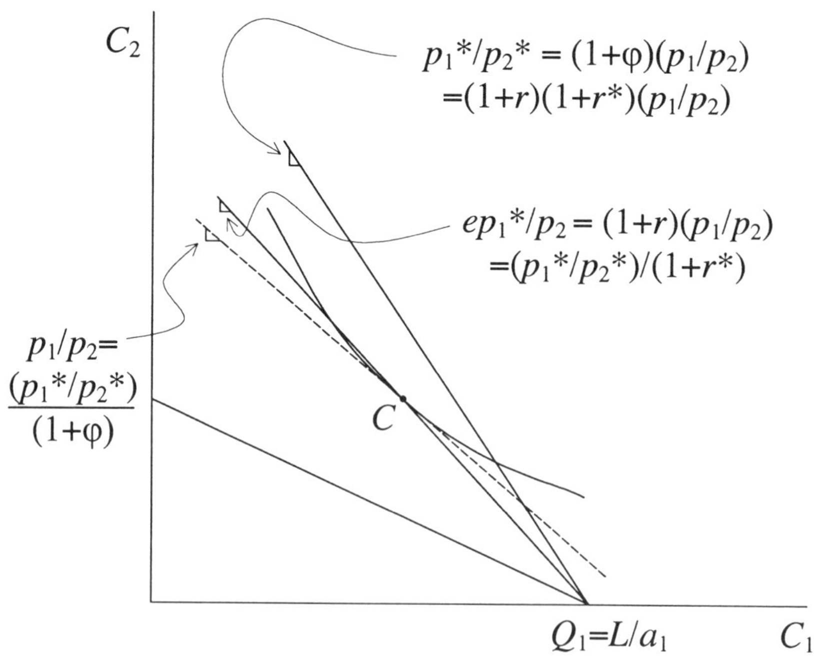

Figure 1: *Trading Equilibrium with Financing Costs of Trade.*

anything new would be learned. More interesting would be to consider more complex dynamics, allowing exogenous variables to change over time, and also to explore the dynamic approach to steady state. I do not have time to do any of this, except to note how the model will behave with constant rates of monetary expansion.

Suppose, contrary to the above, that the two money supplies grow at constant rates p and p^* per period. To minimize the effects on the real economy, let this growth be accomplished by governments that print money and give it to consumers, in proportion to what they already have, at the start of each period (before they lend it to firms to pay workers).

In the rest of the world, which is effectively the closed economy of section 1, this monetary growth will simply cause prices to grow at the rate p^*. In a steady state for the domestic economy, prices and money will both grow at the rate p, while the exchange rate will rise (depreciate) at the rate $p - p^*$. This depreciation will hold all relative prices constant over time, at the same levels that clear the markets in the stationary state seen above.

3. CRISIS

I will model a financial crisis as a loss of confidence in a currency. That means that market participants suddenly come to expect the currency to fall in value. They may or may not have good reason to think this, but I will start with the assumption that nothing else has happened, and therefore they perhaps do not.

Suppose then that we begin in the stationary equilibrium discussed above, with variables now labeled with a superscript S, for stationary. Then suddenly, within the current period after firms have borrowed money and hired labor, but before they have sold their output, expectations change. The public comes to believe that the exchange rate *next* period (not necessarily this period; that will be determined) will be some number, $\vec{e}$, that is larger than the equilibrium that has prevailed in the past in the stationary state: $\vec{e} > e^S$. Since e is the price of foreign currency, this means that the domestic currency is expected to depreciate.

What will happen now, in response to this expectation? Firms have already borrowed the available domestic money stock and paid all of it to their employees, so they can do nothing about that, even if they want to. However, some firms are also traders, and they have goods in transit in both directions between the countries. Exporters shipped a portion of their output to the foreign market in the preceding period, and they are waiting to sell it there and bring the proceeds home. Likewise, importers have bought foreign goods and have brought them to the domestic market for sale here. Both groups have the option, if they choose, of delaying either the sale of their good or the sale of the proceeds on the foreign exchange market until next period, when they expect the exchange rate to be $\vec{e}$. Will they do this? If today's exchange rate were to remain unchanged at its previous equilibrium value, then importers—firms that have brought products to the domestic market—would have an even greater incentive to sell now, since they will lose not only time but also money if they wait. But exporters—domestic firms that have taken their product abroad—will have a strong incentive to delay the return of their proceeds to the domestic economy. Recall that in equilibrium these exporters were indifferent between selling in the domestic market and exporting, since the slightly higher price abroad was just enough to pay the interest on their loan. Now, however, if today's exchange rate remains e^S, they can add to their profit the full percentage of the expected depreciation simply by waiting one more period to return their funds at the expected exchange rate $\vec{e}$. If this depreciation is greater than the time cost of waiting—the interest cost r of extending their loan one more period, as I now assume it is—then they will wait.

Therefore the previous equilibrium exchange rate e^S can no longer clear the exchange market *this period*, because at e^S there would be supply of domestic currency from importers but no demand from exporters. Indeed, this would be true for any exchange rate below $\vec{e}/(1+r)$. At that point the expected depreciation is only r, which still adds to the profit of exporters, but only enough to compensate them for the interest cost of waiting an extra period. Therefore the effect of the expectation that next period's exchange rate will be $\vec{e}$ is to drive this period's exchange rate to e^C (C is for crisis), where $e^C = \vec{e}/(1+r)$.

Is this the end of the story? It can't be, since the new exchange rate is higher than what would clear the market in equation (16). That is, we know that in the previous stationary equilibrium

$$e^S p_1^* X_1^S = p_2^S X_2^{*S} \tag{27}$$

Now the foreign price p_1^* is unchanged, and the quantity of imported good 2, X_2^{*S}, does not change, since its importers, as just noted, want to sell as quickly as they can. Domestic consumers have the same amount of money to spend as they did before, and they also have no incentive not to spend it, since they too face a depreciation in its value. Therefore the same domestic prices p_1^S and p_2^S must clear the domestic market. That leaves only X_1^S to adjust in equation (27) when the currency depreciates to $e^C > e^S$, and that is what happens. Exporters of good 1, who, because $e^C = \vec{e}/(1+r)$, are indifferent at e^C between selling now and selling next period, postpone a portion of their sales equal to the depreciation. That is, they repatriate their earnings from sales abroad of only $X_1^C < X_1^S$, such that the exchange market now clears at the new exchange rate e^C:

$$e^C p_1^* X_1^C = p_2^S X_2^{*S} \tag{28}$$

Does this matter? Not right away. Since the home country is small, the fact that its exporters are either selling less abroad or are holding onto foreign currency for an extra period will not matter for the world. And because of the depreciation, these exporters acquire the same amount of domestic currency from their sale of foreign currency as they would have before. So they are able to repay their loans and re-borrow as they enter the next period, thus starting again with the same amount of money, M, available to hire labor, which they do. The only difference for them is that they still have either some unsold goods or some unconverted currency sitting abroad, waiting to be exchanged next period.[5]

However, importers have taken a hit. The currency they earn from selling their imports on the domestic market is now worth less than they expected on the exchange market, and it is no longer enough to pay off the loans they obtained abroad last period. If the model included a better cushion in the form of some sort of wealth, they might be able to repay their loan out of that, with subsequent effects on wealth and whatever these imply. But in the current model this is not possible, and I assume that importers simply default on their loans. This has no further effects that I can see in the current period.

But in the next period it will have what I take to be the main adverse effect of such a crisis: borrowers find their access to credit either cut off entirely or made available only on much less attractive terms. If credit is cut off entirely, then there is not much to say. Without credit, countries cannot import, and without imports trade will break down completely. The country will return to autarky and will give up its gains from trade.

Less drastically, I will assume that credit remains available, but that the default in the crisis causes the risk premium on foreign loans to rise substantially. Thus r^* goes up, raising the domestic price at which imports must sell in order to cover the increased cost of financing. Since world relative prices are fixed, this means that the domestic relative price of the export good falls, as does the hybrid price ratio (ep_1^*/p_2) at which trade is balanced. This is illustrated in Figure 2. Starting from the initial equilibrium at C, the foreign interest rate that is charged to domestic borrowers rises from r^* to $r^{*\prime}$. This lowers the relative price of the country's export good at which trade must balance from $(p_1^*/p_2^*)/(1 + r^*)$ to $(p_1^*/p_2^*)/(1 + r^{*\prime})$, and it lowers the domestic price faced by consumers correspondingly, moving the equilibrium to C'.

This, then, is my story of a financial crisis. When all is said and done, it does not really amount to much. Expectation of a currency depreciation undermines an otherwise satisfactory and stationary equilibrium, forcing a depreciation even before it is expected. This causes those who have financed their trade by borrowing abroad expecting to repay with domestic currency in my model, these are importers) to default on their loans. And this raises the financing costs for borrowing in the next period. Because financing costs

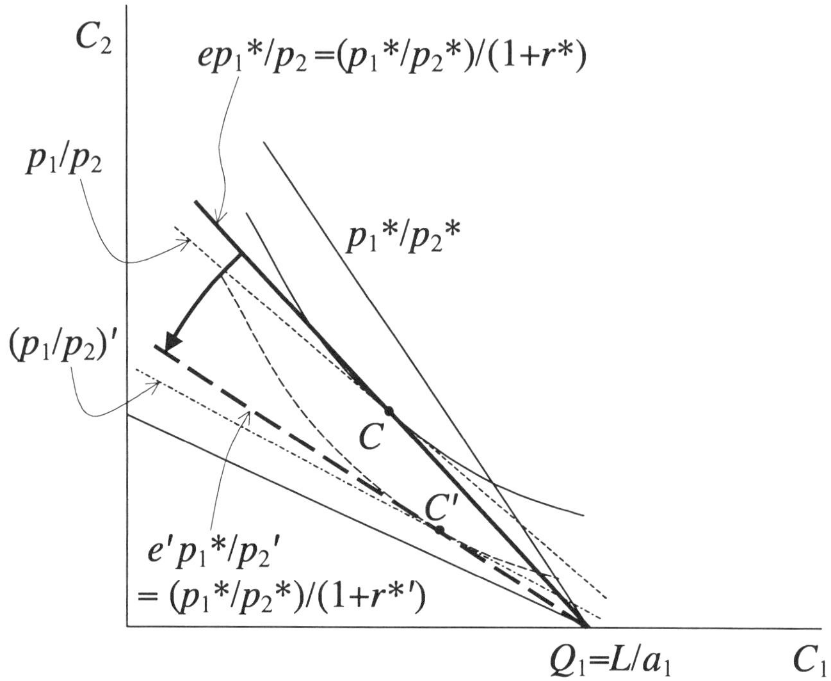

Figure 2: *Effects of a Crisis that Raises Financing Costs.*

enter the model in a manner analogous to other costs of trade, this increase in the risk premium on foreign borrowing worsens the country's terms of trade and causes a drop in its welfare.

This is a simple story, and it captures at best only a small part of what really goes on in a financial crisis. One extension that comes immediately to mind would be to include a banking sector as an intermediary between firms and both domestic and foreign lenders. If these banks were to be the ones that either defaulted on loans or suffered the adverse effects of such default, their costs might rise on loans in both foreign and domestic currencies, increasing the cost of exports as well as imports. Qualitatively, the effect would be the same as above, but this might be more plausible, and quantitatively it could be even larger.

4. ONE CAUSE OF CRISIS: A PEGGED EXCHANGE RATE

So far, my story does not attempt to explain why the crisis occurs, except to say that there is an unexplained loss of confidence in the country's currency. Many reasons for such a loss of confidence exist, but the one that is perhaps the most common arises in the presence of an attempt to peg an exchange rate by exchange market intervention. In this section I explore that possibility.

Suppose, then, that the government/central bank of a country attempts to peg its exchange rate at a rate, $\bar{e}$. It could easily do this by setting its money supply appropriately, but I will assume that it does not, and that instead it finds itself with an overvalued currency. To make this happen, I will start in a stationary equilibrium like those above without intervention, based upon an initially fixed money supply M^S and other variables that correspond to it. Then, in the current period, the government expands the money supply by a one-time percentage p_1 (which will turn out to be the domestic rate of inflation in the current period $t = 1$) to $M_1 = (1 + p_1)M^S$ that would yield, if there were no exchange-market intervention, an equilibrium exchange rate $e_1 = (1+\pi_1)e^S > e^S$. Yet rather than letting the currency depreciate, the central bank chooses to intervene pegging at $\bar{e} = e^S$.

How can the central bank enforce its pegged rate $\bar{e}$? I will assume that it has reserves of foreign currency, R, perhaps accumulated during past exchange-market intervention or perhaps acquired from a swap or a bail-out from a foreign institution. It now uses these reserves to peg its exchange rate. That is, it stands ready to buy its own currency at the price $\bar{e}$ in whatever amounts are forthcoming from the market.

Thus we start the current period with quantities of both goods in transit, X_1^S and X_2^{*S}, that were shipped in the previous period as exports and imports, respectively. The world price to be received for exports, p_1^*, remains the same, but the domestic price that will clear the domestic market for good 2 in the current period, $t = 1$, is increased by the monetary expansion to $p_2(1) = (1+\pi_1)p_2^S$.[6] Since it was previously true that

$$e^S p_1^* X_1^S = p_2^S X_2^{*S}, \tag{29}$$

the equilibrium exchange rate in the current period without intervention would indeed be $e_1 = (1+\pi_1)e^S > e^S$, since this would increase both sides of the equation by $(1+ p_1)$. Pegging the exchange rate instead at $\bar{e} = e^S < e_1$ will therefore leave the left-hand-side of equation (29), which is the demand for domestic currency on the foreign exchange market arising from (last period's) exports, unchanged. But there is still an increase in the right-handside, the supply of domestic currency arising from imports, by the fraction p_1. The central bank will have to buy this excess domestic currency, using up a part of its reserves.

The change in reserves, $R = R_2 - R_1$, will therefore be

$$\begin{aligned} \Delta R &= p_1^* X_1^S - \frac{p_2(1) X_2^{*S}}{\bar{e}} = p_1^* X_1^S - \frac{(1+\pi_1) p_2^S X_2^{*S}}{e^S} \\ &= \frac{1}{e^S}\left[e^S p_1^* X_1^S - p_2^S X_2^{*S} - \pi_1 p_2^S X_2^{*S}\right] = -\pi_1 p_2^S X_2^{*S} / e^S \end{aligned} \tag{30}$$

That is, the central bank loses reserves during the current period in proportion to the value of imports and the extent of the overvaluation of domestic currency.

Now, it is well-known in international finance that this sort of problem solves itself eventually if the central bank lets that happen, and this is true here as well. When the central bank sells foreign currency out of its reserves, it does this in exchange for domestic currency that it buys and therefore retires from circulation. That is, the domestic currency value of equation (30) equals the change in the domestic money supply from period 1 to period 2:

$$\Delta M = M_2 - M_1 = \bar{e}\left(-\pi_1 p_2^S X_2^{*S} / e^S\right) = -\pi_1 p_2^S X_2^{*S} \tag{31}$$

If this decline in the domestic money supply is allowed to occur, then it will gradually, over time, cause domestic prices to fall and the overvaluation to be corrected.

There are two problems with this, however. One is that the central bank may resist this deflationary change, offsetting (sterilizing) it with an additional monetary infusion to keep the money supply at its new level. Not knowing why the money supply was increased in the first place, we cannot know whether this will be done, but we also cannot rule it out. If that happens, then next period's equilibrium will be just like this one, with the same loss of reserves repeated.

The other problem is simply that there may not be enough reserves. Even without sterilization, the overvaluation will be fully corrected only when the money supply has fallen to the original level at which it began in the old stationary state. That is, we need the money supply to fall by the amount that it recently increased, $p_1 M^S$, and the value of this at $\bar{e}$ may be greater than R. Note in particular that the central bank's ability to defend the peg during the first period is not enough, since first-period nonsterilized intervention cannot be enough to bring the money supply back to its old level. This is so because the intervention in (30) is p_1 times the level of imports only, while the money supply has grown by p_1 times the level of total income per period.

Therefore it is quite possible that the central bank will run out of reserves eventually. This will happen either necessarily, if it sterilizes the effects of its intervention on the domestic money supply, or it may happen if it does not sterilize, if the value at the pegged exchange rate of the initial monetary expansion was greater than the level of reserves. Furthermore, if we can know this, then the market can know it too. Suppose we know that, in some time period down the road, the remaining reserves will be insufficient to finance the necessary intervention. The intervention will then have to stop, and the exchange rate will be determined by the market at a level that is depreciated relative to $\bar{e}$. But in the period before that, market participants will know it is coming, and therefore we will be in the situation of the previous section, with a depreciated exchange rate expected during the next period.

Of course, there is now no reason for this expectation to exist only in the period just before the central bank would run out of reserves if the expectation had not arisen. As seen in the previous section, the expectation of a depreciation will induce those who have already exported to delay buying domestic currency until next period. In the previous section this caused the current exchange rate to depreciate. With a pegged rate, on the other hand, it will instead cause the central bank to have to intervene more heavily. The excess demand for foreign currency will now, in the period before the peg collapses, be the whole value of imports, not just p_l times it as in equation (30). This in turn means that the peg will actually collapse earlier.

The point of this discussion, however, is not just to flag the unsustainability of a pegged exchange rate. Rather it is to connect that with the adverse real effects of a financial crisis that were seen in section 4. Whenever the expectation of a depreciation first arrives, it will cause an actual depreciation then or shortly thereafter as the peg collapses, and this in turn will cause defaults on loans and an increase in the financing costs of trade that deprive the country of part or all of the gains from trade. Thus a financial crisis that originates in nominal misbehavior by the central bank can have significant adverse real effects.

5. FRAGMENTATION

The point I wish to make in this section about fragmentation is simply this: fragmentation can enhance the gains from trade, but it does so by increasing the volume of trade that must be financed. This increases a country's vulnerability to financial crisis, both because there is more trade being financed and because there are greater gains from trade to be lost if a crisis undermines that financing.

To illustrate that point I will work through just one example. Starting from the model considered so far in this paper, suppose that fragmentation becomes possible in the technology for producing good 1. That is, in addition to the unified technique of producing good 1 using a_1 units of labor per unit of output, it now becomes possible, everywhere in the world, to produce good 1 in two stages. In the home country, the first stage uses a_{11} units of labor to produce a unit of an intermediate input, while the second stage uses one unit of the intermediate input (call it good 11) together with a_{12} units of labor to produce a unit of final good 1. I assume that the fragmented technology would actually be inferior to the unified technology if both stages were done at home, in the sense that $a_{11} + a_{12} > a_1$. These two stages can be carried out in different locations, and the intermediate good 11 is freely tradable, subject only to the same requirements of financing assumed above for goods 1 and 2. Both stages can therefore be undertaken abroad, but because this is a Ricardian model their unit labor requirements abroad are different. All that matters for the home country, however, is the prices abroad of the intermediate input, p_{11}^*, and of the final goods p_1^* and p_2^*.

With these assumptions, the home country may specialize in one or the other fragment of production, or it may not, depending on how the labor requirements of both stages compare to world prices and financing costs. For my example, I will assume that a_{11} is small enough relative to other parameters that the country specializes in fragment 11, producing only the intermediate input and exporting all of it in return for imports of both final goods, 1 and 2. It turns out that what is needed for this to be the outcome is the following pair of conditions:

$$\frac{a_{11}}{a_1} < \frac{1}{1+\varphi}\frac{p_{11}^*}{p_1^*} \tag{32}$$

$$\frac{a_{11}}{a_2} < \frac{1}{1+\varphi}\frac{p_{11}^*}{p_2^*} \tag{33}$$

where, as before, $(1 + \varphi) = (1 + r)(1 + r^*)$. These say, in effect that it must be possible to produce fragment 11, and to export it in exchange for final goods 1 and 2, respectively, more cheaply than those goods can be produced at home.

With these conditions satisfied, the situation is as shown in Figure 3. The stationary equilibrium from Figure 1, without fragmentation, is shown with lightly shaded lines and output and consumption levels Q_1^0 and C^0. With fragmentation, the relevant restrictions on stationary equilibrium prices and trade are given by the following equations:

$$(1+r)p_{11} = ep_{11}^* \tag{34}$$

$$p_1 = (1+r^*)ep_1^* \tag{35}$$

$$p_2 = (1+r^*)ep_2^* \tag{36}$$

$$p_1X_1^* + p_2X_2^* = ep_{11}^*X_{11} \tag{37}$$

Equation (34) requires that the foreign price of exports of fragment 11 be high enough to pay the domestic interest on its financing. Equations (35) and (36) do the same for imports of final goods, 1 and 2, both of which must now be financed by borrowing abroad. Equation (37) is the balance-of-payments or exchange-market-equilibrium condition, requiring that the domestic currency receipts of importers (with which they must buy foreign currency in order to repay their loans) equal the domestic currency value of exports of the fragment 11.

It is still technologically possible for the country to produce goods 1 and 2 on the transformation curve of Figure 1, and thus to produce an output Q_1^0 of good 1. However, with the assumed comparative advantage in fragment 11, it

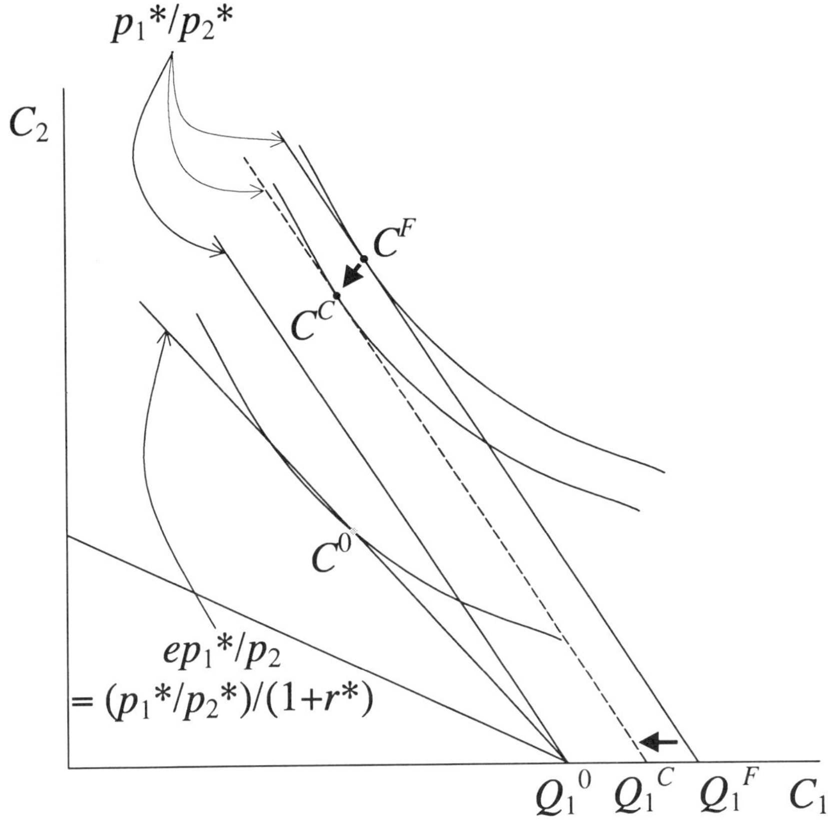

Figure 3: *Trade with Fragmentation and Effects of a Financial Crisis.*

can do better by producing the fragment and trading it for good 1. Setting $X_2^* = 0$ in (37) we can solve for maximum obtainable import of good 1,

$$Q_1^F = \frac{ep_{11}^* X_{11}}{p_1} = \frac{1}{1+r^*} \frac{p_{11}^*}{p_1^*} \frac{L}{a_{11}} \tag{38}$$

which from (32) is larger than the maximum output of final good 1 produced directly:

$$Q_1^0 = \frac{L}{a_1} < \frac{L}{a_{11}} \frac{1}{1+\varphi} \frac{p_{11}^*}{p_1^*} < \frac{1}{1+r^*} \frac{p_{11}^*}{p_1^*} \frac{L}{a_{11}} = Q_1^F \tag{39}$$

Therefore, in Figure 3, the consumer's budget line starts from Q_1^F to the right of Q_1^0. Dividing equation (35) by equation (36), the ratio of domestic prices is now the same as the ratio of foreign prices of the final goods, p_1^*/p_2^*. The consumption point with fragmentation, C^F, is therefore shown as a tangency to a budget line with this slope extending from Q_1^F.

It is notable that fragmentation has permitted a considerable increase in economic welfare compared to the original equilibrium C^0. In effect, fragmentation permits a country to expand its consumption possibilities outward, in addition to giving it access to an imported good at lower price, tilting these possibilities in its favor. Of course, this is all still just the effects of comparative advantage, but fragmentation seems to focus comparative advantage more narrowly and thus permit greater gains from specialization and trade.

But, like other trade in this model, fragmentation-based trade requires financing, and the costs of financing (like other trading costs not considered here) can reduce these gains from trade. In particular, following the argument of the preceding sections, if a financial crisis raises the risk premium on borrowing from abroad and thus raises r^*, then the gains from fragmentation and trade will be reduced or even eliminated.

This is shown in Figure 3, where the effect of a crisis that increases r^* is shown as reducing the intercept of the budget line from Q_1^F to Q_1^C, as we know it must from equation (37). Consumption and welfare are therefore reduced, as is the volume of trade. However, if r^* were to increase by even more, then this would raise φ in equation (32) and might eliminate the country's comparative advantage in fragment 11 completely. If so, then the economy would return to the equilibrium we had without fragmentation at C^0, and the welfare loss would be that much greater.

Thus, although fragmentation-based trade is not intrinsically any different from any other trade that is based on comparative advantage, it does appear to make an economy more sensitive to trade costs at the same time that it expands the gains from trade. Therefore a financial crisis that increases these trade costs can have a more devastating impact.

6. CONCLUSION

The purpose of this paper is to point out and model one link between the international financial performance of a country and its international trade. Economists have not routinely paid much attention to such links, in part because the difficulty of the subject has led most of us to specialize in studying either trade or finance, not both. But the links between trade and finance have become painfully evident to residents of several Asian economies in the last three years, as a financial crisis has caused real suffering there, at least some of which seems to have been transmitted to them by trade. This paper has tried to show one simple mechanism by which this may have happened, working through the need to finance international trade in international credit markets.

Recent research has also directed attention to international fragmentation as a means by which the world economy has become more integrated. It seems plausible in particular that the most successful of the Asian economies may have achieved their success in part by exploiting opportunities for fragmentation. This paper has suggested, however, that by doing so they may also have made themselves more vulnerable to financial crisis, with repercussions that have only recently been felt.

The message should not, however, be that increased trade and fragmentation was a mistake, in spite of that vulnerability. For what has been lost, according to this analysis, has been primarily only the gains from trade and fragmentation themselves, or a portion of them. The message should instead be that the world economy needs to find better ways to prevent crises like these. And that is a task that I will gladly outsource to my colleagues in international finance.

References

Deardorff, Alan V. (1998a). "Fragmentation across Cones," Discussion Paper No. 427, Research Seminar in International Economics, University of Michigan, August 7.

Deardorff, Alan V. (1998b). "Fragmentation in Simple Trade Models," Discussion Paper No. 422, Research Seminar in International Economics, University of Michigan, January 7.

Lucas, Robert E., Jr. (1982). "Interest Rates and Currency Prices in a Two-Country World," *Journal of Monetary Economics* 10, pp. 335–360.

Endnotes

[1]Table 1 excludes one other country that was hit hard by the crisis, Malaysia, because its balance of payments data in *International Financial Statistics* (*IFS*) stop in 1997. *IFS* also does not report any data for Taiwan, or balance of payments data for Hong Kong, two other countries that were affected by the crisis.

[2]See Deardorff (1998a,b).

[3]With these consumers lending money to firms to use throughout the period of production, you might think that they should be paid an interest rate equal to the rate of time preference on the loans. That will indeed be the case when we allow them to lend across periods. However, within the period the consumers have nothing else they can do with the money, since newly produced goods are not yet available for consumption, and the markets for last period's goods have closed. So the equilibrium interest rate is zero, at which consumers are indifferent between lending the money to firms and not.

[4]If (10) were to hold with equality, then the pattern of trade would be determinate, but the volume of trade would not be.

[5]In the stationary world I am looking at here, it doesn't matter whether they wait to sell goods or wait to sell currency. With inflation it might, depending on where and at what interest rate they refinance their loan.

[6]Similar to the argument above, this occurs because consumers have been paid $(1+\pi_1)M^S$, and they are offered the two goods in the same quantities as before, so that both equilibrium prices rise by the same proportion.

Chapter 3

HORIZONTAL ASPECTS OF VERTICAL FRAGMENTATION

Ronald W. Jones and Henryk Kierzkowski

1. INTRODUCTION

The volume of world trade relative to the volume of world gross domestic product has gone through a severe cycle this century, with a comparative value now that is much higher than it was a few decades ago but not much different from what it was at the beginning of the century. Recent academic interest in the current degree of globalization has, however, been stimulated by more than the sizeable ratios of international trade in goods and services to levels of gross national product. No longer is it accurate to think of countries as engaged in the production of commodities that may be traded on world markets but to which value added at most stages is contributed in a vertically integrated process. Instead, the composition of international trade has made increasingly prominent the role of trade in goods-in-process, intermediate goods, raw materials, and other "middle products." It is becoming apparent that conditions in world markets have encouraged international outsourcing or *vertical fragmentation* of the production process. For example, Alexander Yeats (2001b) recently estimated that about 30% of world trade in machinery and transport equipment takes the form of trade in components. (At least US$800 billion of world trade in manufactures is based on global fragmentation of production.) In a series of papers (1990, 2001a, 2001b) we described and analyzed this phenomenon; in the latter two papers we focused particularly on the consequences of such fragmentation for a country's internal distribution of income. In this paper we turn our attention to various horizontal aspects of the process of international fragmentation. These aspects include both the ways in which changes in the costs of service links spill over to encourage vertical fragmentation in a horizontal array of industries and the two-way connection between frag-mentation and technological change. We also illustrate some of these con-nections for the photo-imaging and pharmaceutical sectors of the world economy, as well as the spread in the use of cellular batteries.

2. VERTICAL FRAGMENTATION

The key concept introduced in our earlier treatment of the process of vertical fragmentation is the distinction between production blocks and service links. Adam Smith highlighted the advantages of increasing the scale of production whereby a finer and finer division of labor is possible, leading to increasing returns to scale. This suggests that at small scales of output a firm manufactures a product in a single plant, in a single location. As the output of the firm expands an alternative scenario is possible—one in which production is separated into a pair of production blocks, each of which is the site of some part or fragment of the entire process. The advantage of this alternative is that marginal costs are lowered with increased degrees of specialization. The disadvantage is that there are now extra costs of service links between the two production blocks—links in the form of comm-unication, transportation, coordination, and accounting. However, these are costs of a relatively fixed kind, so if scale is sufficiently large it pays to "outsource" part of the production process.

Figure 1 illustrates the behavior of total costs in relation to output. Line 1 admits some degree of increasing returns even within a single production block. Some internal services are required, and they provide the bulk of the fixed cost component, 0*A*. Line 2 illustrates an alternative scenario in which output is broken down into a pair of production blocks in a manner that serves to lower marginal costs but to raise the service costs of linking the blocks in a coordinated fashion. If output reaches the level shown by point *D*, vertical fragmentation becomes worthwhile. Line 3 illustrates costs associated with a finer degree of fragmentation into three production blocks, albeit with an even higher cost of required service links, shown by distance 0*C*.

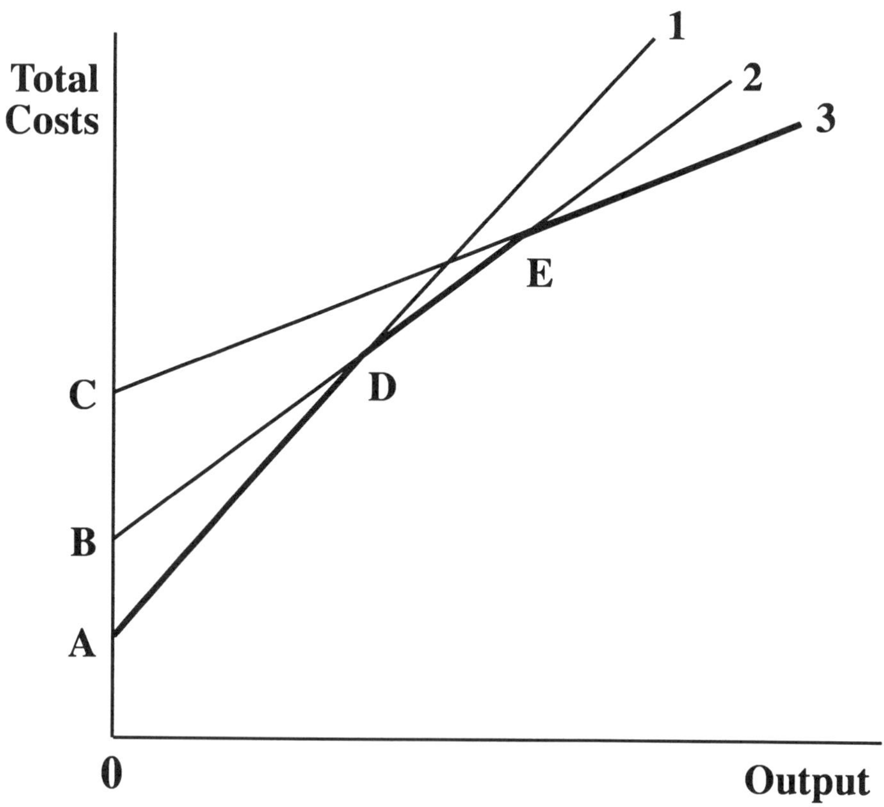

Figure 1: *Vertical Fragmentation.*

This process of vertical fragmentation may take place solely within a country and completely within the firm. However, for two possible reasons, outsourcing to foreign locales is potentially attractive. On the one hand, the relationship between factor productivities and factor prices may be different abroad—a la the Ricardian emphasis on technologically based differences in comparative advantage. On the other hand, various production blocks typically require factors in different proportions, and countries vary in their relative factor prices. This is the Heckscher-Ohlin basis for trade. Thus it is attractive to move more unskilled labor-intensive fragments to countries in which such labor is relatively inexpensive compared with its productivity. To be balanced against this are the possible added service costs of attempting to coordinate activities across national borders, as well as the possible higher costs of transportation and communication. The horizontal nature of service links and technology discussed in this paper points to a crucial role played by multinationals in the process of fragmentation.[1] Developing new production blocks is expensive. When a new production block can potentially be used in a number of sectors or products, it is likely to cost more to redesign and standardize the component. High research and development (R&D) costs act as a deterrent, and frequently only large firms, which often operate internationally, have access to the resources needed to produce a new production block.

The size of the potential market is one of the factors determining whether investments in product development will be made. With the application of a production block in a number of products, market potential expands accordingly. In Section 6 we discuss the use of long-lasting cellular batteries in laptops, cellular phones, and electronic organizers. With the world market for cellular phones approaching 1 billion and the other two markets measured in hundred of millions, the rewards for designing a standardized long-lasting cellular battery will be huge. These possible cross-product applications aid and abet the process of fragmentation.

Figure 1 thus suggests that in an environment in which world demand for a firm's product is expanding, it is reasonable to expect both greater reliance on international trade for parts and intermediate goods and shifts in resources from manufacturing to service-sector activities. However, such growth is by no means the sole, or indeed the most important, impetus behind greater vertical fragmentation. In recent decades technology has made great strides toward lowering the costs of many service links, especially costs associated with communication and transportation. Instant communication among widely dispersed agents is now virtually costless, and transport costs have continued the downward drift that has been evident for the past two centuries. Lowered barriers to international trade and the pressures in many countries to lessen the degree of regulation of economic activities have also directly lowered the costs associated with vertical fragmentation. Finally, less government regulation is usually associated with more market competition, and costs are lower on this account.[2]

3. THE HORIZONTAL NATURE OF SERVICE LINKS AND TECHNOLOGY

An improvement in the technology of providing service links is rarely sector specific. The introduction of fax technology proved a boon to all sorts of economic activity in which written documents had to be transferred from country to country at low cost and high speed. Telephone service has benefited from technical improvements, the use of satellites, and, not least, greater competition unleashed by deregulation. Such cost reductions encourage vertical fragmentation in a wide variety of industrial pursuits. Firms now also have access to much more information about other countries' legal systems, so they are better protected in case of contract violations. Increased competition abroad, stemming from deregulation and an increase in the number of potential supplying firms and in the number of foreign countries able to provide supplies and host fragmented production blocks, serves to lessen purchasing firms' fear that they may be held hostage to monopolistic practices. This, in our view, will encourage more reliance on arms-length transactions between firms in different countries and will lessen, relatively, the importance of keeping transactions under the umbrella of multinational firms.

Technical progress can alter the costs of service links. In addition, some types of progress can affect a wide variety of industries by introducing new products and materials that are useful inputs into many sectors. A recent issue of *Business Week* (summer 1999) cites a number of these that have been introduced in this past century. For example, nylon was first introduced to replace silk in stockings but was soon being used in parachutes and other items made of fabric, and in brush bristles. Other innovative materials whose use has been widespread include synthetic rubber, Bakelite (leading to plastics), and vacuum tubes. The invention of vacuum tubes was followed by the invention of transistors in 1947; these are now used not only in radios but also in PC's and missiles. (Intel's Pentium III chip contains 9.5 million transistors.) Lasers are used in surgery as well as at supermarket checkout counters. Zippers and transparent tape have countless uses.

In addition to providing new materials, technical progress has also resulted in new methods of combining inputs to produce outputs. Henry Ford exploited mass-production techniques in making automobiles, and such techniques spread fairly rapidly to other sectors of the economy. In more recent times the use of "just-in-time" inventory management and new accounting techniques have served to lower production costs, again revealing a horizontal appeal throughout the economy.

In earlier academic literature the assumption of "technical progress in the industry" was often made, as if each sector of the economy had unique technological features to match the unique character of the final item it produced. Rather, changes in technology often find applications in industries whose final products are quite dissimilar. Ohyama and Jones (1995) argue that such horizontal spread helps to account for the phenomenon of "leapfrogging," whereby today's industry leaders can become tomorrow's followers because they possess a comparative advantage in today's soon-to-be outmoded techniques. In what follows we argue that one consequence of vertical fragmentation is that it may serve as an inducement for further technological change affecting a horizontal spread of industries.

4. FRAGMENTATION AND THE INDUCEMENT FOR TECHNOLOGICAL INNOVATION

In a preceding section we describe several possible causes of international fragmentation of production processes, such as a technological improvement that lowers the costs of service links. Here we start by assuming that some such change has indeed taken place. In Figure 2 we illustrate an initial situation in which the reduction in service link costs encourages fragmentation in a series of industries, *A*, *B*, and *C*. In each of these sectors there is now a pair of production blocks. The basic assumption that we now introduce is that the fragmented production blocks, say in stage 2, may have more in common with each other than do the vertically integrated blocks. Just as different commodities with very different characteristics may nonetheless share some aspects of technology, so some fragmented production blocks may be more similar to fragments from other sectors than they are to anything within the vertically integrated processes. This is what Figure 2 suggests for the second production block in industries *A*, *B*, and *C*. This is not to say that such blocks are identical but rather that they may be similar enough to induce entrepreneurs to attempt to modify them so as to come closer to the "one size fits all" state. In Figure 2 the final segment suggests that after R&D expenditures have been made it becomes possible to alter the nature of a common production block so that it can be used in all three industries.

This process, whereby vertical fragmentation encourages R&D efforts to create a production block or material that a number of sectors can use, is not the end of the story of how fragmentation and technical progress are intertwined. Suppose that these efforts are successful, as is illustrated in the third stage in Figure 2. Then the volume of production of the second type of production block has expanded enough to encourage further fragmentation. The new industry producing the second kind of production block may itself become fragmented, with most production taking place in labor-abundant countries but headquarter activities and R&D taking place in a more capital-abundant country. This serves to change the industrial landscape from that found in the earlier stage in which each industry had its own separate production block to one in which separation again takes place. This time, however, the separation is dictated by the measure of factor intensity instead of by industry. The most widespread illustration of this kind of phenomenon in recent times has probably been in the computer chip industry. Chips originally designed for use in computers, and produced by the computer sector, are now produced in a separate sector for use in a wide variety of applications, from automobiles to toasters.

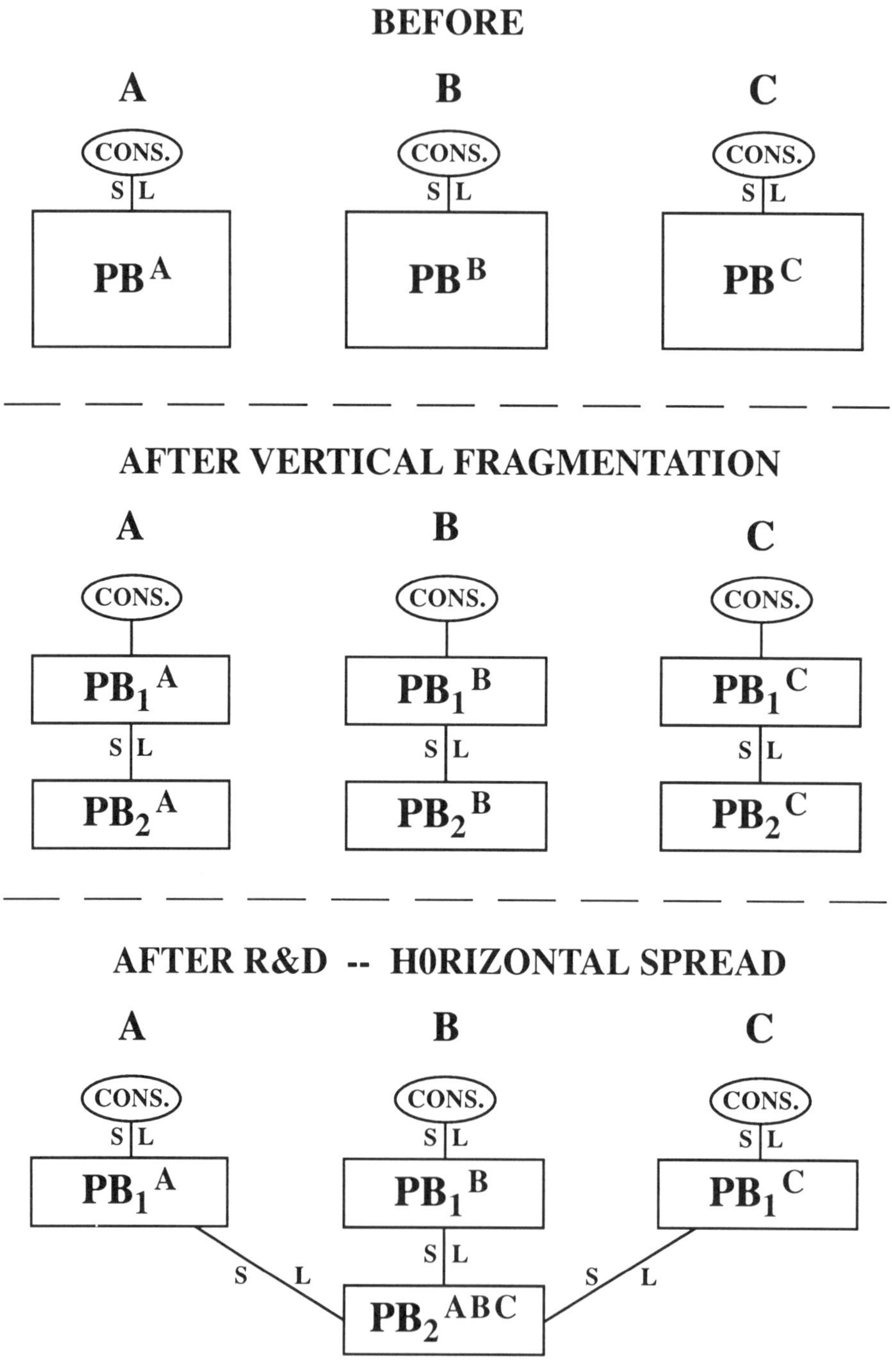

Figure 2: *Fragmentation Induces Technical Progress.*

5. FRAGMENTATION AND THE ADOPTION OF NEW TECHNOLOGIES

At any point in time the technology that is actually used in a production process may not be up to the standards of state-of-the-art technology at the frontiers of knowledge. Resistance to change is a common human characteristic. Eventually, of course, the benefits to be derived by change outweigh the costs of change, which are expressed, for example, by the resistance of workers or of researchers who are collecting rents on previously acquired knowledge. This process of change, whereby modern technology is adopted to replace older and more costly methods of production, is aided if other shocks to established procedures are encountered. One such shock is evident when mergers take place between firms. In such cases there is frequently a change in personnel or location of production that allows a simultaneous change to more modern productive techniques with new personnel to take place. Here we focus on a different kind of shock—one in which international fragmentation has occurred.

Figure 3 helps isolate the various costs and benefits whose comparison determines the rate at which new technology is actually adopted. Costs and benefits are shown on the vertical axis, whereas the horizontal axis, labeled *t*, represents time elapsed since the existing (vertically integrated) process has been installed. With the passage of time, new technologies emerge. Following

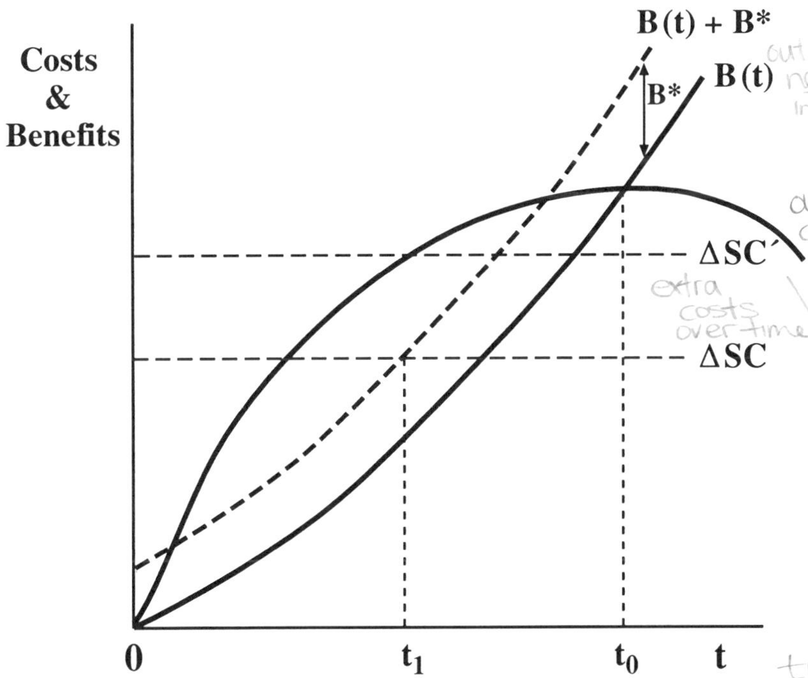

Figure 3: *Advantages of Fragmentation.*

the treatment in Ohyama and Jones, we ignore the direct R&D costs that may be undertaken by this firm in order to improve technology. Instead we appeal to the notion that, with the passage of time, technological progress is taking place in various other sectors of the economy and that this firm can use some of these new ideas to lower costs. This is a kind of "horizontal aspect" described earlier. The potential benefits of these new developments accumulate with time and are illustrated in Figure 3 by the upward-sloping $B(t)$ schedule. The $C(t)$ schedule captures the costs of adopting the new technology, which comprise the resistance of factors trained in the existing techniques. If these techniques become too "old-fashioned," resistance may weaken, but we are not concerned with this section of the $C(t)$ schedule. If no fragmentation or other shock takes place, these two schedules suggest that at time t_0 a switch will be made to the new techniques.

As already suggested, the possibility of international fragmentation entails extra costs represented primarily by higher expenditures on service links, but it also helps to reduce marginal costs of the entire operation. For a *given level of output* let the extra costs of the service links involved in furthering the extent of fragmentation be represented by horizontal line ΔSC and the improvement in benefits by B^*, a constant reflecting the benefits associated only with fragmentation at that level of output. These benefits should be added to the $B(t)$ schedule, as shown by the dashed $\{B(t) + B^*\}$ rising curve. The argument is that with fragmentation the new technology is adopted at time t_1, where the increased costs of service links, ΔSC, is balanced by the combined benefits of fragmentation coupled with the replacement of the old technology by the new levels appropriate at time t_1. Fragmentation provides the kind of shock that allows the firm to avoid the costs associated with convincing factors to accept the new technology. Figure 3 illustrates the situation in which fragmentation has caused an earlier adoption of already existing new techniques. Even higher levels of increased service link costs (e.g., those represented by dotted line $\Delta SC'$), would still imply that vertical fragmentation of the production process encourages a lower time lag in adopting the new technology.

6. THE PHOTO-IMAGING, COMPUTER, AND PHARMACEUTICAL SECTORS

We have identified the photo-imaging, computer, and pharmaceutical sectors as good examples of fragmentation that spills across industries and affects the organization of production, changes the market structure, and even redefines industry.

Technological change, broadly defined, is a driving force behind the evolution of these three industries. As we have already argued, technological change is rarely confined to a single sector number, but it has to start somewhere. As Staffan Linder (1961) argued 40 years ago, production and technological changes and decisions often come about in response to specific problems arising in entrepreneurs' environment. The environment means here a specific geographic location and a specific industry. Thus Linder attached a great deal of importance to the needs of the national market in the creation of a new industry and in its technological evolution.

New technology spreads across industries. How far can it spread? Some authors take the position that there are general-purpose technologies that are bound to affect every single sector in the economy.[3] Although examples such as electricity and chemicals are very telling, we suggest that there are also technological changes that are less all-embracing.

In what follows, we first look at the photo-imaging sector and show how a highly integrated product (in the beginning it was just a black box with a hole instead of a lens) has undergone fragmentation. We next turn to the pharmaceutical industry—a very complex sector that is in a perpetual state of flux. One could concentrate on fragmentation in the production of drugs, and on such fragmentation's international implications. Fragmentation in this industry is indeed taking place. But there is also fragmentation of R&D in the pharmaceutical industry. On the one hand, some medical breakthroughs, such as laboratory-grown skin or intestinal cells, change the R&D process in the industry by eliminating or at least reducing the need for animal tests. On the other hand, pharmacologists' enormous knowledge of compounds and molecules finds uses in agriculture and food sectors. As the market responds to these new technological possibilities, the structure of the industry changes as well. Big mergers within the sector and the creation of life-science firms will doubtless have additional effects on fragmentation and its horizontal spread. Finally, we turn to the very important recent changes in the computer industry.

The Photo-Imaging Sector

The French physicist Joseph Nicéphore Niépce is credited with developing, in 1831, the earliest photograph on record. At that time images were recorded on glass plates covered with light-sensitive compounds. The photographer had to prepare his own plates in a darkroom; the wet plates would be introduced into a camera. As plates had to be developed at once after they were exposed, the photographer would return to the darkroom to fix the image and transfer it to paper. George Eastman made a crucial technological advance by replacing plates with long strips of paper. He also invented the first box camera. Eastman's company would not only produce the entire camera but would preload it with film, ask the consumer to send the camera with the exposed film inside to Kodak laboratories for processing and, finally, return the reloaded camera to the consumer, together with black-and-white pictures.

Although highly integrated technology dominated the new industry in its initial stages of development, fragmentation of production in pursuit of a finer division of labor, and resulting gains from specialization, soon emerged, a process that continues to this day. Changes in technology and in the organization of production have produced new trade patterns. Innovations in electronics, telecommunications, and data storage, though external to the industry in question, have in recent years been exerting additional influences on fragmentation of production and international specialization. These complex relationships can be illustrated with a number of episodes from the history of the photographic industry.

The names Zeiss and Leica are widely known. Carl Zeiss, a German industrialist, set up a workshop in Jena in 1864. The location was not incidental, since the glass produced in Jena was internationally acclaimed for its quality and resistance to heat and shocks. When Zeiss opened his workshop it specialized in repairs of microscopes for the University of Jena.[4] The appointment of German mathematician and physicist Ernst Karl Abbe as Zeiss's director of research moved the young company in the direction of camera and microscope design and manufacture. In short order about 100 kinds of glass were developed to go with newly designed camera bosdies and microscopes. The company, together with another German producer, Leica, dominated the domestic and international market for many decades.

Fragmentation of production was already evident by the late 1940s. When the Swedish company Hasselblad was launched, its owner realized that his comparative advantage was in his superior clockmaking skills and that he could not rival the quality of Zeiss lenses. The German company was willing to help Hasselblad by developing an exclusive series of lenses, which made Hasselblad possibly the best medium-format professional camera in the world. This production arrangement continues to this day. Zeiss demonstrated that the best location for the production of a fragment or a superior product may have been Germany, not Sweden. Zeiss was soon to find out that the best location for such production was actually Japan.

When the electronics revolution permeated the photo-imaging industry in the 1960s, cameras were being continually redesigned to take advantage of new technological opportunities. The industry needed workers with experience and new skills. Japanese workers had both. In addition, Japanese electronics firms could meet huge R&D costs that a small European company producing very sophisticated cameras for a limited market could not bear. Lower labor costs provided an additional impetus for shifting the production of components to Japan. Kyocera Corporation has become Zeiss's strategic partner, and one of its divisions, Yashica, supplies the German firm with just about everything but lenses. In fact, some standard lenses are now also produced by Yashica according to German design, specifications, and quality standards, and they bear Zeiss's name.

The recent introduction of digital photo imaging has further fragmented the production process and has brought new industries into the picture. The so-called Charge Couple Device (CCD) is an important new innovation. It was first presented by Bell Laboratories in 1969 and was capable of converting optical images into electronic pictures. The CCD consists of a very fine grid of sensors. The most advanced models have more that 6 millions pixels packed into a 24- by 36-millimeter area. These sensors are extremely sensitive to light, about 15 times more sensitive than photographic film. When light reaches an individual sensor, electric charges are created; these charges can be described by numbers between 0, representing darkness, to 65,536, signifying very intensive light. Once the pixels translate an image into numbers, a digital image is born. Computers can reconstruct digital images, process them, and send them from one station to another.

Actually, the point of entry for the new technology was outside the photographic industry. NASA was quick to recognize the possibilities offered by the CCDs and invested heavily in the new technology. The Galileo mission to Jupiter and the Hubble Space Telescope were early beneficiaries of the digitalization of images. But it was apparent from the beginning that there would be numerous applications outside the space industry.

Computer and electronic firms were quick to spot profitable commercial applications. Indeed, Casio Computer, one of the biggest producers of watches and calculators, launched the first digital camera in 1995. It hoped to sell 3,000 units a month but was soon producing 80,000. Computer and electronic firms enjoy a clear advantage in designing and refining the new product. Only the biggest traditional producers of cameras have the necessary skills and can afford the high R&D costs associated with digital photography. Kodak has invested heavily in the new type of camera. It now produces a range of its own digital cameras, and it also supplies Nikon and Canon with specially designed camera backs that replace traditional film and digitalize images. Some of the digital cameras presently available are produced by traditional suppliers such as Canon, Kodak, Konica, Olympus, and Polaroid. But there are also newcomers to the market—Casio, Epson, Hewlett Packard, Panasonic, Sanyo, Sony, and Toshiba. It will take some time for the dust to settle, but it is quite likely that standardized fragments will emerge that will be used across a range of industries.

Computers, Cellular Phones, and Electronic Organizers: Convergence of Technologies

In a famous misjudgment of the market, Thomas J. Watson, Sr., founder of IBM, remarked in 1943: "I think there is a world market for maybe five computers." More recently, Bill Gates also mispredicted badly, saying "640K ought to be enough for anybody."

It seems that a new high-tech product cannot develop without key players misjudging, at least at the initial stages, the latent demand and the scope of the market. In the early 1980s AT&T entertained the idea of developing cellular phones, but it pulled away from the new product when McKinsey, commissioned to estimate future market potential, predicted that the global market for the new device would not exceed 900,000 units. Today there are about 1 billion cellular phones in use worldwide, and this number could easily double or triple in the next couple of years.

The three episodes referred to above show how hard it is for even the most qualified experts to judge a market's future size. It is equally difficult to predict the technological evolution of new products, their characteristics and functions, and their possible applications in other industries.

Personal computers (PCs), cellular phones, and electronic organizers have changed our private and business lives. They were developed as distinct products with well-defined characteristics and non-overlapping markets. Increasingly, however, technological convergence has been making information technology products perform similar functions, use similar designs, and share the same production blocks.

With the introduction of laptops, computers' portability has increased dramatically. And computers in general can perform many more functions than they could only a few years ago. We use them to buy books, make theater reservations, and check weather conditions or our bank accounts. It has also become possible for a computer to double as a telephone; once a voice message is stored and digitalized, it can be sent at great speed over any distance at negligible cost. And a digital message received on the other end of the line can readily be transformed into sound.

Cellular phones, on their part, are no longer used only to transmit voice messages; they are now capable of receiving, storing, and transmitting other types of information, including images. And their capabilities are increasing. Most important, advanced cellular phones, such as the *i*-mode model introduced recently by Japan's NTT, can interact with the Internet.

Although it once seemed that the computer would be able to replace the telephone, the latter is now coming back with a vengeance.[5] Technological advances in this field are nothing short of spectacular, even compared with progress made in computer technologies. And it is worth keeping in mind that the global market for cellular phones is far larger than that for PCs.

The first generation of cellular phones was based on analog technology, which only allowed the user to transmit voice messages.[6] The second generation was based on digital technology; this markedly improved reception quality and allowed the phones to do more than receive and send voice messages. But it is the most recent third generation of cellular phones that gave the user the ability to surf the Internet.

The electronic organizer has also changed within the last couple of years. From being an address book, it has become a hybrid computer and cellular phone. The most recent Palm model can be used to send and receive e-mail messages, engage in electronic shopping, or surf the Internet.

The convergence of the laptop computer, the cellular phone, and the electronic organizer has created the possibility of developing and sharing common "production blocks." The need to coordinate the development of future generations of the three products manifested itself first in the area of a common operating system. The ability of PCs, cellular phones, and organizers to interact and communicate easily with each other enhances their usefulness and market potential.

The introduction of a common operating system became only a question of time. Such a system would have to be very simple to be of use in cellular phones and electronic organizers. Electronic organizers' and cellular phones' screens and keyboards are small and often awkward compared to PCs', so it is important to simplify and speed up operating instructions. It is precisely these considerations that led Microsoft to introduce a streamlined operating system called Windows CE for use in desktop computers, laptop computers, cellular phones, electronic organizers, and even cable and on-demand television.

Microsoft has recently introduced the Pocket PC in a direct challenge to electronic organizers such as Palm V. The new product has a simplified Web browser and an updated and further streamlined operating system, Windows CE3.0.[7] Oracle, a perennial competitor of Microsoft, has also introduced a new technology, Portal-To-Go, that allows Web pages to be quickly reformatted and shown on the screens of portable computers, cellular phones, and electronic organizers.

It is not necessary to spell out the consequences of the emergence of an operating system with multiple applications. As it has been for some time, the battle is between Bill Gates and the rest of the world. About 1,000 companies including Nokia, Ericsson, Psion, Motorola, Sun Microsystems, and Japan's NTT and Matsushita have created an alliance under the name Symbian, whose purpose is to develop a common operating system. In commenting on this effort *The Economist* remarked that "a ferocious battle is raging over who controls this operating system that will drive these new devices. One of the reasons for founding Symbian was to prevent Bill Gates from turning the wireless world into another part of its empire."[8]

The convergence of the high-tech products discussed here creates other commonly shared production blocks. Much about future progress depends on improvements in battery technology. A one-size-fits-all battery that would operate for many hours or even days without having to be recharged is around the corner. Only a few years ago a typical battery used in cellular phones had only one hour of talk time and six hours of standby power. The most advanced batteries today are ten times as efficient. But, of course, it would be even better for users of PCs, cellular phones, and electronic organizers to have batteries that would last for weeks or even months without being recharged. If radically new batteries come on the market one day, there will be more opportunities to use this common production block across a range of products, including vacuum cleaners and perhaps even electric automobiles.

Display screens are another example of a common production block that is increasingly shared by small laptops, cellular phones, and electronic organizers. As laptops' screens get smaller and smaller, those used in cellular phones and electronic organizers become larger. And of course all users desire screens producing crisper images and a wide range of colors. The screen as a common production block is used in such diverse activities as digital cameras, electronic games, or modern automobile dashboards, which display road maps, images, e-mail messages, or Web sites.

Research and Development Fragmentation in the Pharmaceutical Industry[9]

The pharmaceutical industry is a giant, with annual global sales of prescription drugs exceeding $250 billion a year. Not surprisingly, large firms dominate, and they get even bigger as time goes on. Reports of new negotiations that could create yet bigger firms through mergers appear regularly in the press. Upjohn and Pharmacia joined forces in 1995 through a transaction valued at US$13.0 billion. Beecham and SmithKlineBeckman created a US$8.9 billion company in 1989. And in 1996 Sandoz and Ciba engineered a merger that produced Novartis in a transaction worth US$62.1 billion. One of the reasons these mergers are taking place has to do with sharp increases in R&D costs. Two decades ago it took about US$50 million in R&D costs to develop and place a new drug on the market; today this cost is estimated to exceed US$300 million. In a winner-takes-all race even very big players can hardly afford a string of misses. Just as Hollywood stars are valued by how much their last movie made, global pharmaceutical companies live by star products still protected by patents and new promising medicines coming out of the R&D pipeline. The average development time was about 8 years in the 1960s; it now takes a bit longer than 15 years successfully to shepherd a new product through three development stages—discovery, clinical testing, and approval. Discovering a new, safe, and powerful drug is not the end of the story: "Having produced a successful drug, a company has no time to rest on its laurels. The period of patent protection (normally 20 years) starts when the compound is registered with the patent office, not when it first appears on the pharmacist's shelves (which can be as much as a decade later). So drug companies have only a few years to recoup their research and development costs (including the cost of testing all the compounds that did not make it to the end of the pipeline), and to earn profits for shareholders, before their rivals are permitted to see how much of the market they can snaffle. For an average drug, every day of delay after a patent has been applied for costs $lm in protected sales."[10]

Developing new drugs is a risky, expensive, and drawn-out process. It is also very fragmented, with complex arrangement of R&D "production blocks." Figure 4 gives a schematic representation of a typical drug discovery. The R&D pipeline starts with finding new molecules. They can be created in the company's laboratory or extracted from living organisms such as various fungi. However, only 1 in 10,000 discovered molecules gets to the

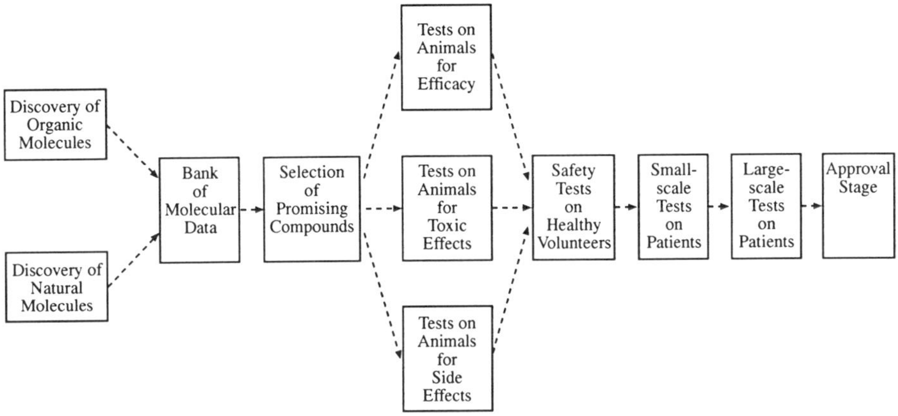

Figure 4: *R&D Fragmentation in the Pharmaceutical Industry.*

finish line in the R&D race. For this reason a pharmaceutical company must create a large pool of molecules from which to choose. Indeed, the molecular "library" of a large firm contains hundreds of thousands or even millions of "volumes." The number of organic molecules chemists could possibly create exceeds 10^{60}, so the tricky trick is to pick up a large but manageable number of compounds that at least promise to be medically beneficial.

Once a suitable package of molecules is selected, preclinical testing on animals can begin. This stage is represented by a number of vertically arranged blocks to reflect the fact that several characteristics of a promising drug must be taken into account—efficacy, toxicity, side effects, and so on. After that, the clinical stage of drug development can get under way. This stage consists of three successive blocks—a safety test on *healthy* volunteers, small-scale tests for efficacy and safety on patients and, finally, large-scale tests. The approval process is the final hurdle to be overcome before a drug can be marketed. It goes without saying that a new drug-development project can be aborted at any point; the later this happens, the larger the investment that must be written off.

The configuration of R&D blocks presented in Figure 4 evolves over time in response to new developments in the pharmaceutical industry and elsewhere. Development of so-called predictive medicine could add another block to the beginning of the R&D structure. It may be much more effective to establish early in a person's life whether he or she is likely to develop, say, high blood pressure later on and to design a preventive package of drugs, diet, and even genetic intervention to prevent the problem from actually occurring, rather than developing drugs to fight the disease after it has already set in. Predictive medicine requires enormous amounts of information concerning a specific individual, his or her family history, and medical knowledge. One can also envisage new blocks appearing toward the end of the R&D chain in Figure 4. One such a block could be called the individualization of drugs. About half of prescription drugs given to patients are ineffective because what works for one individual may not work for another with an identical medical problem. A slightly different drug or dose could perhaps do the trick. Designing a drug for an individual rather than for an average patient would again require a great deal of knowledge and information. The appearance of combination drugs (for example, the drugs used against AIDS) reinforces the need for "custom production."

Having put the R&D process in the pharmaceutical industry in the context of fragmentation, we can now illustrate some of the horizontal aspects discussed earlier. Testing of promising compounds in the preclinical stage used to be done on animals. However, new developments in the medical field offer new possibilities, which could greatly speed up the process, reduce development cost and, last but not least, reduce the use of animals and thus their suffering. Human skin can now be grown in laboratories; this is very useful for testing burn drugs. Other human and animal cells (e.g., intestinal cells) can now be produced in a tissue culture. Feeding a drug to an animal and then taking and analyzing blood samples to see how much of it the intestines have absorbed is the traditional way of testing for bioavailability. Since membranes created from laboratory-grown intestinal cells are now available; they are increasingly used by pharmaceutical companies in bioavailability tests. This example suggests that the ongoing research on artificial organs can find applications in pharmaceutical research. Furthermore, the awareness that medical research can have crossover applications in the pharmaceutical industry may well affect the direction and intensity with which medical researchers look for new solutions.

It turns out that the use of animals as well as humans can be further reduced if virtual organisms, created with powerful computers, are used instead. While computer models of animals, of the human body, or of parts of the human body are too crude at present to eliminate preclinical and clinical testing, future developments in computer science will transform R&D in the pharmaceutical industry.

Fragmented technology used in pharmaceutical R&D lends itself to the international spread of research and testing tasks. Some R&D work on new molecules has been shifted from pharmaceutical laboratories to biotechnology companies. The latter have developed new ways of looking for drugs. Instead of synthesizing new molecules, they use proteins with established medical potency. Thus they no longer need to construct a protein from numerous molecules to start preclinical testing Over 1,000 American, British, French, Swiss, German, Swedish, Italian, and other biotechnology companies specializing in therapeutic proteins have been established in the last decade.

Testing new drugs is another task that can be transferred from traditional pharmaceutical companies to specializing firms. Designing an optimal test that yields statistically meaningful results and convinces drug-approval authorities is no easy task. A new type of company has found a niche in the market. *The Economist* describes one such a company: "Behind closed doors in an office in Princeton, some 60 computers hum quietly to themselves. Each is running a clinical trial of an aspiring drug. Some of these trials are taking place in several countries simultaneously, so the computers are multilingual. Their task is to tell the doctors taking part in the trials, in each participant's own language, which patients to dose when with what, to ensure a statistically meaningful result will emerge from the trial. The computers belong to Covalence, a leading contract research organization (CRO), and thus an example of a type of pharmaceutical company that scarcely existed 15 years ago."

In spite of mergers, big pharmaceutical firms increasingly rely on outside companies specializing in individual "production blocks" that are useful in more than one sector. "Thanks to (new) . . . technology, companies specializing in individual stages of the research and development process—from designing libraries to applying for regulatory approval—are springing up like midnight mushrooms. The traditional drug firms (which the newcomers, in a mixture of awe and contempt, tag as "big pharma") are thus able to outsource any part of the research and developing process, and increasingly do so. . . . In America, for example, more that half of the substances currently undergoing clinical trials originate outside the laboratories of big pharma."[11]

There are other areas in which common production or research blocks can be shared by a number of industries such as pharmaceuticals, agriculture, and biotechnology. For example, consumers can already buy margarine containing compounds designed to lower cholesterol. Vitamin supplements can be easily introduced into many foods. The creation of life-science firms is reinforcing the wave of mergers among pharmaceutical firms. A recent merger between an agricultural giant, Monsanto, and one of the big players in the pharmaceutical industry, American Home Products, serves as just one example of how borders between industries are being redefined. Sharing common or overlapping R&D blocks is only one reason this process is taking place. Creating joint distribution networks also creates economies of scale. When new life-science firms are formed, opportunities arise to restructure the production process.

7. CONCLUDING REMARKS

The process of vertical fragmentation is not confined to industry by industry. Instead, reduced costs of service links and advances in technology tend to spread horizontally over other sectors of the economy. Often this results in a two-way interplay between technological change and fragmentation. Technological improvement encourage fragmentation, and one consequence of fragmentation is that new inducements for further improving technology arise. Furthermore, international fragmentation is often associated with the adoption of newer techniques that have been resisted by rent-preserving agents in the original vertically integrated structures. Our paper has illustrated some of these features in three industries—the photo-imaging industry, the computer industry, and the pharmaceutical industry.

Southeast Asian countries play a major role in the fragmentation of the photo-imaging industry. Fifty years ago they had already started eroding the comparative advantage of traditional European and American producers in manufacturing camera bodies if not the whole camera. The introduction of electronics into cameras has given first Japan and then other regional producers new niches to exploit. The most recent developments involving digital photography have created new production blocks and given rise to new opportunities. The notion of a traditional integrated camera producer makes little sense today; electronics, computer, and even software producers can take advantage of the ever-fragmenting photo-imaging industry.

By contrast, Southeast Asian producers have had little involvement in the fragmentation of the pharmaceutical industry. Perhaps the types of medicine practiced in Asia are less suitable for North American and European needs and, as a result, pharmaceutical companies from the Far East have production profiles that do not allow them easy access into markets in the North. But this may change. India is emerging as a major regional drug producer and exporter, although its success is based largely on imitation. Given the growing importance of the software industry in the R&D of new drugs, India may well find a place in the global pharmaceutical sector. Perhaps the greatest potential for East Asian countries, especially Taiwan, South Korea and Japan, lies in computers, cellular phones and electronic organizers. In fact, producers in those three countries have already contributed to the rapid process of fragmentation of production and convergence of technologies.

References

Business Week (1999): Special Summer Issue

Helpman, Elhanan (1998), ed., *General Purpose Technologies and Economic Growth* (Cambridge, MA: MIT Press).

Jones, Ronald W. and Henryk Kierzkowski (1990): "The Role of Services in Production and International Trade: A Theoretical Framework," in R. W. Jones and A. Krueger, eds., *The Political Economy of International Trade* (Oxford: Basil Blackwell).

---------- (2001a): "Globalization and the Consequences of International Fragmentation," forthcoming in R. Dornbusch, G. Calvo, and M. Obstfeld (eds.), *Money, Factor Mobility and Trade; The Festschrift in Honor of Robert A. Mundell* (Cambridge, MA: MIT Press).

---------- (2001b): "A Framework for Fragmentation," in S. Arndt and H. Kierzkowski, *Fragmentation: New Production Patterns in the World Economy* (Oxford: Oxford University Press), forthcoming.

Linder, Staffan Burenstam (1961), *An Essay on Trade and Transformation* (New York: Wiley).

Ohyama, Michihiro and Ronald W. Jones (1995), "Technology Choice, Overtaking and Comparative Advantage," *Review of International Economics,* June, pp. 224–34.

The Economist (1998) "The Alchemists," February 21.

Yeats, Alexander J. (2000), "Just How Big is Global Production Sharing?" in S. Arndt and H. Kierzkowski, eds., op. cit.

Endnotes

[1]Although a large volume of international fragmentation may take place under the rubric of a multinational enterprise, it is also the case that a significant number of firms find it less costly to deal in arms-length transactions across borders with other firms.

[2]We should add, however, that other types of government policies and regulations may encourage international fragmentation, although not necessarily in the best direction. Pharmaceutical firms' or aircraft companies' access to a national market sometimes depends on their initiating some local production. To penetrate the Chinese market, Boeing may be persuaded to set up production of certain parts and components in China, although not of the entire aircraft. Although local content requirements have encouraged fragmentation, these are not developments that would necessarily have transpired in free markets in China.

[3]This view is presented in Elhanan Helpman, ed., *General Purpose Technologies and Economic Growth,* MIT Press, Cambridge, Massachusetts, 1998.

[4]Most of the information concerning Zeiss can be obtained from The Zeiss Historica Society at http://www.cmaeraquest.com/zeis.htm.

[5]What gives the cellular phone an edge over the laptop is its greater portability and connectivity.

[6]For a review of technological advances in telecommunications and an analysis of their impact on the whole industry the reader is referred to *The Economist's* survey "The World in Your Pocket," October 9, 1999.

[7]In advertising the new product over the Internet, Microsoft stresses the virtues of the new operating system installed in the Pocket PC: "We can't let all the desktop PC users have all of the fun, now can we?... When you connect your Pocket PC to the Internet and browse to PocketPC.com, you will be presented with a version of our site that is formatted to Microsoft Pocket Internet Explorer. You will get the same information as the desktop version, but in a nicely designed Web site tailored just to suit your Pocket PC's screen." See http://www.microsoft.com/pocketpc.

[8]"The World in Your Pocket," October 9, 1999, p. 9.

[9]Much of the factual information presented here comes from an excellent review of the pharmaceutical industry published by *The Economist,* February 21, 1998.

[10]Quoted from "The Alchemists," *The Economist,* February 21, 1998, p. 4.

[11]*The Economist*, February 21, 1998, p. 4

Chapter 4

FUNCTION VS. FORM IN THE FRAGMENTED INDUSTRIAL STRUCTURE: THREE EXAMPLES FROM ASIA PACIFIC EXPERIENCE

Henry Wan, Jr.[*i]

1. INTRODUCTION

The study of fragmentation pioneered by Jones and Kierzkowski (1990) is among the most stimulating topics in the trade literature. New elements of realism are continually being incorporated into the discussion. All evidences suggest that the fragmentation process causes certain predictable changes to the aggregate production set. It appears that a full-fledged new theory is waiting to be born. With many rapid changes taking place in its economies, the Asia Pacific region is clearly an economists' Constellation of Cassiopeia, rich in grist for future theorists' mill.

2. CLOSED VS. OPEN NETWORKS

When fragmentation takes place, firms in an economy may form networks of different types. The present note contains three Asia-Pacific examples pertaining to a single theme: what is known as *closed* and *open* networks (Naughton 1997, p. 20) have different economic implications. To give precise definitions to such networks, we shall first present four cases for illustration. Firms are divided into subcontractors and assemblers. A network is a set of links, each of which joins an ordered assembler-subcontractor pair. Consider two classes A and B of similar firms forming networks. The difference between the two is their permissible collections of networks, in which the links in any network must obey some association rule.

Case 1

Class A has two subclasses of subcontractors A' and assemblers A", just as class B has subcontractors B' and assemblers B". In A, the rule of association R_A is that each subcontractor a' in A' must only trade with one particular member a" of A. In B, rule R_B means that there is no such restriction. How these notions work is illustrated in Chart 1.

Chart 1: Closed versus Open Networks

The set of members, A $=$ The set of subcontractors, A' $\cup$ The set of subcontractors, A''

A 'closed' network: $N^A{}_1$

The set of members, B $=$ The set of subcontractors, B' $\cup$ The set of subcontractors, B''

An 'open' network: N^B

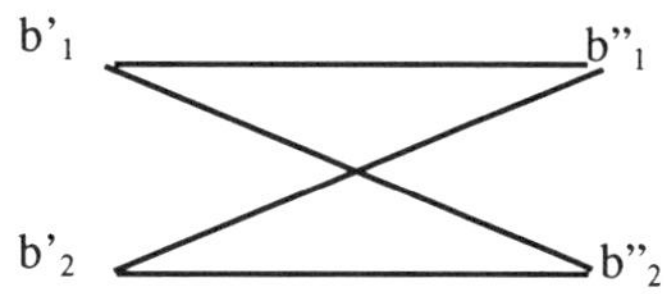

There are actually other "closed" networks for A, such as $N^A{}_2$, which is equivalent to $N^A{}_1$, as well as $N^A{}_3$ and $N^A{}_4$, all of which are displayed in Chart 2:

Chart 2: Alternative Closed Networks

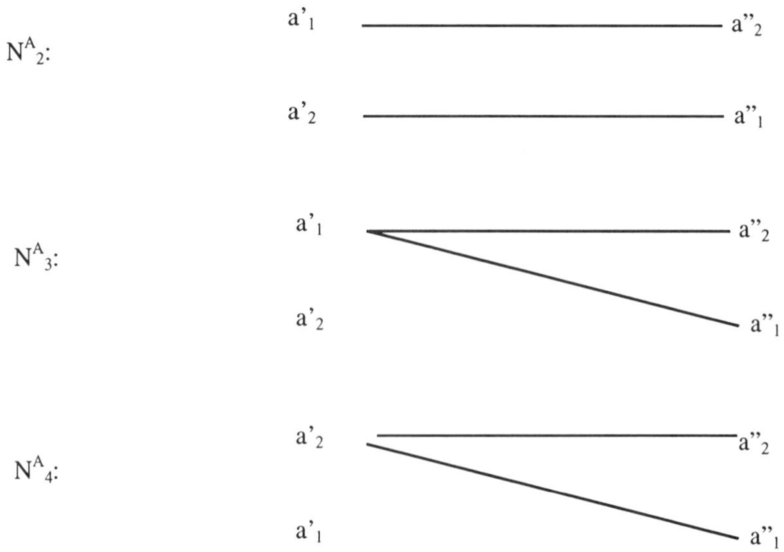

In the economic context what matters is the entire collection of networks, where each collection is associated with specific rules of association and not with specific networks. This choice of classification is important for our discussion, just like in botany, trees have to be organized into a typology by species, and not by the number of branches. We next consider three more cases that are directly relevant to the Asia Pacific context.

Case 2

Everything here is the same as Case 1, except that here each a' in A' can trade only with some particular subset A''_0 of A'', say, with A''_0 being domestic assemblers. Note that to be fully general, A and B should have the same number of members, by the definition of similarity (formally, each member in A may be placed into a 1-1 correspondence with a member in B), but they need not be distinct. In particular, A'' is the nonintersecting union of A''_0 and C; B'' is the nonintersecting union of B''_0 and C, where A''_0 is similar to B''_0 with different economy-specific assemblers, but C is the same, representing foreign assemblers. Thus R_A grants domestic assemblers the privilege of being the sole client for its subcontractors, while R_B has no such restriction.

Case 3

Everything here is the same as Case 1, except that here the set A' (B') is the nonintersecting union of A'_1 and A'_2 (B'_1 and B'_2), where subscript 1 denotes fabrication only, while subscript 2 denotes fabrication and product design. Again A_i (B_i) is the nonintersecting union of A_{i0} and C_I (B_{i0} and C_I). A_{i0} (B_{i0}) is the domestic pool, while C_i is the common foreign pool, i = 1, or 2. The closed rule R_A restricts product design to domestic subcontractors only.

Case 4

Everything here is the same as Case 2, with a particular restriction. A'', the permissible assemblers under R_A, contains none of A''_0 (!) and only a particular *proper* subset C* of C, with C* having the interpretation of *favored* foreign assemblers.

The point to be made here is that the fragmented industrial structure is a basic concept necessary for policy analysis. It means different operations like fabrication (or product design, or assembly) are under separate management. Yet that concept in itself still concerns only the 'form' of industrial organization. In influencing the performance of an economy, sometimes industrial policy works through what association rule is explicitly or implicitly in force, which deals with how the structure 'functions'.

3. ASIA PACIFIC EXAMPLES

Three examples given below illustrate the relationship between the architecture of subcontracting networks and three issues in trade development:

(1) Technology development and acquisition (bicycle industry)
(2) Competition under rapid innovation (electronic products)
(3) The threat of "hollowing out" (general manufacturing)

Example 1: Competition in the Bicycle Industry—Taiwan, Korea, and Japan

Formally, this example corresponds to Case 2 presented in Section 2 above. According to Chu and Li (1996) the success of Taiwanese bicycle producers as compared to their Korean rivals is due to the way the subcontracting systems function in these two economies. Bicycles are produced within a fragmented industrial structure in both economies, where parts are produced by subcontractors. In Korea subcontractors are dominated by their clients, which belong to various business groups. Each subcontractor is only allowed to serve one assembler, which is also the source of technology. In Taiwan both assembled bicycles and bicycles parts are exported all over the world. Subcontractors can directly acquire information and experience through their dealings with foreign firms. This has allowed Taiwan to increase its level of technological expertise quickly; as a result it has become the largest exporter of bicycles in the world, in value terms.

It would be simplistic to claim that under all circumstances, the open network *in* Taiwan is superior to the closed network *in* Korea. One may more accurately say that the open network *for* Taiwan is superior to the closed network *for* Korea. This is because neither country has much proprietary technology to support product development, and the two economies compete against each other for the same world market.

Korea did not invent the closed network; it emulated the Japanese model. Japan has developed world-class proprietary technology for bicycle production, so its closed network helps it retain the exclusive character of its product design (which is firm-specific human capital). According to Crown and Coleman (1996, p. 179), although their labor cost is much higher than their Taiwanese counterparts', Japanese producers of high-quality bicycles (e.g., National Bicycle of Matsushita) are able to keep Taiwanese competitors at bay for decades while charging 25% to 30% higher prices. (Lower-quality Japanese brands like Bridgestone have been displaced by Taiwanese brands, however.) As an indication of how Japanese technology has progressed, Shimano has become the dominant firm in the world for derailleurs because of its innovative product's quality.

Example 2: Competition in Electronics—The United States, Japan, and Overseas Chinese Firms

Formally, this case corresponds to Case 3 presented in Section 2 above. According to Borrus (1997), until the 1980s Japanese firms often served both as subcontractors (supplying U.S. firms with key parts and components) and as U.S. firms' rivals. The Japanese electronics industry exploited their dual position: got American technology free in their role as suppliers, then improved upon it sometimes, and in their role as rivals outcompeted American firms for market share. But Japanese prosperity raised the dollar cost of Japanese labor. To save on labor costs, both U.S. and Japanese firms have relied on overseas Chinese in Taiwan, Hong Kong, and Singapore, as well as in the southern part of the People's Republic of China [PRC]) as subcontractors. By the 1990s the situation had changed dramatically. The revival of U.S. competitiveness in this industry can be attributed to (1) U.S. and Japanese electronics firms' different strategies for dealing with their overseas Chinese suppliers, and (2) the rapid introduction of new products. Borrus (1997, p. 140) states that "the recent success of U.S.-owned firms has rested in large part on the growing technical sophistication and competitive strength of Asia-based producers in the China Circle, Singapore, and South Korea."

Under the highly fragmented industrial structure of the bicycle-production industry, the United States entrusts overseas Chinese subcontractors with both fabrication and product design, but the Japanese sign subcontracts for production only. Compared to Japanese firms, U.S. firms engage in more extensive international division of labor, using small overseas Chinese firms in a flexible and highly fragmented industrial structure. As a result of this difference, U.S. firms can ship their products much faster than can Japanese firms, even to Japan itself. In addition to shortening the production cycle, this trend clearly favors the reemergence of U.S. firms, which have renewed their competitive advantage.

Example 3: The Threat of Deindustrialization—Singapore vs. Hong Kong

Formally, this case corresponds to Case 4 presented in Section 2 above. The living standard in Hong Kong has always been much higher than that in the PRC. The same can be said about Singapore relative to its hinterland, Malaysia.

Hong Kong has never had an industrial policy. After the PRC's economic reform went into effect in 1978 industries migrated from Hong Kong to the southern part of the PRC, which is near Hong Kong. But this increased profitability to capitalists also brought the threat of unemployment to Hong Kong workers. Contrary to the optimistic assessment of Naughton (1997, p. 13) that the loss of well paid jobs in the manufacturing sector can be fully offset automatically by the expansion of service industries, the unemployment peak in the recent recession in Hong Kong suggests that de-industrialization is not something that can be taken lightly.

Singapore could follow in Hong Kong's footsteps and lose most of its industries with their well paid jobs to nearby Malaysia, where labor is less expensive. However, Singapore's government has successfully attracted foreign plants with high value added to Singapore, so much so that a labor shortage has developed in the island economy, even though this government policy has helped multinational firms on the island dominate their local subcontractors.

One may note that without granting as much favor to multinational firms as Singapore has, Taiwan has not become "hollowed out" by the PRC economy, as Hong Kong has, either. Addressing the reasons for this is beyond the scope of this paper.

4. POLICY ASSESSMENT

The three examples provided above show that neither the open nor the closed network is superior to its alternative under all circumstances. Example 1 shows that for bicycle production the closed network works better than the open network for Japan, which is technologically advanced, but not for Korea, which is technologically less developed. Example 2 shows that for electronics the open network of U.S. firms and their overseas Chinese associates gives the United States a competitive edge over Japanese firms. In both Example 1 and Example 2 the particular network type emerged as a result of market forces, including the influence of dominant private firms.

Example 3 shows that a farsighted, active industrial policy appears to be superior to laissez faire and that corresponding types of networks arose as a consequence of the choice of that industrial policy by the government of Singapore.

One ought not to conclude that each economy (Korea and Japan) and each industry (bicycle and electronics) are so different that no lesson can be learned and perhaps it does not matter which type of network is in place. It matters tremendously, but the way in which it matters differs depending on the context and on other contributing factors, particularly the level of technology for innovation (illustrated in Example 1) and the pace of product development (illustrated in Example 2).

In all our three examples the world economy is externality ridden, so that laissez faire is by no means always the best policy. On the other hand, the presence of externalities does not necessarily mean that a closed system is superior to an open one.

For instance, in Example 1 we see that the informational externality of the North-South technology transfer is the by-product of trade and is, therefore, a "free lunch" for the south (Korea, Taiwan). The open network provides more contact and hence more gain for the less-informed agents in the developing world. Crown and Coleman (1996) note the instances when Shimano (the Japanese producer of derailleurs) gained valuable information through business contacts and when Giant (the Taiwanese firm that is the largest worldwide supplier of bicycles today) received free advice on occupational safety and plant layout.

Because the Southern firms gained from their association with the Northern firms, one would think that quid pro quo the Northern firms should also be able to extract some business advantages in return. According to Crown and Coleman (1996, p. 214), in fact, that was the view of some former employees at the American firm Schwinn vis-à-vis the Japanese firm Shimano, "Shimano should have helped Schwinn. Helped us because we helped them." But then "They [Shimano] were getting so popular, they didn't have to play the game." Translation: the externality-ridden trading equilibrium was such that the informed party (Schwinn) could not extract *all* the gain from the uninformed party (Shimano), over the benchmark equilibrium of 'no trade at all.'

According to Crown and Coleman, in the dealings between Schwinn and Shimano, the latter gained from externalities; the former did not. But at least Shimano never became Schwinn's competitor. In contrast, in the dealings between Schwinn and Giant and China Bicycle, Schwinn directly helped its own future rivals by providing trade information for free. So perhaps Schwinn could have bargained for a better deal. Nevertheless, in order to extract some gains from the information one party possesses that the other party does not, it takes both parties to foretell the information's ultimate worth. And that is not easy to do.

The moral of all this is that to conclude that one of two alternative systems is superior one needs to scrutinize the situation, case by case. The same open network that may help Taiwanese firms acquire technology and information quickly and cheaply may also threaten Japanese firms, which are struggling to hold on to the invisible asset of their distinct product image. A closed network may appear alluring to Korean firms, whose technology level is not much superior to that of their Taiwanese counterparts, as well as to the ambition of their Japanese rivals to prolong their dominance.

5. A HISTORICAL NOTE

How did Japan's, Korea's, and Singapore's closed systems come into being? The answer to this question is multifaceted. In Japan the system came about by circumstance, presumably because of the growing oligopolistic power of large firms. In Korea it came about by design: Korea's government policy favors large firms because it hopes that by doing so it will make them more "competitive" against foreign rivals. In Singapore the system again came about by circumstance, that is, as a result of the dominant market position gained by multinational firms as a result of government favor.

6. IS A NEW THEORY REALLY NEEDED?

One might be tempted to assume that a closed network is a form of intervention and that the deterministic, competitive market model may thus be adequate for policy analysis once we formally introduce such an intervention into the analysis. A closed network is in fact an intervention on the market; it raises the barrier to entry. But this explanation misses a great deal. As we see below, dynamic factors are at work.

One often-noted fact is that a more fragmented system is advantageous for industries with rapidly changing market conditions or technology. The reason given for this is that the shorter the chain of command, the faster the institutional response. This has been mentioned as the reason that in certain industries Korean business groups are less efficient than firms in Hong Kong or Taiwan. (See a survey of such views in chapters 18 and 19 of Thorbecke and Wan [1999].) Likewise, as can be seen in our Example 2, the alliance of Americans and Overseas Chinese has gained an advantage over the alliance led by Japan. In short, a reformulation of the theory is needed to incorporate some dynamic or stochastic elements into the picture.

7. CONCLUSION

Fragmentation is a concept that allows economists to study many trade and development issues, including the types of network considered here. One important lesson to take away from our three examples is that oftentimes a situation is more intricate than the "single-cause models" economists have found so valuable in pedagogy. In reality the type of network may matter either because of the channels of information externalities it allows (Example 1), the speed of an industry's response to a changing environment (Example 2), or an entire economy's adaptation to the global dynamics of a production cycle (Example 3).

References

Borrus, Michael (1997), "Left for Dead: Asian Production Networks and the Revival of the US Electronics," in Naughton, Barry (ed.), *The China Circle: Economics and Electronics in China, Taiwan and Hong Kong* (Washington, DC: Brookings Institutions Press).

Crown, Judith and Glenn Coleman (1996), *No Hands: The Rise and Fall of The Schwinn Bicycle Company, An American Institution* (New York, NY: Henry Holt).

Chu, W. W. and J. J. Li (1996), "Growth and Industrial Organization: a Comparative Study of the Bicycle Industry in Taiwan and South Korea," *Journal of Industrial Studies* 3/1, 35–52.

Jones, Ronald and Henryk Kierzkowski (1990), "The Role of Services in Production and International Trade: A Theoretical Framework," in Jones, R. W. and A. O. Krueger (eds.), *The Political Economy of International Trade: Essays in Honor of Robert E. Baldwin* (Oxford: Blackwell).

Naughton, Barry (1997), "The Emergence of the China Circle," in Naughton, Barry (ed.), *The China Circle: Economics and Electronics in China, Taiwan and Hong Kong* (Washington, DC: Brookings Institutions Press).

Thorbecke, Erik and Henry Y. Wan, Jr. (eds.) (1999), *Taiwan's Development Experience: Roles of Government and Market* (Boston, MA: Kluwer Academic Publishers).

Endnote

[i]Professor Leonard Cheng's comments are deeply appreciated.

Chapter 5

PRODUCTION SHARING IN EAST ASIA: WHO DOES WHAT FOR WHOM, AND WHY?

Francis Ng and Alexander Yeats

1. INTRODUCTION

Published studies that use trade statistics from at least as early as the 1960s document the remarkable increase in international production sharing that is reflected in unusually high growth rates for the exchange of components or partially assembled manufactured goods.[1] In its most recent form, production sharing involves the development of specialized and (often) labor-intensive activities within vertically integrated international manufacturing industries. As an example, electronic semiconductors, valves, tuners, and other components are now assembled for multinational firms in places like Mexico, Malaysia, or the Philippines. Parts of clothing and leather goods are assembled in Jamaica and the Dominican Republic for reexport to the U.S. market; one estimate by the World Bank (1994b) puts the value of assembly exports from the Caribbean at over US$3 billion. Among the many other industries in which major parts of a production process were transferred abroad are the television and radio receiver industry, the sewing machine industry, the calculator and other office equipment industry, the electrical machinery industry, the power and machine tools industry, the typewriter industry, and the camera and watch industries (USITC 1996).

Theoretical analyses that examine the motivating factors behind international production sharing generally agree on two key points: (1) manufacturing processes for some goods can be separated into distinct stages (like the cutting of cloth for eventual sewing into clothing), and (2) owing to differences in the nature of the processes at each stage, significant differences may exist in the countries comparative advantages' over a manufacturing chain.[2] Much of this literature presumes that manufacturing stages that are labor intensive in nature (e.g., many assembly operations) would be transferred from richer countries to poorer developing countries where wage costs are relatively low. However, recent statistics published by the USITC (1996, p. B-4) showing that Japan and Germany are among the major suppliers of finished goods that use U.S. parts and components assembled abroad are seemingly at odds with this assumption.

Using a heretofore largely untapped data source on production sharing, this study provides new empirical information on the magnitude, direction, and motivation for this operation in East Asia. East Asia was chosen for study because the available evidence indicates that regional production sharing is expanding at a far faster pace there than it is in either North America or Europe.

The paper is organized as follows. First, it describes the nature of the newer data sources on production sharing and assesses their strengths and limitations for empirical analysis. Second, it employs the available statistics to analyze the magnitude, composition, and direction of production sharing in East Asia. Third, it describes a methodological procedure that determines which countries have a comparative advantage in either the assembly or the production of components of a specific manufactured good and also indicates how a country's potential for further development of these operations might be assessed. This new methodological approach is then applied to statistics on East Asian trade in components to determine what motivates the regional pattern of this exchange, and also to determine how Asian comparative advantage in production sharing is evolving. The paper closes with an evaluation of how well the observed empirical results "square" with much of the theoretical discussion, and what their implications are for future research.

2. DATA SOURCES ON PRODUCTION SHARING

Revisions to the original Standard International Trade Classification (SITC) system provide extensive new information on international trade in parts and components. The SITC Revision 2 system distinguishes between 60 broad groups of component products, but in the recently adopted harmonized system (HS), over 200 such product groups are identified. As more countries begin to report trade data in terms of the HS, the potential for empirical research on production sharing will greatly increase.

In its original form, the SITC system did an inadequate job of distinguishing between trade in final goods and trade in parts and components. At the lowest (five-digit) level the SITC identified about 800 individual products—only 10 of which were "parts." However, in the late 1970s and early 1980s many countries also adopted the more detailed SITC Revision 2 system, which expanded the number of product groups composed solely of components. The coverage of these groups was most extensive within the machinery and transport equipment sector (SITC 7), where about 60 individual three-, four-, and five-digit groups consisting solely of components of manufactured equipment to be assembled were identified.[3]

Outside this sector the SITC Revision 2 still fails to differentiate sufficiently between assembled goods and components, so tabulations of the true magnitude of trade in parts are clearly downward biased. The fact that some imported components may be processed and then reexported as components, and not necessarily as assembled final products, may also cause the data to understate the true importance of production sharing. Furthermore, some developing countries did not shift to the SITC Revision 2 system until the early or mid-1980s, so it is not yet possible to fully monitor longer-term trends in this exchange.[4] It should, however, be noted that many countries' recent shift toward reporting trade data in terms of the HS has the potential to greatly expand future opportunities for empirical research on international production sharing.

Given the potential importance of these newer data sources, we undertook several checks aimed at verifying the underlying quality of the Revision 2 trade statistics compiled by the United Nations (UN). These tests, which included partner-country comparisons, revealed three anomalies that should be noted:

- From 1984 through 1995 the Philippines classified approximately 20% of all imports and 30% of all exports as SITC 931 "special transactions." Very few countries reporting to UN COMTRADE have trade shares for this item close to these levels. Partner country data suggest that some exchange reported in SITC 931 involves assembly operations that are not recorded correctly. As a result, the available data probably understate the importance of the Philippines assembly operations.[5]
- Hong Kong's trade data consistently incorporate some of the largest negative trade balances for components observed in any UN country's statistics. For example, in 1996 component imports of US$18.7 billion were reported, as opposed to US$3.1 billion in exports. We suspect that some transit trade with the People's Republic of China (PRC), Macau, and other East Asian countries may be incorrectly included in the data reported to the UN.
- Singapore stopped reporting any trade with Indonesia in 1964. We contacted the UN Statistical Office about this problem, but it was unable to provide an explanation for why it was occurring. However, in 1995 Indonesia reported exports of US$3.7 billion to, and imports of US$2.4 billion from, Singapore. Singapore's reporting practices will bias downward some of our estimates of intra-regional components trade.

In spite of these problems, the available UN Revision 2 data reflect the relative importance of the East Asian countries' trade in components. For example, in 1996, the United States and Germany were (according to this source) the 2 largest exporters and importers of components, but Singapore, Hong Kong, Japan, Malaysia, and the PRC were among the 10 largest exporters or importers of these goods (see Appendix Table 1). Furthermore, the Revision 2 statistics show that recent growth rates for East Asian trade in components have exceeded corresponding rates for any other major regional group of countries.

3. EAST ASIAN TRADE IN PARTS AND COMPONENTS

There is more East Asian trade in components than is often recognized. Regional exports of parts and components totaled US$178 billion in 1996, while Asian imports of these products stood at about US$12 billion less. Components now constitute one-fifth of East Asian exports of manufactures.

Table 1 uses UN statistics to examine the composition and relative importance of the individual component product groups in East Asian trade.

The table identifies each product by SITC (Revision 2) number; it provides a short description of the item and also indicates its value and share in East Asian global imports and exports of all components. (These totals include the combined trade of Japan, Hong Kong, the Republic of Korea, the PRC, Singapore, Taiwan, Malaysia, Thailand, Indonesia, and the Philippines).

One interesting point evident from these statistics is that East Asian component trade is concentrated in a relatively few items. Specifically, 5 of the 60 SITC groups account jointly for almost 75% of all imports (79% of exports), with parts of telecommunications equipment (SITC 764) alone accounting for about one-quarter of this exchange. An analysis of underlying individual East Asian countries' trade statistics reveals little variation in the relative importance of these products across countries. A similarly high level of product concentration has also been observed in global statistics on components trade, except that motor vehicle parts were the largest single traded product group on all world markets (Yeats 1997). In East Asia these items are the fourth-largest product group.

Table 1: *The 1996 Value and Share of East Asian Countries' Imports and Exports of Parts and Components Identified in the SITC Revision 2 Classification System. (Values in US$ Million)*

SITC Rev. 2 – Description*		Imports		Exports	
		Value	Share (%)	Value	(Share (%)
711.9	Parts of Steam boilers and auxiliary plants	426.8	0.26	495.5	0.28
713.19	Parts of aircraft internal combustion engines	139.9	0.08	142.8	0.08
713.31	Parts of outboard motors	99.1	0.06	483.0	0.27
713.32	Parts of outboard motors, nes	388.7	0.23	462.9	0.26
713.9	Parts of internal combustion engines, nes	3,341.3	2.02	5,107.1	2.87
714.9	Parts of engines and motors, nes	2,362.8	1.43	889.0	0.50
716.9	Parts of rotating electric motors	1,748.3	1.06	1,385.4	0.78
718.89	Parts of water turbines and hydraulic motors	40.1	0.02	42.0	0.02
721.19	Parts of cultivating equipment	32.2	0.02	38.6	0.02
721.29	Parts of harvesting machinery	60.7	0.04	64.3	0.04
721.39	Parts of dairy machinery	27.0	0.02	4.8	0.00
721.98	Parts of wine making machinery	9.0	0.01	4.3	0.00
721.99	Parts of other agricultural machinery, nes	54.3	0.03	17.6	0.01
723.9	Parts of construction machinery	1,245.6	0.75	705.3	0.39
724.49	Parts of spinning and extruding machinery	420.3	0.25	291.7	0.16
724.69	Parts of looms and knitting machinery	469.7	0.28	293.2	0.16
724.79	Parts of textile machinery, nes	201.8	0.12	147.2	0.08
725.9	Parts of paper making machinery	704.7	0.43	201.2	0.11
726.89	Parts of bookbinding machinery	16.7	0.01	6.3	0.00
726.9	Parts of printing and typesetting machinery	399.0	0.24	151.5	0.09
727.19	Parts of grain milling machinery	30.8	0.02	29.1	0.02
727.29	Parts of food processing machinery	24.9	0.02	1.5	0.00
728.19	Parts of machine tools for special industries	237.1	0.14	167.9	0.09
728.39	Parts of mineral working machinery	648.7	0.39	132.0	0.07
728.49	Parts of machines for special industries, nes	4,009.3	2.42	2,065.4	1.16
736.9	Parts of machine tools for metal working	1,149.6	0.69	810.6	0.45
737.19	Parts of foundry equipment	182.3	0.11	186.7	0.10
737.29	Rolling mill parts	615.7	0.37	573.5	0.32
741.49	Parts of refrigerating equipment	350.4	0.21	350.4	0.19
742.9	Parts of pumps for liquids	686.1	0.41	473.3	0.27

Table 1: (*Continued*)

SITC Rev. 2 – Description*	Imports		Exports	
	Value	Share (%)	Value	Share (%)
743.9 Parts of centrifuges and filters	1,772.0	1.07	1,260.4	0.71
744.19 Parts of fork lift trucks	111.3	0.07	59.6	0.03
744.9 Parts of lifting and loading machines	2,993.2	1.81	2.027.5	1.14
745.19 Parts of power hand tools	104.6	0.06	125.1	0.07
745.23 Parts for packing machinery	279.2	0.17	72.0	0.04
749.99 Parts of non-electric machinery, nes	715.1	0.43	738.7	0.42
759 Parts of office and adding machinery	30,726.8	18.56	47,130.9	26.10
764 Parts of telecommunications equipment	41,013.1	24.77	42,459.9	23.86
771.29 Parts of electric power machinery	2,178.9	1.32	1,673.3	0.93
772 Parts of switchgear	19,003.0	11.48	21,625.0	12.18
775.79 Parts of domestic electrical equipment	113.7	0.07	85.0	0.05
775.89 Parts of electrothermic appliances	472.7	0.29	606.3	0.35
776.89 Parts of electronic components, nes	19,849.7	11.99	7,868.6	4.46
778.19 Parts of electronic accumulators	184.7	0.11	161.8	0.09
778.29 Parts of electric lamps and bulbs	136.1	0.08	79.9	0.04
778.89 Parts of electrical machinery, nes	451.0	0.27	4,228.7	2.42
784 Parts of motor vehicles and accessories	11,983.6	7.24	21,373.6	12.03
785.39 Parts of carriages and cycles	2,419.8	1.46	4,012.3	2.26
786.89 Parts of trailers and non-motor vehicles	141.0	0.09	169.9	0.09
791.99 Parts of railroad equipment and vehicles	391.9	0.24	126.7	0.07
792.9 Parts of aircraft and helicopters	4,514.5	2.73	2,360.9	0.98
821.19 Parts of chairs	421.6	0.25	367.9	0.20
821.99 Other furniture parts	731.6	0.44	976.5	0.53
874.29 Parts of measuring or drawing machines	719.7	0.43	307.0	0.18
881.19 Parts of still cameras, nes	1,799.6	1.09	1,144.8	0.62
881.21 Parts for cameras under 16mm	40.6	0.02	12.2	0.01
881.29 Parts of cameras under 16mm, nes	43.4	0.03	19.6	0.01
884.11 Parts of unmounted optical elements	262.5	0.16	174.2	0.09
885.29 Parts of clocks and watches	1,683.4	1.02	1,420.1	0.68
899.49 Parts of umbrellas and canes	167.4	0.10	154.2	0.09
ALL ABOVE ITEMS	165,548.5	100.0	178,546.7	100.0
Memo Item: East Asia Trade				
All Above Parts and Components	165,548.5	18.2	178,546.7	20.6
All Manufactures Excluding Chemicals	907,522.5	100.0	944,083.5	100.0

**The trade statistics represent the combined imports and exports of the following East Asian Countries: Japan, Hong Kong, Republic of Korea, China, Singapore, Taiwan (China), Malaysia, Thailand, Indonesia, Philippines.*

Source: *Computed from UN COMTRADE statistics.*

With regard to the relative importance of parts and components in total East Asian trade, the table's memo item shows that these goods account for approximately one-fifth of the region's total exports and imports of manufactures. There is, however, considerable variation in this share across East Asian countries. For example, components consist of almost 30% of the Philippines' imports of manufactures, as opposed to about 7% of Japan's. In contrast, exports of components account for about 20% of Japan's exports, as opposed to under 4% of Indonesia's.

Asian imports of components, measured as a share of all manufactures, is growing considerably faster than imports in the Organization for Economic

Cooperation and Development (OECD) Europe or North America (Canada, Mexico, and the United States). The value of Asian global imports of components rose more than ninefold over the 1985–96 period. Almost three-quarters of all Asian imports of telecommunications equipment are components for further assembly.

Table 2 provides a different perspective on trade, in parts by showing the value of all components imports within two-digit SITC groups, along with their share of all group imports in 1985, 1990, and 1996. For comparison, similar statistics are shown for OECD Europe and North America.

Table 2: *The Relative Importance of Parts and Components Imports in Selected Regional Groups*

Regional and Product Group (SITC)	Value of Parts and Components Imports ($million)			Share of Parts and Components In Product Group Imports (%)		
	1985	1990	1996	1985	1990	1996
EAST ASIA						
Power Generating Equipment (71)	1,478	4,816	8,547	35.3	39.1	35.9
Special Industry Machinery (72)	1,637	4,527	8,592	14.9	17.0	16.1
Metal Working Machinery (73)	171	656	1,948	10.3	12.0	13.9
General Industrial Machinery (74)	637	2,471	7,012	8.7	11.9	13.3
Office Machinery (75)	1,866	8,515	30,727	32.5	46.5	47.2
Telecommunications (76)	3,150	13,707	41,013	43.0	61.8	71.8
Electric Machinery (77)	5,338	13,575	42,390	33.3	27.6	25.2
Road Vehicles (78)	1,759	10,020	14,544	25.1	42.0	35.6
Other Transport Machinery (79)	854	2,271	4,906	8.4	15.8	19.3
Misc. Manufactures (8)	803	3,269	6,870	4.0	4.6	3.7
All Above Products	17,693	63,827	165,549	19.5	24.1	25.1
OECD EUROPE						
Power Generating Equipment (71)	5,221	14,660	19,548	37.8	43.0	40.7
Special Industry Machinery (72)	4,620	10,893	12,298	25.3	22.2	24.2
Metal Working Machinery (73)	905	2,898	2,929	20.2	19.7	22.1
General Industrial Machinery (74)	3,089	11,312	15,765	13.9	17.6	19.2
Office Machinery (75)	10,237	24,316	35,871	35.7	33.9	33.3
Telecommunications (76)	6,455	21,567	38,088	47.4	51.8	64.1
Electrical Machinery (77)	7,518	19,836	29,480	24.2	24.1	22.0
Road Vehicles (78)	14,581	37,345	61,005	29.8	25.0	28.8
Other Transport Machinery (79)	3,911	8,399	15,863	32.6	22.7	44.5
Misc. Manufactures (8)	1,590	5,284	7,937	2.1	2.4	2.9
All Above Products	58,128	156,510	238,783	21.5	20.6	23.4
NORTH AMERICA						
Power Generating Equipment (71)	5,092	7,739	13,655	37.7	34.4	39.5
Special Industrial Machinery (72)	3,162	4,098	5,678	24.1	21.5	21.4
Metal Working Machinery (73)	738	979	2,040	19.1	19.4	22.3
General Industrial Machinery (74)	1,581	4,313	6,628	13.7	18.5	16.6
Office Machinery (75)	6,549	11,004	26,655	41.9	32.6	33.7
Telecommunications (76)	9,625	14,532	23,902	43.9	53.7	54.9
Electrical Machinery (77)	5,407	10,642	19,902	23.3	23.6	18.3
Road Vehicles (78)	23,277	30,014	42,792	28.4	29.8	30.0
Other Transport Equipment (79)	2,965	4,767	5,647	45.4	45.7	43.1
Misc. Manufactures (8)	408	2,504	5,175	0.7	2.5	3.1
All Above Products	58,084	90,591	152,073	23.5	23.4	22.9

Source: *Computed from UN COMTRADE statistic.*

One striking point reflected in these statistics concerns the "dynamism" in East Asian parts trade, as reflected in its changing relative importance in all imports. From 1985 to 1996 the share of components in all SITC 7 and 8 product imports rose by about six percentage points, and it now accounts for 25% of all trade in these goods. As such, East Asian trade in parts went from levels that were relatively less to relatively more important than those in OECD Europe and North America (both had a share of about 23%) over this interval. The striking growth in East Asian trade in parts is most evident in the telecommunications group (SITC 76), where imports rose from US$3 billion in 1985 to over US$41 billion in 1996, while the value of "parts trade" in electronic machinery experienced a slightly smaller (eightfold) increase over this same interval.[6] Table 2 also shows that the growth in the relative importance of East Asia's components trade, as reflected in its increased share of all imports, occurred for all but two SITC groups (electrical machinery and miscellaneous manufactures).

East Asian global exports of components grew at an annual rate of 15% over the 1984–96 period—more than four points above the growth rate for all goods (10.6%). However, exports of components to other East Asian markets grew at a considerably faster than average pace (about 20.9%). As a result, the share of all parts and components exports destined for regional markets almost doubled, rising from 25% to 46%. Table 3 provides additional information on the extraordinary expansion of Asian trade in components. The top two rows show the value of East Asian exports of these goods, to the region and globally, for selected years during the 1984–96 period. These rows are followed by two rows of statistics showing the corresponding annual growth rates of regional and global exports. Finally, the last (fifth) row of the table shows the share of global exports destined for East Asian intratrade. To help assess the implications of these statistics, the table provides similar information for other broad product groups like foodstuffs, agricultural raw materials, chemicals, and transport and machinery equipment. Several important points emerge from these statistics:

- Asian global exports of components increased more than fivefold over the 1984–96 period, while total exports of all goods rose by a factor of approximately 3. However, the value of component exports to the region grew by a factor of about 10, which was roughly double that for all regional trade.
- Over the 1984–96 period components recorded the fastest annual growth rates for all the major product groups in both regional (20.9%) and global exports (15%). These rates were approximately three points per year higher than those for all manufactures and five to six points higher than for all goods. In short, trade in parts and components was increasing dramatically in relative importance.

Table 3: *East Asian Global and Regional Trade Trends for Parts and Components and Other Major Product Groups*

Year	All Items	Foods & Feeds	Agricultural Materials	Mineral Fuels	Ores & Metals	All Manufactures	Of which: Chemicals	Transport & Machinery	Other Manufactures	Parts & Components
					Value of exports to East Asia in terms of US$ million					
1984	110,992	8,888	5,655	25,081	3,201	66,391	6,516	28,595	31,280	8,457
1988	197,831	15,920	9,383	17,207	6,832	146,253	14,513	64,616	67,124	23,940
1992	332,906	21,883	8,968	27,540	7,603	263,444	22,944	120,991	119,509	42,258
1996	557,338	30,899	11,889	35,521	13,174	459,287	41,057	248,477	169,753	82,487
					Value of exports to the world in terms of US$ million					
1984	347,247	20,916	10,442	34,480	5,485	268,948	12,853	138,958	117,138	32,983
1988	563,168	30,172	15,694	21,505	9,275	476,607	25,570	253,374	197,662	73,106
1992	792,603	39,063	14,310	34,213	10,010	684,126	39,071	372,509	272,546	108,885
1996	1,157,622	51,710	19,170	43,275	17,313	1,010,745	67,289	591,553	351,880	178,547
					Annual growth rate of exports to East Asia (%)					
1992–96	13.75	9.00	7.30	6.57	14.73	14.91	15.66	19.71	9.18	18.20
1988–96	13.82	8.64	3.00	9.48	8.55	15.47	13.88	18.34	12.29	16.72
1984–96	14.39	10.94	6.39	2.94	12.51	17.49	16.58	20.43	15.14	20.90
					Annual growth rate of exports to the world (%)					
1992–96	9.93	7.26	7.58	6.05	14.68	10.25	14.56	12.26	6.60	13.16
1988–96	9.42	6.97	2.53	9.13	8.11	9.85	12.86	11.18	7.48	11.81
1984–96	10.55	7.83	5.19	1.91	10.05	11.66	14.79	12.83	9.60	15.11
					Share of East Asian intra-trade in total exports of the group (%)					
1984	32.0	42.5	54.2	72.7	58.4	24.7	50.7	19.2	26.7	25.6
1988	35.1	52.8	59.8	80.0	73.7	30.5	56.8	25.5	34.0	32.7
1992	42.0	56.0	62.7	80.5	76.0	38.5	58.7	32.5	43.8	38.8
1996	48.1	59.8	62.0	82.1	76.1	45.4	61.0	42.0	48.2	46.2

Source: *Computed from UN COMTRADE statistics.*

- Sizeable shifts occurred in the share of all 10 product group exports going to the region over the 1984–96 period, but this shift was most pronounced for transport and machinery equipment, other manufactures, and components. In 1984, 26% of all component exports went to other regional countries, but 12 years later this share had increased to 46%.

As Table 4 further illustrates, the dramatic expansion of East Asian trade in components was largely an intra-regional phenomenon that was unmatched by any other geographic groups of countries. For example, over the 1985–96 period the share of East Asian imports of components from Europe held relatively stable in the 14% to 16% range, while North America's share of both imports and exports fell by 14 percentage points.

Table 4: *The Origins and Destinations of East Asian Countries Trade in Parts and Components*

	Imports			Exports		
Region	**1985**	**1990**	**1996**	**1985**	**1990**	**1996**
	Value of Trade in US$ million					
WORLD	17,693	63,827	165,549	33,152	89,469	178,547
	Share of Trade Destined for or Originating in					
WORLD	100.0	100.0	100.0	100.0	100.0	100.0
ASIA	43.5	59.4	58.5	31.9	41.2	49.6
Of which:						
Japan	28.7	29.1	25.6	1.7	3.2	5.3
Other East Asia	14.2	29.3	32.0	24.2	34.9	41.3
South Asia	0.2	0.2	0.3	1.5	1.0	1.0
Oceania	0.5	0.6	0.6	3.8	1.9	1.5
Former Soviet Union	--	0.1	0.1	0.8	0.3	0.4
NORTH AMERICA	37.1	25.1	23.3	44.9	35.3	29.5
Of which:						
United States	36.2	24.3	22.3	41.8	32.4	27.2
EUROPE	16.6	15.0	17.0	14.0	19.2	16.2
Of which:						
European Union	15.5	13.8	16.0	13.1	18.3	15.3
Eastern & Other Europe	1.1	1.3	1.0	0.9	0.9	0.9
LATIN AMERICA	0.2	0.2	0.4	3.1	2.3	2.8
Of which:						
Mercosur	0.1	0.1	0.2	0.8	0.6	1.1
MIDDLE EAST	0.1	0.1	0.3	4.0	1.1	1.0
AFRICA	--	--	--	1.6	0.6	0.4
Of which:						
Sub-Saharan Africa	--	--	--	0.9	0.3	0.2
UNSPECIFIED COUNTRIES	2.4	0.2	0.7	0.5	0.2	0.4

Source: *Computed from UN COMTRADE statistics.*

The Asian group of countries appears to be the big gainer, but the table shows that other sub-regional Asian groups (e.g., South Asia, Oceania) did not participate in the expansion (i.e., their East Asian import shares remained low and static). In contrast, the share of East Asian imports originating in other (non-Japanese) East Asian countries more than doubled, to about 32%.

A similar pattern is evident in the export statistics. The 1985–96 shares of East Asia's components exports going to Europe remained essentially stable, while North America's shares experienced a sizeable decline. Within the Asian regional subgroups the relative importance of South Asia's share was low and remained virtually unchanged, while the share of East Asian exports destined for Oceania declined by more than 2 points from its 1985 level (3.8%). As with imports, it was in the other (non-Japanese) East Asian countries that regional exporters registered their greatest relative trade gains. In short, as measured by either exports or imports, the driving force behind the expansion of production sharing in East Asia came largely from other East Asian countries.

Table 5 provides additional information on the pattern of intratrade by showing the 1996 value of component exports from each East Asian country to its specific regional destination. In addition, the middle third of the table shows bilateral regional percentage trade balances for components (i.e., the value of exports less imports expressed as a percentage of total exports), while the lower third of the table shows trade intensity indices for components. These latter statistics measure the tendency for two countries to trade more or less heavily than expected given their global importance in trade. If the intensity index is above unity, the countries are said to have a greater-than-expected bilateral trade, based on their share in world trade. If the index is below unity, the intensity of trade is lower than expected.[7]

Three interesting points emerge from the regional trade matrix:

- Even though, as Table 4 shows, exports from other regional countries are rapidly expanding, Japan is still the largest components exporter in Asia, accounting for US$32 billion (or approximately 40%) of intratrade in these goods. Japan also has a positive trade balance with every East Asian country that appears to specialize in the assembly of Japanese-produced components. This pattern conforms to expectations, since Japanese average annual wage costs (US$40,404, as reported by the World Bank 1999) are more than 10 times higher than Malaysia's and 16 times higher than the Philippines'.
- Again, as expected, low-wage countries like Thailand, the PRC, Malaysia, and the Philippines generally have sizeable negative intra-regional trade balances, while the results are mixed for higher-wage countries like Korea, Singapore, and Taiwan.[8] We have not been able to find a satisfactory explanation for Hong Kong's persistent large negative trade balances with every one of the other Asian countries, but (as noted) we suspect that there may be a problem with the proper recording of transit trade.

Table 5: *The Matrix of 1996 Intra-Trade and Trade Balances in Parts and Components among East Asian and Other Major Countries*

	Exporting Country									
Partner	**China**	**Hong Kong**	**Indonesia**	**Korea**	**Malaysia**	**Philippines**	**Singapore**	**Taiwan**	**Thailand**	**Japan**
	Values of Exports of Components in US$ Million									
China	--	2,152	4	828	84	10	671	110	32	3,548
Hong Kong	2,686	--	62	535	564	154	1,872	4,154	287	3,500
Indonesia	131	18	--	334	124	19	--	305	64	2,135
Korea Rep.	494	58	28	--	151	42	465	372	79	4,445
Malaysia	205	77	128	314	--	94	5,436	632	363	3,959
Philippines	64	68	6	250	154	--	603	425	49	2,197
Singapore	475	331	615	524	4,201	175	--	863	1,597	3,635
Taiwan	345	95	24	166	292	104	644	--	156	4,514
Thailand	83	94	39	226	732	611	1,612	623	--	4,157
Japan	2,126	158	310	1,053	1,186	686	1,230	1,740	945	--
All East Asia	6,609	3,051	1,216	4,230	7,488	1,895	12,533	9,244	3,572	32,090
World	10,706	4,227	1,767	11,917	12,504	3,570	23,558	21,521	7,071	80,939
	Trade Balance in Components as a Share of Component Exports (%)									
China	--	−222.0	−2,155.2	29.2	−213.5	−496.7	−29.3	−159.3	−760.5	25.4
Hong Kong	51.4	--	70.3	78.6	−11.5	−189.2	39.7	94.7	54.7	92.4
Indonesia	911.3	−117.1	--	91.5	−88.4	12.5	--	88.6	50.2	82.8
Korea Rep.	−94.2	−1,056.7	−308.3	--	−197.1	−646.7	−0.1	46.7	−142.4	71.4
Malaysia	21.8	−570.9	66.9	63.2	--	−129.9	21.2	56.3	−138.7	74.0
Philippines	37.4	−136.8	−37.8	89.2	−20.1	--	60.1	87.3	−607.6	56.9
Singapore	−38.3	−466.6	63.8	50.4	36.4	−252.9	--	62.5	57.2	78.1
Taiwan	−310.3	2,750.9	−543.9	−109.1	−151.9	−174.3	−82.1	--	−198.9	55.1
Thailand	−135.0	−116.8	−38.5	64.1	−11.5	80.6	6.2	74.3	--	77.3
Japan	−153.0	−3,437.1	−741.4	−376.1	−313.0	−369.5	−356.6	−120.1	−552.3	--
All East Asia	−53.1	−513.2	−172.5	−55.5	−45.3	−179.2	−22.0	−120.6	41.6	67.9
World	−60.4	−517.7	−276.0	−16.5	−53.5	−186.7	−12.6	64.2	−101.5	70.7

(Continued)

Table 5: (*Continued*)

Partner	Exporting Country									
	China	Hong Kong	Indonesia	Korea	Malaysia	Philippines	Singapore	Taiwan	Thailand	Japan
	Trade Intensity Index for Components									
China	--	17.7	0.1	2.5	0.2	0.1	1.0	0.2	0.2	1.5
Hong Kong	5.8	--	0.8	1.1	1.0	1.0	1.8	4.6	0.9	1.0
Indonesia	1.1	0.4	--	2.6	0.9	0.5	--	1.3	0.8	2.4
Korea Rep.	2.0	0.6	0.7	--	0.5	0.5	0.9	0.8	0.5	2.4
Malaysia	0.6	0.6	2.3	0.8	--	0.8	7.2	0.9	1.6	1.5
Philippines	0.4	0.9	0.2	1.2	0.7	--	1.5	1.2	0.4	1.6
Singapore	1.0	1.8	8.1	1.0	7.7	1.1	--	0.9	5.2	1.0
Taiwan	2.5	1.8	1.1	1.1	1.8	2.3	2.1	--	1.7	4.3
Thailand	0.3	0.9	1.0	0.8	2.5	7.4	2.9	1.3	--	2.2
Japan	5.1	0.9	4.6	2.3	2.4	5.0	1.3	2.1	3.4	--

Source: *Computed from UN COMTRADE Statistics.*

- Japan's trade intensity indices are at or above unity for every East Asian country, which indicates that Japan's production-sharing operations are quite widely dispersed in the region. Similarly, with the exception of Hong Kong, trade intensity indices for other East Asian exports to Japan are also above unity. The fact that the indices for Indonesia and the Philippines are between 4.6 and 5 suggests that some component products may be manufactured using labor-intensive production processes.[9]

Available statistics (through 1997) indicate that the recent Asian economic slowdown has not had a major detrimental impact on regional trade in parts and components. Although Japanese exports declined slightly in 1997, shipments from most other East Asian countries increased by between 9% and 16%. Since Asian trade in components has been such a dynamic leading sector in regional imports and exports, there is an obvious interest in monitoring future changes in this exchange.

A question of obvious interest concerns the influence of the recent Asian economic recession on intra-regional trade in components. Here, two alternative scenarios exist. First, if assembled goods that use imported components are destined primarily for other Asian countries' markets, a general decline in effective demand would probably cause a marked reduction in the trade in parts. However, if the assembled goods were largely destined for non-Asian markets, the reduction would likely be less severe. Indeed, the economic situation within East Asia may increase the importance of expanding these and other types of exports to build up foreign exchange reserves. This scenario could lead to an increase in intratrade in components.

Unfortunately, at the time this report was completed (March 1999), 1996 was the last full year for which UN COMTRADE statistics were available for all the East Asian countries. However, 1997 export statistics were available for Japan, Thailand, Singapore, the Republic of Korea, Hong Kong and the PRC, and these data should provide some information on the initial impact of the slowdown. In general, there is little evidence of an extensive adverse impact on trade in components. Regional exports of the PRC, Indonesia, and Thailand all increased by at least 16% from 1996 to 1997, while those of Singapore and Malaysia rose by 9% to 10%. In contrast, component exports from Japan and Hong Kong fell slightly (by 2% to 3%) from their 1996 levels.[10] Given that East Asian trade in components has been such a dynamic leading sector in regional imports and exports, there is an obvious interest in monitoring future changes in this exchange.

4. COMPARATIVE ADVANTAGE IN PRODUCTION SHARING

Traditionally, "revealed" comparative advantage (RCA) indices have been computed using export statistics. These results show whether a country has a comparative advantage in the manufacture of a product. When RCA indices are computed using import statistics for a given component product the results should indicate whether a country has a comparative advantage or disadvantage in assembly operations.

Economists often use the RCA concept to identify countries whose factor endowments make it advantageous for them to specialize in the production of a good. The RCA index for country i in the production of product j has been defined as

$$\mathrm{RCA}_{ij}^{p} = [x_{ij}/X_{j}]/[x_{wj}/X_{w}] * 100 \tag{1}$$

where x_{ij} and x_{wj} represent the value of j exported by country i and the world, while X_j and X_w are total exports by j and the world, respectively. The index has a relatively simple interpretation. If its value exceeds unity the country is said to have a comparative advantage in the *production* of product *j*; in contrast, if the RCA index is below one, the country is at a comparative disadvantage in the good.[11] However, equation 2 must be used with some caution, since domestic measures (e.g., local subsidies) that have nothing to do with comparative advantage or foreign trade barriers can impart a bias in the index.[12]

This study employs a variant of equation 2 to also identify countries that appear to have a comparative advantage in further upstream operations—that is, the *assembly* of the next stage of the manufacturing process. Specifically, the revealed comparative advantage of country *i* in the assembly of product *j* is

$$\mathrm{RCA}_{ij}^{a} = [m_{ij}/M_{j}]/[m_{w}j/M_{w}] * 100 \tag{2}$$

where the ms represent imports but otherwise correspond to the terms in equation 2. The reasoning behind this proposition is relatively straightforward. Parts and components typically have no general end use in themselves but are exchanged for further assembly into a product that does.[13] Therefore, it follows that countries with above-average import shares for components have a comparative advantage in the assembly operation.

The structure of a country's overall RCA profile can sometimes provide useful information about its potential for the development of future assembly or manufacturing operations when examined in a "stages" context. Relatively "immature" countries whose potential for future development may be large typically have highly skewed RCA profiles with few indices above unity. These profiles tend to flatten out as the capacity of the country to engage in assembly or production increases.

In addition to using import statistics for estimating RCA indices, we propose to further expand the "informational content" of this measure by analyzing the structure of index values across different component products. This extension could convey information on the "stage of development" or "maturity" of a country in the production of components, or in their upstream assembly. Several previous studies show that countries that are just moving into the production of manufactures have different RCA profiles for these goods than nations where these operations are broadly based and expanding.[14] We propose to extend this concept to trade in components.

This line of analysis argues that useful information on a country's comparative advantage may be observed in a profile of its RCA indices for each component product if they are ranked from their highest to their lowest value. Several different conceptual profiles are illustrated in Figure 1. A country that either (1) is an early stage of development as either a manufacturer or an assembly point for components, or (2) has economically advanced (grown sufficiently high cost and wealthy) to the point that it is "exiting" assembly operations may have a pronounced comparative advantage in a limited number of component products, but practically no advantage in most others.[15] Countries falling in this category would have a profile similar to (*f*) in Figure 1. As the capacity for production or assembly improves, the RCA indices for other component products, which previously were low, begin to rise and gradually level out. That is, the country's RCA profiles might shift from, say, *e* or *f* toward *a*. This schema can convey useful information about the *potential* for broadening future operations. However, we recognize that adverse developments like unanticipated wage inflation might prevent a country from achieving its potential capacity.

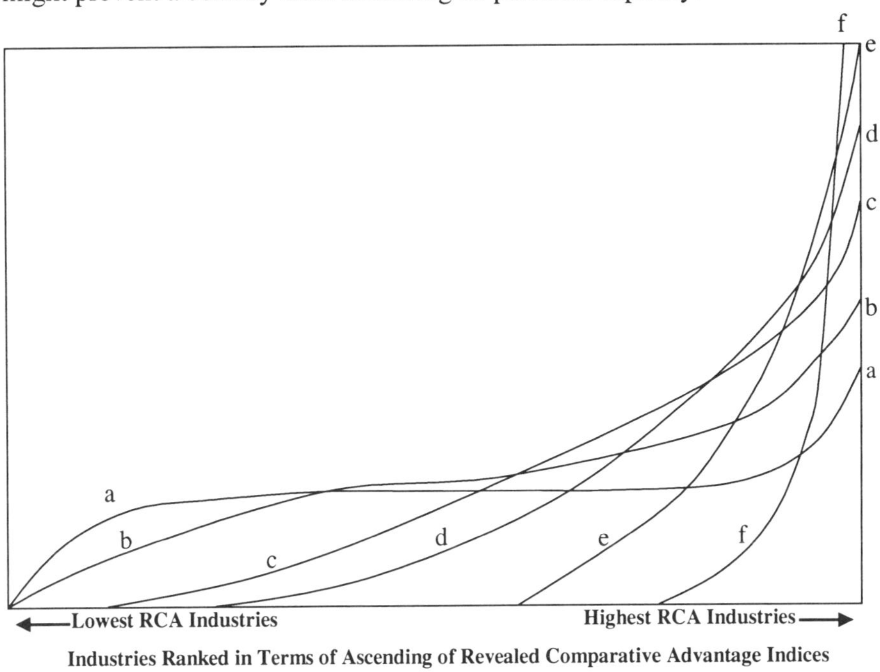

Figure 1: *A Stages Approach for the Analysis of Revealed Comparative Advantage Index Profiles.*

Our intention is to empirically examine the structure of East Asian countries' RCA profiles for the production and assembly of parts and com-onents within this "stages" approach To assist in this effort we compare the East Asian profiles with those for Mexico, Hungary and Poland, all countries with extensive assembly operations, and with the United States, which is the world's largest exporter of components.

Assembly operations are tending to migrate to the relatively low-wage Asian countries, while countries like Japan, Singapore, and Taiwan have increased their specialization in the manufacture of components. Overall, the revealed comparative advantage profiles conform to predictions based on factor proportions theory. Low-wage countries have a disproportionately high revealed comparative advantages in assembly operations (most of which are assumed to be labor intensive in nature), while the wealthier Asian countries have high RCAs for the production of components.

Table 6 shows each East Asian country's RCA indices, which were calculated using their 1996 import statistics. As noted, these data should indicate whether or not the country has a comparative advantage in the assembly of each product.[16] Similarly, Table 7 shows 1996 RCA indices computed using export data that indicate whether these countries have a comparative advantage in the production of components. To help summarize the implications of this information, two additional statistics are shown (see the memo item). The first is an aggregate RCA index estimated for all component products taken as a group, while the second is a count of the number of individual components in which each country has a comparative advantage. Finally, the table also includes (as a comparison) similar data for the United States, which is both the world's largest importer and its largest exporter of parts and components (see Appendix Table 1).

The pattern reflected in these statistics generally conforms to expectations based on factor proportions theory. Japan, which has the highest unit-wage costs in the region, only has a comparative advantage in the assembly of 8 of the 60 component product groups, 3 less than the United States. Furthermore, Japan lost a comparative advantage in the assembly of 7 component groups in which its 1985 RCA indices exceeded unity (see Appendix Table 2). This contrasts markedly with the results for low-wage countries like Indonesia, Malaysia, and Thailand, which have a comparative advantage in the assembly of about half of the 60 product groups. With the exception of the Philippines, where the quality of the underlying data is suspect, relative wage costs appear to be an important determinant of where assembly operations have been and will be located.[17]

Table 6: *The Revealed Comparative Advantage of East Asian Countries in Assembly Operations as Reflected in Their 1996 Import Statistics*

Component Product Group*	China	Hong Kong	Indonesia	Rep. of Korea	Malaysia	Philippines	Singapore	Thailand	Taiwan	Japan	USA
Steam boilers	2.09	1.26	8.56	1.72	2.36	1.48	0.86	0.23	0.19	0.63	0.39
Aircraft engines	0.30	0.02	3.83	0.60	4.29	0.46	4.60	0.29	0.06	0.59	1.25
Outboard motors	0.16	0.73	1.30	0.57	1.56	0.26	0.39	0.24	0.35	0.15	1.30
Outboard motors, nes	1.79	0.22	0.90	1.94	1.60	0.41	1.60	0.61	0.27	0.38	0.91
Combustion engines, nes	0.77	0.58	0.91	1.19	0.30	0.39	0.64	1.63	0.59	0.37	0.98
Engines and motors, nes	0.22	0.23	1.18	0.92	0.63	0.16	0.42	0.15	0.06	1.15	1.37
Rotating electric motors	2.77	1.53	1.51	1.21	1.65	0.81	1.11	2.52	0.86	0.56	0.79
Water turbines	0.62	0.01	1.93	0.07	0.22	0.13	0.08	0.01	0.16	0.14	0.27
Cultivating equipment	0.05	0.01	0.11	0.56	0.13	0.27	0.27	0.16	0.07	0.22	1.05
Harvesting machinery	0.17	0.08	0.12	0.83	0.17	0.09	0.12	0.04	0.10	0.18	0.71
Dairy machinery	0.20	0.04	0.53	0.52	0.18	0.11	0.21	0.37	0.04	0.23	0.20
Wine making machinery	7.79	0.13	0.99	0.15	0.63	0.34	0.30	1.79	0.52	0.05	0.35
Agricultural machinery, nes	0.92	0.04	1.87	1.28	0.62	1.14	0.17	0.18	0.34	0.41	0.92
Construction machinery	1.46	0.39	3.65	1.36	2.16	0.64	5.39	1.09	0.18	0.38	0.52
Spinning machinery	2.14	0.28	5.17	1.36	0.91	0.67	0.41	1.80	2.28	0.36	0.79
Looms and knitting machinery	1.40	1.12	2.18	0.96	1.22	0.45	0.30	0.66	1.85	0.77	0.82
Textile machinery, nes	1.59	0.96	2.13	0.48	1.30	1.70	0.62	0.70	0.62	0.30	0.48
Paper making machinery	0.53	0.10	12.12	0.81	0.32	0.85	0.35	2.61	0.15	0.34	0.73
Bookbinding machinery	0.04	0.21	0.30	0.25	0.07	0.08	0.32	1.42	0.08	0.61	1.15
Printing machinery	0.66	0.49	0.54	0.39	0.54	0.70	1.53	0.36	0.30	0.68	0.71
Grain milling machinery	1.06	0.10	4.30	0.48	0.44	3.32	0.45	0.31	0.11	0.19	0.42
Food processing machinery	0.00	2.05	0.00	0.00	0.00	15.61	0.00	0.00	0.00	0.00	0.00
Machine for special industries	1.32	0.48	1.67	0.83	2.36	0.57	1.24	0.63	0.66	0.43	0.80
Mineral working machinery	1.30	0.27	13.47	0.38	2.54	4.81	0.70	1.86	0.38	0.33	0.52
Special industry machines, nes	1.76	0.42	1.04	2.59	1.80	5.22	1.98	1.09	1.18	0.98	0.80
Machines for metal working	1.19	0.30	0.60	1.69	1.38	0.85	1.01	1.04	0.63	1.08	0.93
Foundry equipment	5.38	0.18	1.10	1.49	0.40	0.27	0.67	0.17	0.72	0.23	1.58

(Continued)

Table 6: (*Continued*)

Component Product Group*	China	Hong Kong	Indonesia	Rep. of Korea	Malaysia	Philippines	Singapore	Thailand	Taiwan	Japan	USA
Rolling mill parts	2.18	0.03	1.94	1.79	0.63	0.80	0.08	5.56	1.38	0.51	1.32
Refrigerating equipment	0.67	0.47	3.19	0.19	0.54	0.92	1.27	0.89	0.33	0.40	0.46
Pumps for liquids	0.50	0.16	1.93	1.17	0.27	0.51	0.85	0.38	0.32	0.51	0.87
Centrifuges and filters	0.88	0.50	1.40	1.58	1.26	0.91	0.88	0.77	1.19	0.73	0.80
Fork lift trucks	0.12	0.02	0.93	15.38	0.33	0.64	3.67	1.29	0.03	0.08	0.63
Lifting and loading machines	1.14	0.45	4.41	0.88	0.70	0.92	1.83	0.74	0.34	0.47	0.87
Power hand tools	0.19	0.17	2.50	0.37	0.75	1.33	1.56	0.18	0.25	0.36	1.07
Packing machinery	0.76	0.47	0.77	0.62	0.23	0.74	0.24	1.20	0.23	0.37	0.72
Non-electric machinery, nes	0.88	0.24	1.23	1.49	1.05	0.49	1.60	1.42	1.14	0.73	0.21
Office and adding machinery	0.87	1.04	0.06	0.52	1.63	2.28	2.64	1.57	0.43	1.29	1.32
Telecommunications equipment	1.63	1.79	1.94	1.02	1.41	2.21	1.47	0.99	0.55	1.18	0.84
Electric power machinery	3.21	3.80	0.52	1.75	2.94	2.40	2.09	1.82	0.49	0.56	0.48
Switch gear	1.15	0.84	0.97	1.04	2.25	1.26	1.73	1.70	1.32	0.61	0.84
Domestic electrical equipment	0.89	0.91	0.50	0.32	0.40	0.14	0.66	0.11	0.11	0.40	0.98
Electro thermic appliances	2.14	2.53	0.64	0.39	1.27	0.91	0.60	1.22	0.42	0.66	0.83
Electronic components, nes	0.72	1.16	0.14	2.00	17.38	25.27	3.44	6.40	1.97	0.14	0.32
Electronic accumulators	3.50	1.02	1.63	0.71	0.31	0.69	0.53	0.21	0.27	0.16	0.32
Electric lamps and bulbs	1.04	0.34	2.70	1.94	1.12	2.96	0.22	1.30	0.42	0.36	0.44
Electrical machinery, nes	0.93	0.01	0.23	0.31	1.08	2.17	0.33	1.23	0.55	0.42	0.19
Motor vehicles and accessories	0.33	0.14	1.51	0.41	0.18	0.34	0.25	1.69	0.51	0.27	0.99
Carriages and cycles	1.31	1.23	13.34	0.20	1.00	0.77	0.97	2.59	2.66	0.69	0.62
Non-motor vehicles	0.11	0.04	0.56	0.15	0.72	0.16	0.54	0.61	0.16	0.20	0.36
Railroad vehicles	0.28	0.19	1.67	2.81	0.70	0.30	0.11	0.19	0.47	0.20	0.96
Aircraft	0.64	0.12	0.87	0.99	0.33	0.68	1.58	1.41	0.22	0.43	0.69
Chairs	0.05	0.04	0.06	0.16	0.36	0.12	0.09	0.08	0.39	0.76	1.76
Other furniture	0.10	0.37	0.07	0.22	0.17	0.45	0.52	0.17	0.48	1.61	1.00
Drawing machinery	0.59	0.37	0.28	2.38	2.83	0.23	1.50	0.37	0.72	2.17	0.76

Still cameras, nes	6.38	4.48	0.93	1.25	3.51	8.40	1.02	1.49	3.16	1.75	0.39
Cameras under 16 mm	0.59	0.31	0.84	0.56	0.72	0.97	1.69	0.17	0.92	0.77	0.88
Cameras under 16 mm, nes	0.23	0.53	0.12	0.35	1.29	14.14	1.73	0.04	0.33	0.39	0.57
Unmounted optical elements	0.30	0.53	0.39	0.57	0.36	0.06	1.19	0.19	0.39	2.46	0.56
Clocks and watches	4.28	7.75	0.06	1.52	2.19	4.62	1.69	4.23	0.43	0.62	0.12
Umbrellas and canes	8.70	5.71	0.83	0.46	0.47	1.53	0.16	2.25	0.24	0.71	0.45
Memo Item:											
RCA Index for All Components	1.00	0.86	1.44	0.89	1.73	2.27	1.40	1.54	0.67	0.76	0.93
Number of Product Groups with An RCA Over Unity	26	14	30	24	27	19	25	27	9	8	11

**See Table 1 for the SITC Revision 2 classification number of each component product group.*

Source: *Computed from United Nations COMTRADE Statistics.*

Table 7: *The Revealed Comparative Advantage of East Asian Countries in the Production of Components as Reflected in Their 1996 Export Statistics*

Component Product Group*	China	Hong Kong	Indonesia	Rep. of Korea	Malaysia	Philippines	Singapore	Thailand	Taiwan	Japan	USA
Steam boilers	0.34	0.02	0.03	0.43	0.11	0.46	0.15	0.00	0.15	2.07	1.78
Aircraft engines	0.01	0.00	0.00	0.04	5.06	10.43	3.35	0.01	0.02	0.06	2.83
Outboard motors	0.05	0.78	0.00	0.03	0.06	0.00	0.10	0.01	0.00	5.62	2.43
Outboard motors, nes	0.05	0.00	0.01	0.61	0.08	0.00	0.67	0.68	0.00	1.56	0.95
Combustion engines, nes	0.15	0.15	0.40	0.12	0.05	0.01	0.34	0.11	0.15	1.85	1.41
Engines and motors, nes	0.04	0.00	0.01	0.11	0.02	0.02	0.12	0.05	0.11	0.33	2.69
Rotating electric motors	0.97	0.98	0.11	0.24	0.28	0.02	1.23	0.40	0.77	1.38	1.29
Water turbines	0.07	0.00	0.00	0.03	0.06	0.00	0.04	0.01	0.01	0.68	0.71
Cultivating equipment	0.21	0.00	0.03	0.26	0.02	0.00	0.08	0.03	0.73	0.10	1.17
Harvesting machinery	0.19	0.00	0.01	0.07	0.03	0.00	0.05	0.01	0.60	0.18	1.63
Dairy machinery	0.01	0.25	0.00	0.00	0.01	0.00	0.17	0.00	0.04	0.02	0.88
Wine making machinery	0.18	0.00	0.01	0.00	0.97	0.00	1.75	0.03	0.87	0.02	1.35
Agricultural machinery, nes	0.09	0.01	0.00	0.04	0.12	0.00	0.04	0.14	0.73	0.01	2.72
Construction machinery	0.09	0.12	0.42	0.25	0.38	0.00	2.30	0.10	0.01	0.07	5.29
Spinning machinery	0.18	0.21	0.05	0.85	0.12	0.03	0.22	0.03	0.68	0.84	0.74
Looms and knitting machinery	0.31	0.23	0.00	0.71	0.06	0.00	0.27	0.01	0.92	0.71	0.23
Textile machinery, nes	0.41	1.02	0.02	1.52	0.29	0.51	0.38	0.03	0.55	0.37	0.77
Paper making machinery	0.03	0.03	0.20	0.16	0.04	0.07	0.17	0.07	0.44	0.29	1.06
Bookbinding machinery	0.01	0.17	0.02	0.04	0.02	0.00	0.12	0.00	0.27	0.17	0.48
Printing machinery	0.06	0.40	0.00	0.05	0.08	0.13	0.37	0.12	0.23	0.28	1.17
Grain milling machinery	0.47	0.47	0.33	0.06	0.68	0.02	0.41	0.09	1.37	0.15	1.15
Food processing machinery	0.00	4.71	0.00	0.00	0.00	1.02	0.00	0.00	0.00	0.00	0.00
Machine for special industries	0.10	0.25	0.00	0.37	0.22	0.00	0.50	0.04	1.86	0.49	0.94
Mineral working machinery	0.26	0.09	0.02	0.16	0.29	0.09	0.21	0.04	0.23	0.18	1.44
Special industry machines, nes	0.14	0.41	0.10	0.67	0.38	0.14	0.72	0.05	0.45	0.87	1.83
Machines for metal working	0.27	0.20	0.00	0.18	0.16	0.48	0.33	0.20	0.70	0.90	1.96

Foundry equipment	0.15	0.26	0.17	0.58	0.06	0.00	0.51	0.01	0.66	1.39	0.65
Rolling mill parts	0.37	0.00	0.02	0.21	0.06	0.01	0.03	0.07	0.29	1.76	0.47
Refrigerating equipment	0.19	0.13	1.32	0.55	0.53	0.01	0.76	0.47	0.45	0.77	1.45
Pumps for liquids	0.17	0.21	0.00	0.10	0.08	0.09	0.36	0.02	0.21	0.65	1.48
Centrifuges and filters	0.20	0.23	0.04	0.14	0.18	0.16	0.36	0.20	0.84	0.91	1.44
Fork lift trucks	0.19	0.00	0.01	0.59	0.07	0.00	4.03	0.06	3.16	0.08	2.07
Lifting and loading machines	0.24	0.07	0.24	0.48	0.14	0.07	0.93	0.17	0.15	0.73	1.78
Power hand tools	0.17	0.05	0.17	0.06	0.17	0.00	0.70	0.02	2.18	0.71	1.52
Packing machinery	0.03	0.18	0.00	0.05	0.03	0.01	0.11	0.03	0.28	0.13	0.95
Non-electric machinery, nes	0.19	0.24	0.03	0.25	0.07	0.17	0.94	0.15	1.34	0.81	0.72
Office and adding machinery	0.05	1.65	0.48	0.24	2.94	3.46	3.34	2.06	2.71	1.50	1.54
Telecommunications equipment	1.07	1.20	0.83	1.13	1.90	1.52	1.73	1.07	1.08	1.13	1.24
Electric power machinery	0.93	5.06	0.07	1.02	1.64	0.14	1.70	2.24	2.81	1.05	1.53
Switch gear	0.62	1.35	0.24	0.42	1.36	0.74	1.24	1.22	1.33	1.41	1.09
Domestic electrical equipment	0.28	2.55	0.00	0.73	0.24	0.04	0.47	0.17	0.65	0.09	2.17
Electro thermic appliances	0.70	10.04	0.01	3.12	1.03	0.00	0.64	0.66	1.64	0.45	1.10
Electronic components, nes	0.34	2.49	0.23	1.17	4.17	2.45	4.19	3.33	0.74	3.74	0.91
Electronic accumulators	0.39	7.92	0.65	0.15	0.10	0.05	0.38	0.49	1.01	0.82	1.36
Electric lamps and bulbs	0.35	0.31	0.02	0.27	0.22	0.27	0.44	0.23	1.37	0.30	1.40
Electrical machinery, nes	0.25	0.00	0.01	10.60	0.26	0.00	0.36	1.30	4.10	0.52	1.08
Motor vehicles and accessories	0.08	0.00	0.07	0.26	0.04	0.44	0.13	0.09	0.38	1.26	1.60
Carriages and cycles	1.64	0.00	2.67	0.09	1.18	0.65	1.61	2.95	6.76	2.31	0.42
Non-motor vehicles	0.12	0.01	0.01	0.42	0.16	0.14	0.55	0.07	1.14	0.03	0.71
Railroad vehicles	0.15	0.00	0.00	0.17	0.01	0.01	0.20	0.00	0.21	0.20	1.85

(Continued)

Table 7: (*Continued*)

Component Product Group*	China	Hong Kong	Indonesia	Rep. of Korea	Malaysia	Philippines	Singapore	Thailand	Taiwan	Japan	USA
Aircraft	0.07	0.00	0.14	0.12	0.38	0.67	0.43	1.54	0.00	0.25	3.21
Chairs	0.06	0.00	0.49	0.14	0.16	0.03	0.03	0.36	0.19	0.34	1.69
Other furniture	0.99	0.09	4.59	0.14	1.14	4.26	0.12	1.28	1.38	0.02	0.54
Drawing machinery	0.15	0.10	0.01	0.21	0.25	0.01	0.75	0.06	0.27	0.95	1.47
Still cameras, nes	1.25	10.22	0.29	0.60	2.36	3.46	1.31	2.85	3.52	3.13	0.97
Cameras under 16mm	0.08	0.00	0.06	0.01	0.52	0.29	0.86	0.00	0.02	0.12	1.62
Cameras under 16mm, nes	0.17	0.01	0.25	0.01	1.03	0.00	2.49	0.01	0.32	0.06	2.68
Unmounted optical elements	0.47	1.00	3.98	0.19	0.13	0.03	1.58	2.39	0.28	0.13	2.75
Clocks and watches	2.06	22.07	0.63	0.30	4.40	3.05	3.02	11.97	1.96	1.28	0.20
Umbrellas and canes	2.90	0.08	0.37	0.03	0.03	0.00	0.13	0.63	18.48	0.11	0.06
Memo Item:											
RCA Index for All Components	0.47	0.90	0.39	0.59	1.15	1.12	1.27	0.95	1.11	1.17	1.54
Number of Product Groups with An RCA Index Over Unite	5	12	4	6	12	8	15	12	19	16	40

* *See Table 1 for the SITC Revision 2 classification number of each component product group.*

Source: *Computed from United Nations COMTRADE Statistics.*

The pattern reflected in the RCAs derived from 1996 export statistics (Table 7) also generally conforms to what is expected. Higher-wage countries like Japan, Singapore, Taiwan, and Hong Kong have the largest number of RCAs exceeding unity (but still less than half the number the United States has), while Indonesia and the PRC only have a comparative advantage in the production of 4 or 5 product groups. In fact, the statistics in Appendix Table 3 show that all of Indonesia's 1985 RCA indices for production operations were below unity.

Table 8 summarizes statistics on the 1985 and 1996 levels and changes in East Asian RCA profiles. Shown here is the percentage of all 60 component product groups in which each Asian country had a 1985 or 1996 comparative advantage in the production or assembly of components. For example, in 1996 the PRC only had a comparative advantage in the production of 8.3% of the 60 component groups, as opposed to 26.7% for assembly operations. For comparison, the table also provides similar information for the United States and Mexico (the latter engages in extensive assembly operations geared toward the U.S. market), and for Hungary and Poland, which appear to be developing as assembly centers for the European Union. Several

Table 8: *The Percentage of All Parts and Components Products in Which East Asian Countries Have a Comparative Advantage*

Country	Country Classification*	Exports-Production Operations		Imports-Assembly Operations	
		1985	1996	1985	1996
East Asia					
China	e	6.7	8.3	36.7	26.7
Hong Kong	e	16.7	21.7	28.3	23.3
Indonesia	c	0.0	6.7	63.3	50.0
Rep. of Korea	d	0.0	10.0	26.7	40.0
Malaysia	c	20.0	20.0	46.7	45.0
Philippines	d	10.0	13.3	66.7	31.7
Singapore	d	25.0	26.7	36.7	41.7
Thailand	c	10.0	20.0	36.7	45.0
Taiwan	e	15.0	31.7	16.7	16.7
Japan	e	21.7	26.7	25.0	13.4
Comparators:					
Mexico	b	18.3	16.7	70.0	70.0
Hungary	d	na	38.3	na	36.7
Poland	d	na	20.0	na	43.3
United States	e	58.3	66.7	16.7	16.7

**The letters shown below attempt to classify the 1996 RCA component import profiles as shown in Figure 1. Countries with an "a" or "b" profile have a broad and diverse comparative advantage in the assembly and export of parts and components while those with an "e" or "f" profile have an advantage in the assembly of very few industries. See the Annex figures for graphs of each of these countries ranked RCA assembly profiles.*

Source: *Computed from the RCA statistics in Table 6 and Table 7.*

noteworthy points are evident in these summary statistics. These comparisons reaffirm the earlier observation that assembly operations for components tend to be established in relatively low-wage Asian countries, while countries like Japan, Taiwan, and Singapore have an advantage in production. Collectively, East Asian countries are strengthening their comparative advantage in the production of components. Nine of the 10 countries increased the percentage of component industries in which they have a production comparative advantage, with Thailand's and Taiwan's shares doubling. A similar pattern is not evident in the corresponding shares for assembly operations, which fell or held constant for 7 of the 10 countries.

As reflected in the RCA statistics, the diversity of assembly operations in middle-and-low income Asian countries has expanded to about the same degree as it has in Poland and Hungary, which have a comparative advantage in about one-third of the 60 component groups. Similarly, the United States and Japan now are at a comparative disadvantage in most assembly operations, but the U.S. comparative advantage in production of components is far broader.

A "stages" approach analysis of Asian RCA profiles shows that Indonesia, Thailand, and Malaysia have the broadest and most mature assembly capacity for components. However, these Asian profiles are less developed than Mexico's, which suggests that the former may still have the capacity to expand these operations. RCA profiles for Singapore, Taiwan, and Japan suggest that these countries are in a "sunset" stage and appear ready to exit most assembly operations.

We earlier suggested that the "maturity" of a country's involvement in component production or assembly might be assessed by analysis of its overall RCA profile for these operations (Figure 1). For an initial test, Figure 2 plots the 1996 RCA import indices for Indonesia, Hong Kong, and Japan. In addition, Mexico's profile is included for comparison, given this country's extensive involvement in assembly operations (see Table 8). Component groups with RCAs less than 0.8 were not included, since these distributions "tails" seemingly have a small information content. Adoption of a 0.8 cut-off can provide an indication as to the number of assembly operations in which the country is close to achieving a comparative advantage. Similar comparisons for other East Asian countries are presented in Appendix Figures 1 and 2, while Appendix Figures 3 through 5 provide RCA index distributions derived from export statistics. As Figure 2 shows, the four countries' RCA profiles incorporate important differences. Japan's profile is the shortest of the four, with the ranked indices falling below the 0.8 cut-off well in the figure's rightmost portion. Hong Kong has a similar profile, but its distribution extends a bit farther to the left. A subjective judgement might classify these countries' profiles as corresponding to the "e" (Hong Kong) and "f" (Japan) distributions in Figure 1. Given the level of their wage costs and their

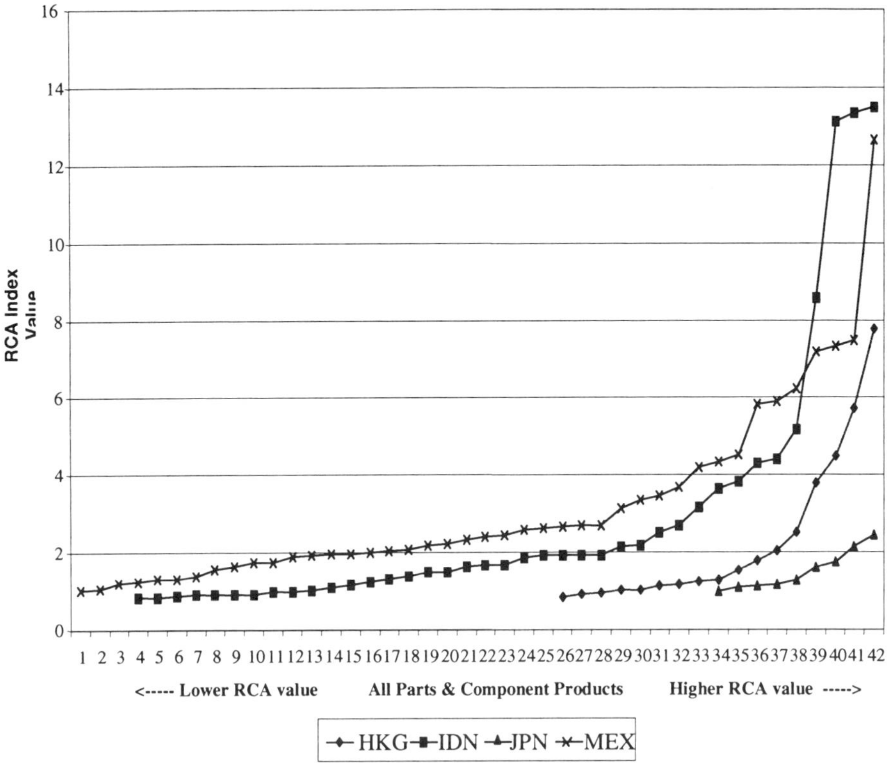

Figure 2: *The Relative Maturity of Asian Countries and Mexico's 1996 Import RCA Indices.*

high level of industrialization, it might be reasonable to expect that the two countries' distributions will shift farther to the right (reflecting a continuation of 1985–96 changes) as their comparative advantage in the few remaining "sunset" assembly operations continues to erode.

In contrast, the RCA profile of Indonesia extends farther to the left and reflects a broader base of operations, but it still falls well short of that for Mexico. This observation has several interesting implications. First, there appear to be similarities between the "stage of development" of Indonesia's RCA profile and that of the Philippines, Malaysia, and Thailand. However, all four Asian countries' profiles truncate well to the right of Mexico's profile—which could be classified as corresponding to the "a" or "b" distributions in Figure 1. If one proposes that Mexico's profile represents an "achievable standard," it appears that these four Asian countries (and the PRC) could potentially further broaden their domestic assembly operations.[18]

Similar conclusions emerge from the RCA profiles derived from export statistics (Appendix Figures 3 through 5). One difference, however, is that these indices often truncate farther to the right than the indices derived from import statistics. This suggests that Asian production operations for components are collectively at a less mature stage than are assembly operations and that they may have more potential for development.[19] The fact that the East Asian RCA export profiles truncate far to the right of the U.S. distribution lends some credence to this view.

Cross-correlations between East Asian countries' RCA indices often fail to achieve statistical significance. One possible explanation is that locational, wage, and communications costs, along with the specific mix of skills and infrastructure required for assembly in specific component industries, are more binding than is often thought. Correlations between specific countries' 1985 and 1996 RCA assembly profiles are often highly significant, indicating that these operations may not be as "footloose" as is sometimes suggested.

A question for consideration is whether the economic factors responsible for East Asian production sharing in specific manufacturing processes are limited to a few locations or are relatively well dispersed across countries. If the former is the case, one could expect relatively few similarities between the RCA profiles of most Asian countries. That is, if the *specific mix* of locational (transport cost), wage rate, communications, and other production requirements needed for assembly of a given component group occur in relatively few Asian countries, and if these requirements differ among types of components, sizeable differences in Asian countries' RCA profiles could result.[20] In contrast, if locational and production requirements are less binding, sizeable differences in these profiles need not occur.

Table 9 tests this proposition by reporting bilateral correlation coefficients between each Asian country's 1996 RCA (assembly) indices for all 60 component product groups. The table also reports results when each country's 1996 RCA indices were correlated with its earlier (1995) profile. The objective here was to assess the stability of comparative advantage over time. This issue is important, since some contend that assembly operations are often footloose, which means that offering substantial financial incentives to attract this activity, or making sizeable investments in any required infrastructure, may not be advisable.[21]

For the most part, the results reported in Table 9 reflect important dissimilarities in the RCA (assembly) profiles of most East Asian countries. For example, all the cross-country correlations involving Indonesia and Japan fail to achieve statistical significance, while the PRC's profile is significantly correlated only with Hong Kong's, and the Republic of Korea's is significantly correlated only with Singapore's. Cross-country correlations involving Malaysia and the Philippines are statistically significant in four instances, but this runs counter to the normal pattern, which suggests that important differences exist in the assembly capacity of Asian countries.[22]

Table 9: *Correlation Coefficients for the 1996 RCA Profiles of East Asian Countries for All Parts and Component Products*

1996 RCA Profile and Country	1996 RCA Profile for All Parts and Components Imports									
	China	Hong Kong	Indonesia	Korea	Malaysia	Philippines	Singapore	Thailand	Taiwan	Japan
China	1.000									
Hong Kong	0.674*	1.000								
Indonesia	–0.028	–0.103	1.000							
Rep. of Korea	–0.056	–0.037	–0.074	1.000						
Malaysia	0.040	0.170	–0.019	0.054	1.000					
Philippines	0.079	0.244	–0.048	0.030	0.825*	1.000				
Singapore	–0.087	0.055	–0.025	0.364*	0.463*	0.304*	1.000			
Thailand	0.284	0.375*	0.142	0.101	0.559*	0.502*	0.161	1.000		
Taiwan	0.289	0.215	0.155	–0.021	0.372*	0.350*	0.012	0.431*	1.000	
Japan	0.047	0.201	–0.151	–0.085	–0.002	–0.044	0.082	–0.104	0.251	1.000
1996 RCA Profile versus 1985 Profile**	0.518*	0.844*	0.131	0.105	0.838*	0.385*	0.604*	0.288	0.864*	0.578*

**Statistically significant at the 99 percent confidence level.*

***China did not report SITC Revision 2 trade data to UN COMTRADE until 1987. Therefore, the correlation for this country compare 1987 and 1996 RCA profiles.*

Source: *Computed from the RCA statistics shown in Table 6*

Somewhat different conclusions result from the 1985 and 1996 RCA profiles of each Asian country. In 7 of the 10 cases strong positive significant correlations only occurred with Indonesia; the Republic of Korea and Thailand failed to achieve significance. Altogether, the results suggest the comparative advantage profiles of the 10 countries were generally relatively stable over this 11-year period, a point indicating that assembly operations are not as footloose as is sometimes thought.

5. CONCLUSION

Two of this study's key findings are that the scope of production sharing in East Asia is considerably greater than is generally recognized, and that these countries' comparative advantages in production or assembly operations conforms to factor-intensity theory. Even more important, however, is that the study demonstrates that data sources (including the newer HS statistics) are now available for empirical research, and that it analyses factors leading to the remarkable growth in, and location of, production sharing.

This analysis of newer statistics on production sharing in East Asia shows that production sharing is very important to regional countries and that trade in components is growing at a far above-average pace. These findings are consistent with those of an earlier study that detected growth rates that were well above average for components and estimated that trade in these goods may now account for as much as 30% of world trade in manufactures (Yeats 1997). The current study, however, extended the earlier investigation to suggest that factor intensities appear to be an important driving force behind the spread of this activity. Countries with relatively low wages generally have a higher and more extensive revealed comparative advantage in assembly operations (which are generally labor intensive in nature), while high-wage countries like Japan generally have a stronger comparative advantage in the production of components. An important related point is that production sharing of this sort has been shown to convey numerous job-creation and other benefits to the participating developed and developing countries (USITC 1991, 1996).

A point that has as yet not been discussed concerns the implications of the existence of these data sources (that is, the Revision 2 of the SITC and newer more detailed HS statistics) for production sharing. Since the Revision 2 data are now available for more than 100 countries, and since they often extend back as far as the mid-1970s, a rich source of empirical information exists for the empirical testing of hypotheses relating to the causes, extent, and location of production sharing. Concerning this latter point, special mention should be made of the research potential of U.S. trade statistics. The United States is one of a relatively small number of countries that tabulates and publishes statistics on freight costs for all imports by product and by country of origin, so it should be possible to shipping costs influence the location of production sharing and the types of goods for which this activity is feasible (see Yeats 1981 for further discussion of this data source and for examples of studies that have employed it in analyses of international shipping costs).

References

Arndt, Sven and Kenryk Kierzkowski, eds., (2001). *Fragmentation: New Production and Trade Patterns in the World Economy* (Oxford: Oxford University Press) forthcoming.

Balassa, Bela (1965a). "Tariff Protection in Industrial Countries: An Evaluation," *Journal of Political Economy* (December).

Balassa, Bela (1965b). Tariff Liberalization and "Revealed" Comparative Advantage," *The Manchester School of Economic and Social Studies* (May).

Balassa, Bela (1977). "A Stages Approach to Comparative Advantage," in Irma Adelman (ed.) *Economic Growth and Resources*, Proceedings of the Fifth World Congress of the International Economic Association held in Tokyo (London: Macmillan Press).

Balassa, Bela (1979). "The Changing Pattern of Comparative Advantage in Manufactured Goods," *The Review of Economics and Statistics* (May).

Finger J. M. (1975). "Tariff Provisions for Offshore Assembly and the Exports of Developing Countries," *Economic Journal* (June).

Finger J. M. (1976). "Trade and Domestic Effects of the Offshore Assembly Provisions in the U.S. Tariff," *American Economic Review* (September).

Gereffi, Gary (1996). "Commodity Chains and Regional Divisions of Labor in East Asia," *Journal of Asian Business*, Volume 12, no. 1.

Lary, Hal (1968). *Imports of Manufactures from Less Developed Countries* (New York: National Bureau of Economic Research).

USITC Publication 2365 (1991). *Production Sharing: U.S. Imports Under Harmonized Tariff Schedule Subheadings 9802.00.60 and 9802.00.80, 1986–1989* (Washington: US International Trade Commission, March).

USITC Publication 2966 (1996), *Production Sharing: Use of U.S. Components and Materials in Foreign Assembly Operations, 1991–1994* (Washington: US International Trade Commission, May).

Wall Street Journal and Heritage Foundation (1997). *1997 Index of Economic Freedom*, (Washington: Wall Street Journal and Heritage Foundation).

Watanabe, Toshio and Hirokazu Kajiwara (1983). *Industrialization in Asian Developing Countries and Emerging Horizontal Economic Integration*, mimeo, paper presented at a conference on Trade and Industrial Cooperation in East and South-East Asia, 8–9 March (Tokyo: Institute of Developing Economies).

World Bank (1984a). *China: Foreign Trade Reform* (Washington: The World Bank).

World Bank (1984b). *Coping with Changes in the External Environment*, (Washington: World Bank Caribbean Division, May).

World Bank (1999). *World Development Indicators 1999* (Washington: World Bank, in press).

Yeats, Alexander (1981). *Shipping and Development Policy: An Integrated Assessment* (New York: Praeger Scientific Publishers).

Yeats, Alexander (1989). "Shifting Patterns of Comparative Advantage: Manufactured Exports of Developing Countries," *The Developing Economies*, Summer.

Yeats (1991). *China's Foreign Trade and Comparative Advantage: Prospects, Problems and Policy Implications* (Washington: World Bank Discussion Paper Series).

Yeats, Alexander (1997). *Just How Big is Global Production Sharing?* (Washington: World Bank Policy Research Paper No. 1871).

BOX 1
THE RELATIVE IMPORTANCE OF RECENT AND "VINTAGE" PRODUCTION SHARING

While most recent analyses of international production sharing focus on the fragmentation of manufacturing processes (see Arndt and Kierzkowski 2001 for an overview), this activity has long been a major and evolving process. One of its earlier forms involved the production of primary commodities in developing (and some developed) countries, shipment of these goods to (largely) industrial nations for further processing, and then the reexport (in part) of the processed product back to the primary commodity producer or third countries. As an example, tin ore might be mined in Thailand or Malaysia and then shipped to Japan for refinement and further manufacture; some of it might then be shipped back to the primary product producers. In part, these "production sharing" trade flows were based on comparative advantage (some processing activities, like the fabrication of metals from ore, is highly capital intensive), but factors such as "escalation of industrial countries'" trade barriers contributed to this exchange pattern.

An interesting question is that of how the magnitude of the trade in these "traditional" production inputs (unprocessed commodities) compares with that now reflected in the international exchange of manufactured parts and components. For answers, East Asian global imports and exports of agricultural raw materials (SITC 2-22-27-28) and ore, minerals, and nonferrous metals (27+28+68) were tabulated along with certain types of unprocessed foodstuffs (e.g., cocoa, coffee beans) that have no discernable end use in their natural form. These goods, like those manufactured components listed in Table 1, are being shipped elsewhere for further processing.

The statistics shown below compare the value of East Asian trade in "traditional inputs" with manufactured components for selected years over the 1984–96 period. The figures attest to the remarkable growth dynamism in the exchange of the latter. In 1984 Asian imports of traditional inputs were more than double those of manufactured components, but by 1996 their relative positions had reversed, and component imports were US$67 billion higher. Over the full 198496 period the annual growth rate for components (20.4%) was about two and one-half times greater than that for traditional products.

Trade Flow/Product	Value of East Asian Trade in US$ billion				Growth Rate (%)	
	1984	1988	1992	1996	1984–96	1988–96
East Asian Imports						
Traditional Production	39.2	68.0	71.4	98.9	8.0	4.8
Inputs	17.9	46.4	80.8	165.6	20.4	17.2
Manufactured Components						
East Asian Exports						
Traditional Production	17.0	26.3	25.1	37.6	6.8	4.6
Inputs	33.0	73.1	108.9	177.8	15.1	11.8
Manufactured Components						

Source: *Computed from UN COMTRADE data.*

While 1984 Asian exports of manufactured components (US$33 billion) were approximately twice as large as traditional inputs, the former was five times as large as the latter by 1996. In short, whether it is measured by exports or by imports, the statistics attest to the remarkable dynamism in and importance of East Asian trade in manufactured parts and components.

BOX 2

CHANGES IN THE GEOGRAPHIC DISPERSION OF COMPARATIVE ADVANTAGE IN PRODUCTION AND ASSEMBLY

The previous analyses (Table 3) demonstrate that, over the last decade, East Asian global exports of parts and components were growing at a considerably faster rate than exports of all goods and at about 30% per year faster than the rate for all manufactures. To what extent is this recent dynamism reflected in the comparative advantage profiles of East Asian countries for assembly or production operations? Also, do these changes give any indication of the potential magnitude and direction of future changes?

The statistics reported below provide summary information on the extent to which the 10 East Asian countries had a comparative advantage in the assembly or production of all 60 parts and component product groups in 1985 and 1996. For example, in 1985 only 2 or 3 countries had a comparative advantage in the assembly of 30% of all the component groups, but by 1996 this had changed to 36.7%. In 1996 only 2 or 3 countries had a comparative advantage in the production of 16.7% of the components; this is down from an average of 20% in 1985. The rightmost column shows the average number of East Asian countries with a comparative advantage in all 60 component groups. In 1996 this average was 3.48 for assembly operations as opposed to 1.77 countries for the production of parts.

	Percentage of Industries Ranked by the Number of Countries with RCAs Exceeding Unity					**Average No. of Countries With RCAs over Unity**
Operation/Year	**0 or 1 Countries**	**2 or 3 Countries**	**4 or 5 Countries**	**6 or 7 Countries**	**8 or 9 Countries**	
Assembly of Parts	20.0	36.7	20.0	16.7	6.6	3.48
1996	16.7	30.0	35.0	8.3	10.0	3.83
1985						
Production of Parts	66.7	16.7	3.3	8.3	5.0	1.77
1996	71.6	20.0	6.7	1.7	0.0	1.12
1985						

Source: *Computed from the data in Tables 6 and 7.*

Several interesting points emerge from these statistics. First, East Asian comparative advantage is considerably broader in assembly operations than in manufacturing components. In 1996 an average of 3.5 countries had a comparative advantage in the assembly of the 60 product groups, which was more than double the average (1.77 countries) for production operations. The slight 1985–96 decline in the overall assembly average is due entirely to the Philippines, where the available data seemingly suggest a loss of comparative advantage in 21 assembly industries. (Note our earlier reservations concerning this nation's data quality.)

Six or more countries had an assembly comparative advantage in about 23% of the 60 industries, as opposed to about 13% for manufacturing. An analysis of the tabulations also suggests that nominal transportation costs may be a determinant of the extent to which Asian countries have developed a comparative advantage in individual industries. Between 6 and 8 Asian countries have developed an assembly comparative advantage in component groups like switchgear, electronic components, or office machinery where low-bulk, high-value products normally should be involved. In contrast, none of the 10 countries had a comparative advantage in products like harvesting machinery, cultivating equipment, or dairy machinery.

Endnotes

[1]Production sharing is here defined as the internationalization of a manufacturing process in which several countries participate in different stages of a specific good's fabrication. The process is of considerable economic importance, since it allows stages of production to be located where they can be undertaken most efficiently. If production sharing is increasing in relative importance, this implies that countries are becoming more economically interdependent.

[2]Quite a few empirical studies examine the size of and motivations for this exchange. Many earlier analyses focus on the effects of U.S. and European tariff provisions on the further assembly abroad of domestically produced components. Under these provisions, components, (say those produced in the United States) could be shipped abroad for further assembly and then reexported to the United States with a tariff applied only to the value-added component of the processed good. Finger (1975, 1976) authored some of the most influential early studies on this activity. More recently, the U.S. Congress mandated that the U.S. International Trade Commission provide annual reports on the magnitude and economic effects of production sharing in North America and Europe. (See, e.g., USITC [1991, 1996].)

[3]This analysis is based exclusively on product groups defined as consisting solely of components. This causes estimates of East Asian production sharing to be downward biased. Specifically, some other SITC 7 product group exports (e.g., television picture tubes) may be used for further assembly operations, but it is not possible to accurately determine whether, or to what extent, these items are used for further assembly. Some empirical studies, like Balassa (1965a), use product-classification schemes that included groups called "intermediate input groups I and II," which were composed of items like yarns, pig iron, leather, and some chemicals. Although we do not do so, other empirical analysts of trade in components might consider using such products in their investigations.

[4]In the early 1990s countries began reporting trade statistics to the UN in terms of the more detailed HS. At its lowest six-digit level the HS distinguishes between approximately 3,600 individual items; of these, some 200 consist solely of parts and components. The current disadvantage of the HS is that a sufficient reporting period has not yet elapsed for adequate time-series information to become available, as is the case with the SITC Revision 2 data.

[5]This "error" has also been incorporated in International Monetary Fund (IMF) data series. Readers may wish to verify that in the 1980s and 1990s the IMF Direction of Trade Statistics reported sizeable Indonesian exports to, and imports from, Singapore. However, the Singapore statistics show no trade with Indonesia.

[6]In contrast to imports, the relative importance of East Asia's exports of parts and components is below that of OECD Europe or North America, a fact that probably reflects Asia's comparative advantage in assembly operations (see Table 6). However, the 1985 to 1996 share of components in total SITC 7 and 8 exports was static in Europe (at 23%), and it declined in North America (from 36% to 31%). Over this period East Asia's share increased by 6 percentage points, to 22 percent.

[7]The trade intensity index (I_{ij}) is defined for country i's exports to country j as the share of i's exports going to j (X_{ij}/X_{iw}) relative to the share of j's imports in world imports ($M_{jw}/(M_w-M_i)$). That is,

$$I_{ij} = (X_{ij}/X_{iw}) / (M_{jw}/(M_w-M_i)).$$

[8]Table 2.6 in the World Bank's 1999 *World Development Indicators* reports statistics on average annual labor costs in manufacturing over the 1990–94 period for East Asian countries and other countries. Average annual wages in the Philippines are US$2,459, as compared to US$15,819 in Korea and US$21,534 in Singapore. Regional annual wages were highest in Japan (US$21,534) and lowest in Indonesia (US$1,008).

[9]At first glance, the fact that low-wage countries like Indonesia and the Philippines have such high trade (export) intensity indices with Japan may appear contrary to reason. These results are largely due to the fact that Japan has relatively low *import* shares for components. As such, the relatively small denominator in the formula for the trade-intensity index (equation 1) can produce what appear to be relatively large index numbers. The shares of Indonesia and Philippines exports of components to Japan are small, but they are relatively larger than Japan's world share of component imports.

[10]The most recent UN trade statistics show 1997 exports from the Philippines falling by over 55% from their 1996 levels. It is difficult to assess the implications of this change, since a similar marked reduction also occurred in trade reported as SITC 931 "special transactions." We have asked officials at the UN Statistical Office to verify the most recent Philippines trade statistics.

[11]Balassa (1965) developed and applied the concept of RCA and also made important extensions in Balassa (1977b and 1979). Where there are simultaneous imports and exports of components of the same sort, use of Balassa's (1965) RCA measure based on net trade might be considered, although this index can be biased by high local trade barriers. Our East Asian trade data show few cases in which a country is a relatively high exporter and importer of the same component.

[12]For example, some governments have actively encouraged the import and assembly of specific electronic components, reasoning that this could be an initial step leading to the development of a full comparative advantage in high-tech industries. See World Bank (1994a) for a discussion of the PRC's policies in this respect. The use of financial and other government incentives to attract offshore assembly operations in the Caribbean is well documented by the World Bank (1994b)

[13]There is one exception when an imported component is used as a replacement for a failed part in an already assembled good. However, this should have a neutral effect across countries unless failure rates for parts in assembled products differ substantially between nations, or unless there are large differences in the intensity of use of a given finished product across countries. The latter might be a factor between, say, very rich and very poor countries, but it should be less important in a region like East Asia.

[14]To the best of our knowledge this line of analysis was first developed by Watanabe and Kajiwara (1983). Yeats (1991) undertook a similar analysis using revealed comparative advantage indices for the PRC.

[15]We envisage a certain "life cycle" in these operations that pushes a country forward or backward through the stages reflected in Figure 1. For example, in the 1950s and 1960s Singapore and Japan were relatively high-skill and low-wage cost-assembly centers, which caused their RCA profiles for these operations to shift in the direction of (f) to (a). However, subsequent domestic wage increases reduced their competitiveness in assembly operations, which pushed their RCA profiles in opposite directions—that is, in the direction of (f) toward (a). As such, position (f) in Figure 1 would be consistent with both: (1) an immature assembler with considerable potential for further development, or (2) a country whose advanced stage of industrialization was causing it to exit assembly operations. We differentiate between the two based on factors such as relative GDP or wage costs.

[16]To help determine how comparative advantage profiles changed, Appendix Table 3 shows similar statistics computed from 1985 import statistics. Appendix Table 6 shows RCA indices

computed using 1985 export statistics and is intended to help assess the 1996 RCA indices shown in Table 7.

[17]The pattern reflected in the statistics on assembly operations parallels that observed for trade in goods. Lary (1968), for example, employed U.S. Census Bureau statistics to estimate the labor intensity of different manufactured goods and then showed that developing countries had a production advantage in the most labor-intensive items. Yeats (1989) extended Lary's analysis through the 1980s and showed that developing countries greatly improved their OECD market shares for the most labor-intensive products.

[18]Some geographic factors could cause important permanent differences between the profiles for Mexico and the East Asian countries. The first is proximity, because Mexico borders the United States and is connected to main U.S. centers of consumption by well-maintained road, rail, and shipping links. Nominal transport costs for components and assembled goods may be significantly lower for Mexico than for East Asia, which could prevent the latter from developing production sharing to the same degree as North America has. The second is that aside from some Caribbean countries, transport and other required infrastructure in Latin American countries, as well as their political instability, may make it unprofitable, or riskier, to establish foreign assembly operations there. In short, Asia may offer more choices as to where to locate assembly operations. If this is so, then both transport costs and the availability of more alternative locational sites could prevent Asian countries from developing RCA profiles that are as broad as Mexico's.

[19]This point is also evident in the data reported in Table 8. On average, the 10 East Asian countries have a 1996 comparative advantage in the production of 19% of the component groups, as opposed to an average of 33% for assembly operations.

[20]Published studies that test "transport cost estimation equations" show that distance has a major impact on shipping costs, particularly after stowage factors and product value are accounted for (see Yeats 1981 for a survey). The sizeable distances between many Asian countries might influence the choice of where to locate assembly operations. If Japan were the intended market for consumption of a relatively bulky final good, shipping costs might dictate that assembly operations be located in a country close to Japan. Some Asian tariffs also appear to be set at levels that would influence the location of production sharing.

[21]In the late 1960s, when international development agencies like the United Nations Council on Trade and Development became increasingly aware of the potential economic importance of foreign assembly operations, concerns were expressed about the construction of "factory ships," which were floating plants that (theoretically) had the capacity to move easily and quickly from country to country to take advantage of the most depressed wages or the most attractive financial incentives offered by governments. It was generally assumed that, if they were viable, factory ships would greatly limit any wage or other economic benefits for developing countries.

[22]Gereffi (1996) suggests that intra-Asian cultural and ethnic factors have played an important role in the spread of production sharing in the region and are at least partially responsible for the differences that exist in the assembly capacity of different countries.

APPENDIX TABLES AND FIGURES

Appendix Table 1: *Major Exporters and Importers of Parts and Components in 1996*

	Trade in Parts and Components			
	Imports		Exports	
Country	Value ($million)	Share of All Manufactures*	Value ($million)	Share of All Manufactures*
United States	102,141	17.4	114,672	29.6
Germany	52,109	18.1	74,161	19.3
United Kingdom	43,886	22.1	39,739	22.2
Canada	33,156	26.4	23,006	20.1
France	30,847	17.6	36,402	19.7
Singapore	26,524	26.3	23,558	24.2
Hong Kong	26,108	16.2	4,227	17.2
Japan	23,724	14.2	80,939	22.4
Malaysia	19,196	32.3	12,504	22.0
China	17,176	18.7	10,706	9.0
Mexico	16,776	25.2	12,878	18.2
Italy	16,663	14.9	27,835	13.5
Netherlands	16,451	17.0	14,258	16.9
Spain	16,089	21.5	9,839	13.9
Belgium	14,620	15.7	10,601	10.3
Thailand	14,252	28.8	7,838	20.7
Korea	13,888	16.6	11,917	11.3
Sweden	10,314	23.3	17,957	29.3
Philippines	10,235	42.6	3,570	21.4
Australia	7,831	17.3	2,585	22.0
Austria	7,738	16.0	6,915	15.0
Taiwan (China)	7,710	12.6	21,521	21.3
Brazil	7,315	23.7	3,351	15.1
Switzerland	7,225	12.9	8,768	16.2
Indonesia	6,644	27.0	1,767	7.4
Ireland	5,566	24.0	6,865	26.2
Saudi Arabia	4,673	26.4	14	1.0
Denmark	4,393	16.7	4,443	18.2
Norway	4,074	17.0	2,091	21.2
Portugal	4,050	18.6	1,193	6.4
Finland	3,721	19.6	6,012	19.2
Poland	3,653	15.9	1,347	8.3

**Excludes chemicals. Manufactures are defined here as SITC 6 through 8 less 68.*

Source: *Computed from UN COMTRADE statistics.*

Appendix Table 2: *The Revealed Comparative Advantage of East Asian Countries in Assembly Operations as Reflected in Their 1985 Import Statistics*

Component Product Group*	China	Hong Kong	Indonesia	Rep. of Korea	Malaysia	Philippines	Singapore	Thailand	Taiwan	Japan	USA
Steam boilers	2.62	2.70	2.17	1.08	0.26	4.27	1.64	2.48	0.22	0.01	0.35
Aircraft engines	90.02	1.19	3.24	0.09	6.41	0.69	4.15	0.29	0.00	0.10	0.11
Outboard motors	0.13	1.06	1.35	0.12	2.42	0.18	0.55	0.00	0.00	0.16	0.82
Outboard motors, nes	0.38	0.72	1.15	13.25	1.28	0.27	2.06	0.00	0.00	0.59	1.08
Combustion engines, nes	0.22	0.23	1.39	0.93	0.94	1.37	1.29	2.45	0.00	0.30	0.79
Engines and motors, nes	0.13	0.05	1.13	1.29	0.07	0.25	0.32	0.30	0.00	3.26	1.45
Rotating electric motors	2.00	0.99	3.06	1.27	1.21	1.58	1.06	0.00	0.00	1.12	0.93
Water turbines	0.39	0.00	22.58	0.29	0.53	1.35	0.12	0.05	0.00	0.32	0.89
Cultivating equipment	0.04	0.06	0.27	0.18	0.22	0.54	0.10	2.42	0.00	0.34	1.25
Harvesting machinery	0.46	0.00	0.12	0.56	0.30	0.16	0.05	0.03	0.00	0.21	1.34
Dairy machinery	0.05	0.12	3.34	0.24	0.16	1.36	0.08	0.02	0.00	0.27	1.74
Wine making machinery	12.57	0.00	10.92	0.19	2.81	6.96	0.26	0.58	0.00	0.26	1.47
Agricultural machinery, nes	0.50	0.13	0.62	0.05	0.39	0.58	0.11	0.07	0.00	0.10	0.91
Construction machinery	0.79	0.61	13.19	0.36	3.80	1.42	3.48	2.94	0.00	0.24	0.62
Spinning machinery	3.70	0.64	4.03	1.97	0.34	1.90	0.30	0.01	5.34	0.84	0.92
Looms and knitting machinery	1.48	1.23	1.77	0.63	0.61	3.36	0.21	3.68	0.97	1.61	1.01
Textile machinery, nes	1.70	1.13	0.84	0.35	0.86	1.04	0.15	0.80	0.00	0.19	1.50
Paper making machinery	1.18	0.08	1.21	0.48	15.02	1.34	0.06	2.59	0.00	0.66	1.14
Bookbinding machinery	0.09	0.17	0.24	0.06	0.79	0.42	0.20	0.04	0.00	0.69	1.27
Printing machinery	0.37	0.30	0.56	0.24	0.43	0.32	0.36	0.10	0.00	0.62	1.35
Grain milling machinery	0.30	0.09	12.06	0.25	0.96	10.67	0.31	0.25	0.00	0.08	0.98
Food processing machinery	0.87	0.30	2.99	0.23	0.45	4.42	0.29	2.96	0.00	0.42	1.34
Machine for special industries	0.43	0.37	3.51	0.42	1.00	3.35	0.65	0.00	0.00	0.61	1.14
Mineral working machinery	0.45	0.35	2.56	0.25	1.74	2.82	0.74	0.00	0.00	0.15	0.92
Special industry machines, nes	1.88	0.52	6.15	0.91	2.74	16.22	0.61	2.63	0.00	0.97	1.19
Machines for metal working	0.67	0.20	1.28	0.68	0.63	1.06	1.07	0.66	0.75	0.79	1.02

(*Continued*)

Appendix Table 2: (*Continued*)

Component Product Group*	China	Hong Kong	Indonesia	Rep. of Korea	Malaysia	Philippines	Singapore	Thailand	Taiwan	Japan	USA
Foundry equipment	6.97	0.19	3.03	0.56	1.02	4.06	0.62	2.73	0.00	0.64	0.94
Rolling mill parts	4.44	0.08	2.63	1.64	0.84	37.98	0.14	0.80	0.00	0.36	1.39
Refrigerating equipment	6.78	0.89	1.83	0.13	1.49	1.03	1.49	0.99	0.00	0.57	1.50
Pumps for liquids	0.57	0.22	5.06	0.82	0.71	2.50	1.34	1.16	0.00	0.86	1.04
Centrifuges and filters	0.52	0.33	1.07	0.72	0.89	2.52	1.31	2.08	0.00	2.21	1.54
Fork lift trucks	0.25	0.15	0.28	0.69	0.41	2.33	0.28	0.12	0.56	0.08	1.11
Lifting and loading machines	1.01	1.34	1.73	0.80	1.11	1.80	0.91	0.00	0.00	0.72	1.17
Power hand tools	0.11	0.10	7.88	0.25	1.75	1.08	0.72	0.22	0.00	0.50	1.07
Packing machinery	0.58	0.34	0.66	0.52	0.66	3.29	0.32	0.58	0.00	0.52	1.58
Non-electric machinery, nes	0.33	0.23	1.10	0.98	1.39	2.75	0.82	0.57	1.03	0.93	0.61
Office and adding machinery	0.34	1.03	0.19	0.44	0.32	0.26	1.68	1.14	1.01	1.04	1.41
Telecommunications equipment	1.64	1.33	1.13	1.51	1.77	1.23	1.43	1.94	1.50	0.77	0.70
Electric power machinery	1.17	2.32	1.24	3.44	3.28	3.27	1.24	0.00	2.29	0.23	0.70
Switch gear	0.43	0.87	1.30	1.45	1.88	1.47	2.18	3.75	2.49	1.16	1.09
Domestic electrical equipment	1.04	2.90	0.85	0.08	0.52	0.84	0.30	1.34	0.00	0.45	1.04
Electro thermic appliances	0.23	1.95	0.22	0.00	1.07	0.33	2.73	0.67	0.00	0.08	1.62
Electronic components, nes	0.37	1.32	0.80	10.46	27.88	23.62	8.77	0.02	0.00	0.26	0.98
Electronic accumulators	0.06	0.74	3.72	0.26	2.28	2.37	0.46	0.37	0.07	0.20	0.00
Electric lamps and bulbs	0.11	0.47	2.62	1.95	0.50	4.24	0.80	1.62	0.41	0.24	0.00
Electrical machinery, nes	1.80	1.73	4.12	2.72	0.25	21.19	0.01	2.79	3.20	1.18	0.00
Motor vehicles and accessories	1.37	0.14	0.40	0.26	0.20	0.41	0.25	1.23	0.45	0.18	1.02
Carriages and cycles	1.05	0.42	3.56	0.50	1.04	1.05	0.63	4.99	5.26	0.38	1.04
Non-motor vehicles	0.07	0.04	0.14	0.11	0.94	0.10	0.34	0.41	0.67	0.07	0.00
Railroad vehicles	1.58	0.47	0.88	0.64	0.25	1.26	0.06	1.00	0.43	0.35	0.93
Aircraft	0.83	0.35	1.55	1.22	0.61	1.48	1.49	0.22	0.09	1.47	0.92
Chairs	0.01	0.07	0.37	0.34	0.46	0.11	0.57	0.32	0.00	1.41	0.00

Other furniture	0.20	0.90	0.30	0.64	0.52	0.05	1.06	0.00	0.20	1.54	0.00
Drawing machinery	0.61	0.07	6.95	0.78	1.69	0.47	0.44	1.03	0.00	1.25	0.00
Still cameras, nes	1.71	4.77	0.34	3.11	4.72	3.26	0.57	0.58	7.93	1.40	0.59
Cameras under 16mm	0.33	0.02	1.48	0.06	1.16	0.09	4.90	0.00	0.00	0.00	0.98
Cameras under 16mm, nes	0.41	0.93	0.95	0.36	0.79	0.39	0.85	0.13	0.00	1.26	0.53
Unmounted optical elements	0.12	1.50	0.09	0.76	0.53	0.14	1.20	0.51	0.90	2.66	0.98
Clocks and watches	19.54	19.87	0.13	6.98	1.79	3.92	1.98	6.14	2.27	0.61	0.20
Umbrellas and canes	2.26	5.22	1.91	2.26	2.67	6.71	2.94	8.00	0.04	3.76	0.37
Memo Item:											
RCA Index for All Components	1.07	0.75	1.38	1.15	1.81	1.99	1.33	1.55	0.87	0.74	0.95
Number of Product Groups with An RCA Index Over Unity	23	17	39	16	26	40	22	23	10	15	29

* *See Table 1 for the SITC Revision 2 classification number of each component product group.*

Source: *Computed from United Nations COMTRADE Statistics.*

Appendix Table 3: *The Revealed Comparative Advantage of East Asian Countries in the Production of Components as Reflected in Their 1985 Export Statistics*

Component Product Group*	China	Hong Kong	Indonesia	Rep. of Korea	Malaysia	Philippines	Singapore	Thailand	Taiwan	Japan	USA
Steam boilers	0.03	0.00	0.00	0.42	0.19	0.03	0.09	0.01	0.02	2.97	1.66
Aircraft engines	0.03	0.00	0.00	0.05	7.35	0.00	3.21	0.19	0.00	0.03	7.00
Outboard motors	0.05	1.32	0.00	0.00	0.03	0.06	0.68	0.00	0.00	5.40	0.00
Outboard motors, nes	0.19	0.00	0.00	0.01	0.11	0.00	1.76	0.00	0.00	2.00	0.00
Combustion engines, nes	0.08	0.13	0.00	0.06	0.04	0.03	0.71	0.10	0.00	0.96	1.97
Engines and motors, nes	0.01	0.00	0.00	0.33	0.13	0.00	0.09	0.00	0.00	0.09	3.36
Rotating electric motors	0.14	0.31	0.03	0.16	0.12	0.00	2.57	0.00	0.00	0.99	1.85
Water turbines	0.63	0.00	0.00	0.00	0.12	0.00	0.17	0.10	0.00	0.74	1.14
Cultivating equipment	0.40	0.01	0.00	0.17	0.06	0.00	0.03	0.02	0.00	0.08	2.26
Harvesting machinery	0.14	0.00	0.00	0.00	0.02	0.02	0.07	0.01	0.00	0.13	2.45
Dairy machinery	0.00	0.00	0.00	0.00	0.02	0.45	0.03	0.00	0.00	0.30	0.93
Wine making machinery	0.68	0.00	0.00	0.00	1.18	0.00	0.03	0.00	0.00	0.52	0.00
Agricultural machinery, nes	0.17	0.00	0.00	0.00	0.03	1.06	0.17	0.66	0.00	0.01	1.01
Construction machinery	0.03	0.00	0.04	0.03	0.99	0.01	3.42	1.57	0.00	0.22	4.78
Spinning machinery	0.52	0.02	0.00	0.03	0.02	0.00	0.88	0.00	0.03	0.68	1.20
Looms and knitting machinery	0.56	0.16	0.00	0.20	0.01	0.00	0.11	0.17	0.34	0.62	0.51
Textile machinery, nes	0.14	0.24	0.00	0.04	0.05	0.03	0.08	0.00	0.00	0.25	0.98
Paper making machinery	0.00	0.00	0.00	0.05	0.08	0.00	0.04	0.00	0.00	0.36	1.31
Bookbinding machinery	0.01	0.00	0.00	0.06	0.01	0.00	0.59	0.00	0.00	0.06	1.65
Printing machinery	0.01	0.05	0.00	0.01	0.09	0.00	0.22	0.04	0.00	0.17	2.25
Grain milling machinery	0.44	0.00	0.00	0.01	0.18	0.00	0.86	0.20	0.00	0.25	0.00
Food processing machinery	0.15	0.00	0.00	0.01	0.14	0.00	0.04	0.07	0.00	0.11	0.00
Machine for special industries	0.01	0.00	0.00	0.13	0.07	0.01	0.21	0.00	0.00	0.16	3.32
Mineral working machinery	0.19	0.00	0.00	0.06	0.43	3.05	0.24	0.00	0.00	0.30	1.28

(Continued)

Special industry machines, nes	0.37	0.11	0.00	0.03	1.05	0.04	0.98	0.22	0.00	0.48	0.57
Machines for metal working	0.15	0.02	0.00	0.04	0.08	0.01	0.65	0.08	0.49	0.39	1.66
Foundry equipment	0.11	0.00	0.00	0.00	0.74	0.00	0.01	0.31	0.00	2.87	0.00
Rolling mill parts	0.07	0.01	0.00	0.02	0.11	0.00	0.62	0.00	0.00	2.03	0.26
Refrigerating equipment	0.07	0.04	0.00	0.13	0.12	0.06	0.37	0.03	0.00	0.86	0.00
Pumps for liquids	0.04	0.01	0.00	0.06	0.28	0.03	0.66	0.03	0.00	0.40	2.48
Centrifuges and filters	0.03	0.05	0.00	0.02	0.13	0.00	1.10	1.14	0.00	0.62	0.44
Fork lift trucks	0.02	0.02	0.00	0.81	0.09	0.00	0.18	0.01	0.01	0.57	2.54
Lifting and loading machines	0.11	0.02	0.00	0.10	0.35	0.09	0.42	0.00	0.00	0.48	1.16
Power hand tools	0.03	0.00	0.00	0.00	0.14	0.00	0.66	0.00	0.00	0.46	2.81
Packing machinery	0.00	0.01	0.00	0.01	0.04	0.02	0.05	0.01	0.00	0.12	1.53
Non-electric machinery, nes	0.10	0.06	0.02	0.05	0.22	0.00	1.61	0.00	0.88	0.82	0.97
Office and adding machinery	0.05	2.10	0.00	0.28	0.24	0.01	1.98	0.93	0.37	0.62	2.87
Telecommunications equipment	0.33	1.47	0.16	0.77	1.40	0.25	1.72	0.05	1.34	1.75	1.16
Electric power machinery	0.13	10.98	0.01	0.47	1.95	0.00	1.30	0.00	2.59	0.96	0.56
Switch gear	0.14	0.56	0.00	0.18	1.17	0.63	4.41	8.21	0.82	0.92	1.16
Domestic electrical equipment	0.62	1.86	0.00	0.12	0.16	1.45	0.51	0.06	0.00	0.43	2.13
Electro thermic appliances	0.16	4.45	0.00	0.00	0.32	0.03	1.05	0.02	0.00	1.79	1.59
Electronic components, nes	0.07	0.88	0.13	0.78	18.40	3.71	9.67	0.00	0.00	1.77	2.27
Electronic accumulators	0.09	0.03	0.17	0.18	0.11	0.19	0.40	1.89	0.67	0.13	1.65
Electric lamps and bulbs	0.15	0.16	0.01	0.10	0.02	0.34	0.04	0.14	1.43	0.31	1.18
Electrical machinery, nes	0.12	1.09	0.00	0.07	0.25	0.00	0.14	0.98	10.55	0.29	0.00
Motor vehicles and accessories	1.59	0.00	0.00	0.08	0.02	0.45	0.17	0.10	0.30	0.76	2.00
Carriages and cycles	1.27	0.00	0.14	0.25	0.05	0.00	0.60	0.36	4.45	2.86	0.10
Non-motor vehicles	0.02	0.00	0.00	0.68	0.14	0.00	0.13	0.00	0.43	0.05	0.62
Railroad vehicles	0.24	0.00	0.00	0.19	0.02	0.00	0.01	0.09	0.01	0.57	2.03

(Continued)

Appendix Table 3: (*Continued*)

Component Product Group*	China	Hong Kong	Indonesia	Rep. of Korea	Malaysia	Philippines	Singapore	Thailand	Taiwan	Japan	USA
Aircraft	0.10	0.00	0.04	0.11	1.72	0.01	0.97	0.05	0.01	0.06	4.50
Chairs	0.22	0.01	0.11	0.07	0.27	0.31	0.22	0.28	0.00	0.09	0.00
Other furniture	0.72	0.60	0.98	0.20	0.51	43.80	0.32	0.00	12.92	0.04	0.00
Drawing machinery	0.04	0.08	0.00	0.12	0.13	0.00	0.28	0.01	0.00	1.06	0.00
Still cameras, nes	0.05	1.75	0.02	0.45	1.36	0.00	0.46	0.05	1.52	3.53	0.84
Cameras under 16mm	0.20	0.00	0.00	0.11	14.51	0.00	8.08	0.00	0.00	2.84	0.24
Cameras under 16mm, nes	0.11	0.01	0.00	0,09	6.17	0.00	0.30	0.63	0.00	0.16	2.61
Unmounted optical elements	0.11	0.28	0.00	0.23	0.54	3.41	2.39	4.91	0.50	0.73	1.84
Clocks and watches	27.41	11.52	0.00	0.36	3.40	0.20	1.10	4.91	1.45	1.62	0.24
Umbrellas and canes	1.76	1.80	0.00	0.13	0.21	0.57	0.52	0.22	20.67	0.13	0.05
Memo Item:											
RCA Index for All Components	0.65	0.69	0.04	0.25	0.84	0.66	1.42	1.01	0.64	0.87	2.05
Number of Product Groups with An RCA Index Over Unity	4	10	0	0	12	6	15	6	9	13	35

**See Table 1 for the SITC Revision 2 classification number of each component product group.*

Source: *Computed from United Nations COMTRADE Statistics.*

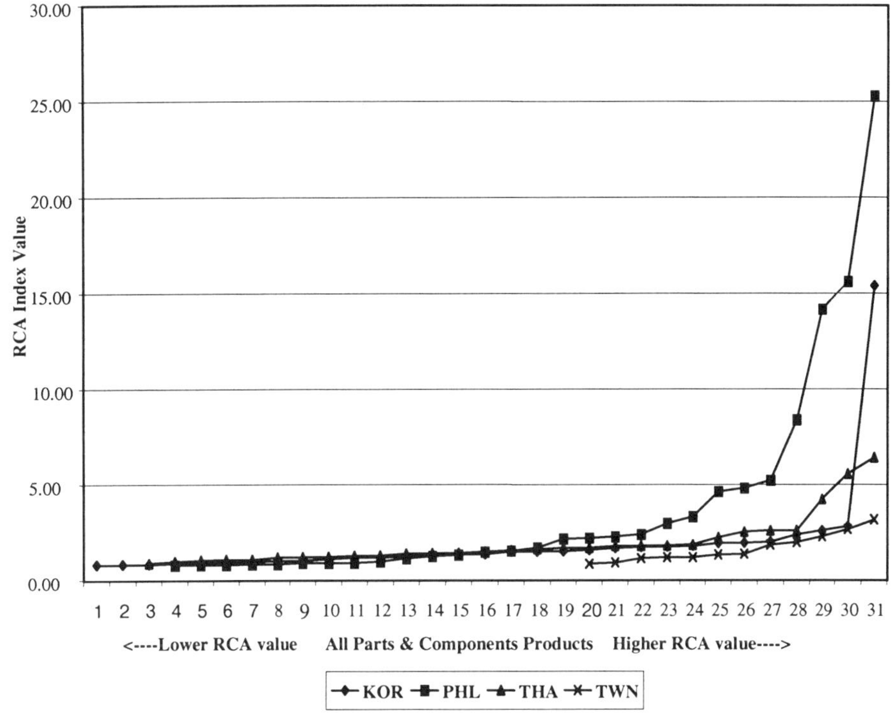

Appendix Figure 1: *The Relative Maturity of Korea, Philippines, Thailand and Taiwan's 1996 Import RCA Indices.*

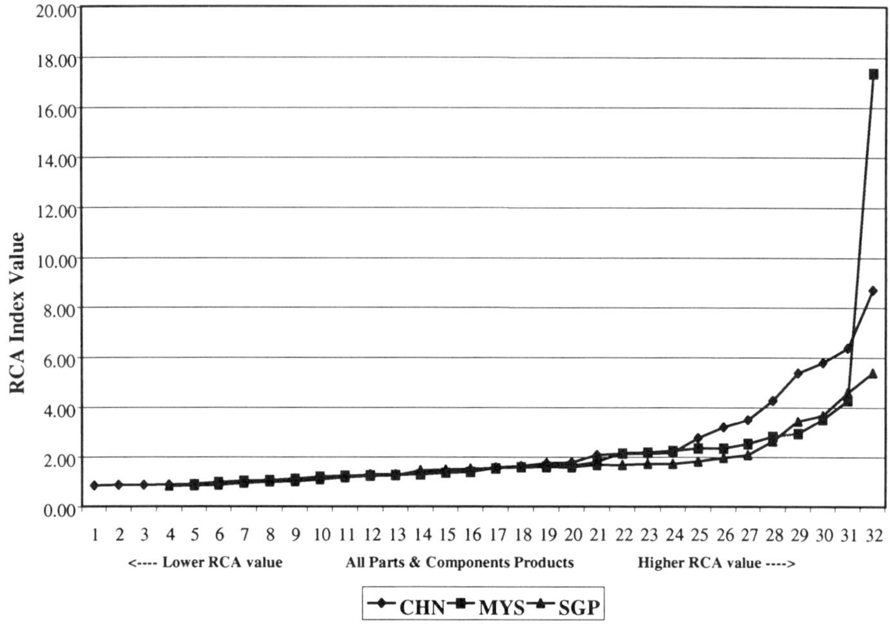

Appendix Figure 2: *The Relative Maturity of China, Malaysia and Singapore's 1996 Import RCA Indices.*

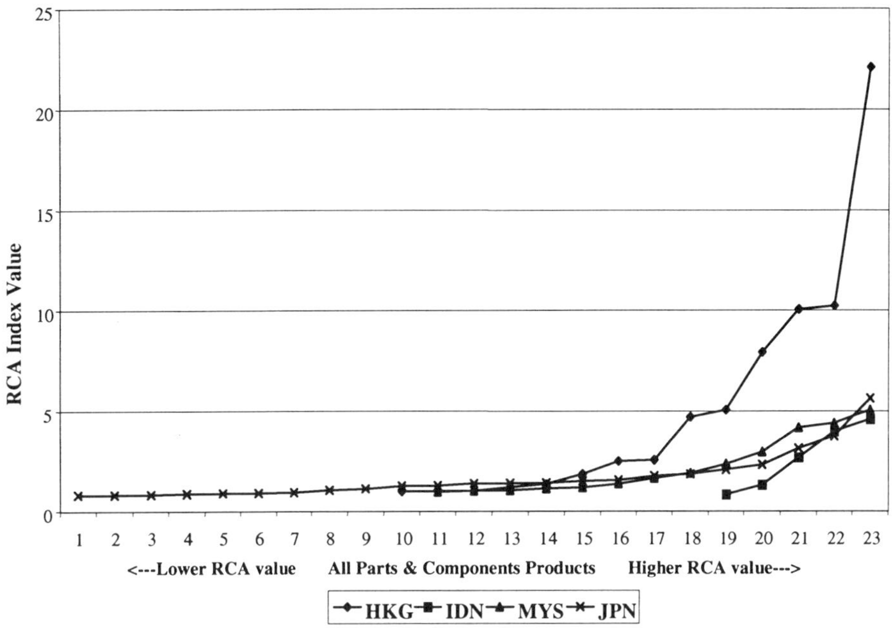

Appendix Figure 3: *The Relative Maturity of Hong Kong, Indonesia, Malaysia and Japan's 1996 Export RCA Indices.*

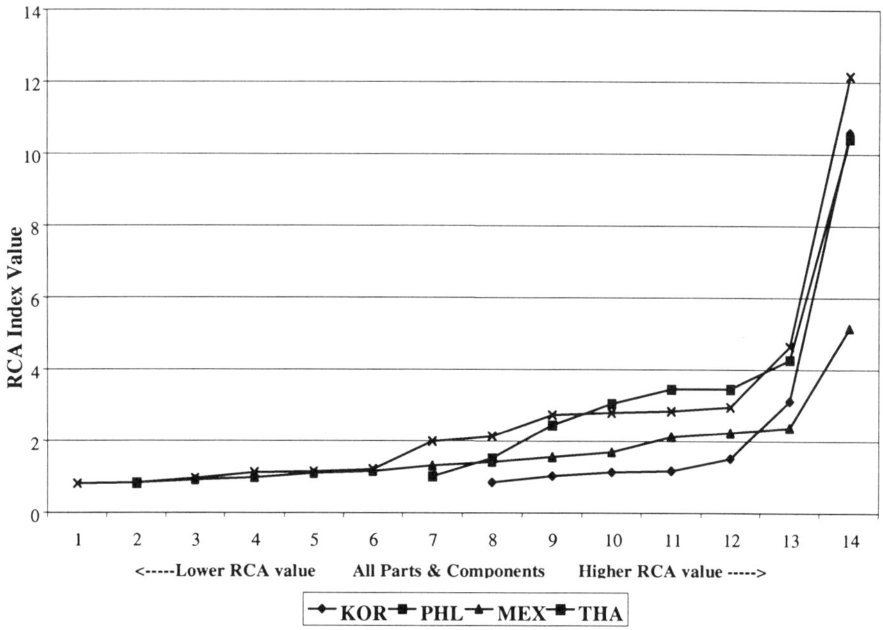

Appendix Figure 4: *The Relative Maturity of Korea, Philippines, Thailand and Mexico's 1996 Export RCA Indices.*

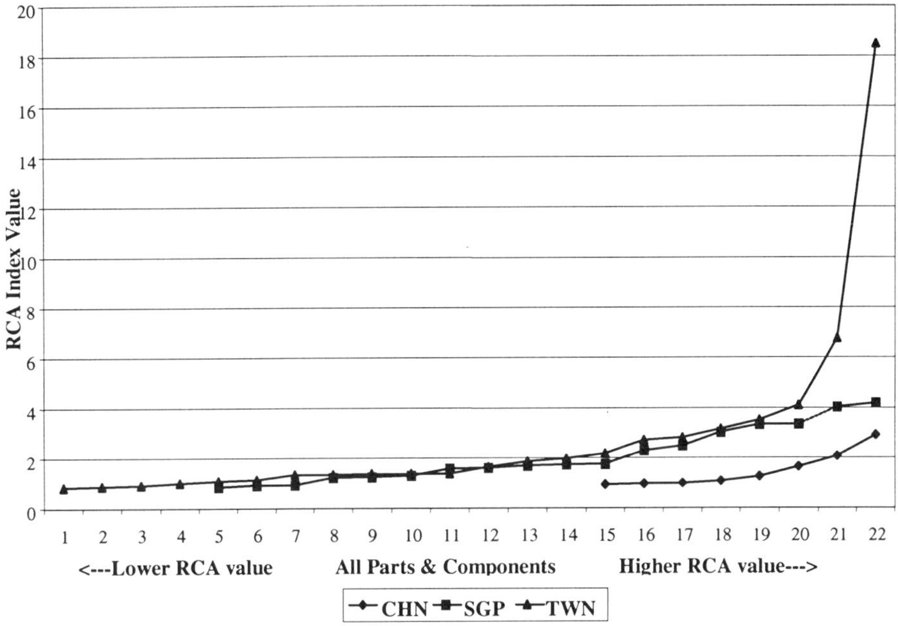

Appendix Figure 5: *The Relative Maturity of China, Singapore and Taiwan's 1996 Export RCA Indices.*

Chapter 6

INTRA-INDUSTRY FOREIGN DIRECT INVESTMENT AND TRADE FLOWS: NEW MEASURES OF GLOBAL COMPETITION

David Greenaway, Peter J. Lloyd, and Chris Milner[*]

1. INTRODUCTION

It has long been recognized that cross-border supply can take two forms: arms-length trade and direct production.[1] In the case of arms-length trade a firm supplies a given foreign market with exports that are then sold on through some supply chain. In the case of direct production the firm establishes (or acquires) an overseas affiliate to produce the commodity locally and then sell it on.

There is now an extensive literature analyzing the choice between supplying a foreign market by exporting or by establishing a foreign affiliate, that is, by foreign direct investment (FDI). This is the analysis of the choice of mode of supply. Much of that literature is concerned with the relationship between trade and foreign direct investment in particular. Following Mundell (1957) it was thought that trade and foreign direct investment were substitutes (crudely, in a world of differential factor endowments, either factors move or goods move). One by-product of the early literature on intra-industry trade, which is dominated by trade between countries with similar factor endowments, is that this particular proposition has been challenged, and many models dating from Agmon (1979) emphasize potential complementarities between trade and FDI. There is now a substantial theoretical literature on trade and foreign direct investment that has been summarized by Ethier (1994, 1996) and Markusen (1995, 1998).

This concept of substitutability/complementarity between trade and foreign direct investment captures an important relationship. However, it misses a rather important aspect, namely the impact of trade *and* foreign direct investment into different markets, irrespective of whether they are substitutes or complements. This paper addresses the fundamental issue of competition between the firms of different countries, recognizing the multiple modes of competition.

The key reason this issue has been missed that there is no widely used measure of "aggregate" supply in a given market (i.e., the aggregation of exports and affiliate production in a given context). We begin by setting out some basic definitions of trade and international production in Section 2 and then use these in Section 3 to develop some new concepts and measures of extended supply. In Section 4 we work through a series of applications to U.S. bilateral intra-industry supply. Section 5 considers how these patterns of extended trade relate to the phenomenon of fragmentation and looks at further extensions and developments.

2. DEFINING INTERNATIONAL TRADE AND INTERNATIONAL PRODUCTION

In the literature on multinational firms, supply through an affiliate is sometimes called "international production." "International production is defined as that production which is located in one country but controlled by a multinational corporation (MNC) based in another" (Cantwell 1994, p. 303). Using this definition we can identify the two modes of supply as international trade and international production, which together comprise "international supply." In its annual *World Investment Report,* the United Nations Council on Trade and Development (UNCTAD) has for some years noted that world production by foreign affiliates exceeds world trade in goods (e.g., UNCTAD 1995).

If we take the case of industry i in one country, we can define the relevant variables as follows:

X_i^j = exports (supply) to the foreign country or countries by mode $j = 1, 2$, and
M_i^j = imports (supply) to the foreign country or countries by mode $j = 1, 2$.

Let mode 1 be direct or arms-length trade in goods and mode 2 be international production. Then, X_i^1 and M_i^1 are exports and imports of goods of country i, and X_i^2 and M_i^2 are international production by country i's foreign affiliates and the international production in country i by the affiliates of foreign corporations. These definitions of exports and imports may be applied either to the total trade of a single country with all countries or to bilateral trade between one country and a second country. Initially we consider the total trade of a country.

There is some data relating to the ratio of production by foreign affiliates to imports and exports of goods and services. On the import side, there are measures of (M_i^2 / M_i^1). In Western Europe the ratio of sales of foreign affiliates to imports of goods and no-factor services in 1994 was 1.22 (UNCTAD 1997, Table I.3).[2] In North America the ratio in 1994 was 1.63. On the export side there are measures of (X_i^2 / X_i^1). In Western Europe the ratio of sales of affiliates abroad to exports of goods and nonfactor services in 1994 was 1.50 (UNCTAD 1997, Table I.3). In North America the ratio in 1994 was 2.07.

We can also add the value of trade by the two modes of supply from and to one country, country i. This gives

$$X_i = X_i^1 + X_i^2 \tag{1}$$

$$M_i = M_i^1 + M_1^2 . \tag{2}$$

That is, we add exports of goods and international production by MNCs of the same country, and the imports of the goods and the international production of foreign MNCs located in the country. This gives us the aggregate international supply from and to a country. The total flows in both directions may be called "extended exports" and "extended imports."

The notions of extended supply of "exports" and "imports" are appealing. They present a new view of competition. As an example, consider the concept of "import penetration." The standard measure of import penetration for an industry (i), in one country is

$$P_i = M_i^1 / S_i \tag{3}$$

where S_i = total domestic sales. When we recognize the existence of two modes of cross-border supply, this measure should be modified to

$$P_i = \left(M_i^1 + M_i^2\right) / S_i \tag{4}$$

For the analysis of structural adjustments in markets in which FDI has been significant, this may be preferable to the traditional measure, M_i^1 / S_i.

But there is a problem with the addition of exports and imports by mode. Statistics of the exports of one country include goods produced in the exporting country by plants of foreign affiliates. If one is including the production abroad of affiliates of the home country in its exports by mode 2 (X_i^2), it is not consistent to include part of the reverse flow by foreign affiliates in the home country as a part of the national supply. Conversely, imports include goods produced abroad by affiliates of the home country that are distinct from exports produced by foreign companies.

One way of overcoming these problems is to define the supply from one country in terms of the supply by companies owned by residents of the country, that is, the *nationality* of the company supplying the goods or services. Julius (1990, chapter 4) suggests a redefinition of world exports and imports of merchandise goods on the basis of the nationality of the producer. Her definition applies to supply by mode 1. Baldwin and Kimura (1998a) extend the concept of nationality of company to international production, and they combine direct cross-border sales with sales of affiliate companies. However, their focus is on a balance-of-payments accounting framework rather than on cross-border supplies and interfirm competition in particular markets. This leads them to measures that differ from those proposed here.[3] For their measures they report the aggregates across all industries, but at the industry level they report only the net trade, that is, the differences between exports and imports by each mode and for the two modes combined. By contrast, our measures take the flows in both directions and examine the extent to which they offset each other in each industry.

In order to compile statistics of cross-border supplies on the basis of the nationality of the firm, the statistics of goods exports should be adjusted by excluding exports produced by foreign companies, and the statistics of imports should be adjusted by excluding goods produced in foreign plants owned by domestic companies of the importing country. One can also measure supplies by mode 2 on the same nationality basis. This too should involve an adjustment for that part of the production in other countries by companies owned by residents of the home country that is exported from the country in which the foreign plant is located and, similarly, for that part of the production in the home country by foreign-owned plants that is exported.

Using the nationality definitions of supply, one can now add the supply by the companies owned by nationals of a country to other countries by means of the two modes. The sum of supplies by companies owned by residents of the home country by means of mode 1 and mode 2 gives the total supply to foreign countries by its companies. Conversely, the supplies received by the home country from other countries are calculated as supplies from companies owned by residents of other countries, either by direct export or by international production.

This definition designates the nationality of all production in the world economy according to the nationality of the firm producing it and then distinguishes between that part of the output of country *i*'s firms in some industry produced at home and that produced abroad. It excludes from exports and from the total supply from one country the value of goods produced in the exporting country by firms owned by residents of another country. This aggregation is desirable in that the "(extended) exports" of one country are equal to the "(extended) imports" of the other country and that world "(extended) exports" are equal to world "(extended) imports."

Most important, in the context of competition among firms and the interpenetration of national markets, aggregation on the basis of the nationality of firms is the logical choice of aggregation.[4] Firms of one country have characteristics that differentiate them from firms of other countries. The literature on international trade frequently assumes that the goods produced by an industry in one country are differentiated from goods produced by the same industry in another country. This is known as the Armington assumption, after the seminal paper by Armington (1969). It is also frequently assumed that the firms of one country have a different technology than those of another. Management literature emphasizes differences across countries in management methods, for example, the Just in Time and Quality Circles of Japanese firms. The classification of international supplies according to the nationality of the firm captures the national characteristics of production for both modes 1 and 2.

3. SOME NEW CONCEPTS OF INTRA-INDUSTRY COMPETITION

In most industries trade in both goods and the outputs of foreign-direct-investing enterprises is two way. We need to incorporate two-way "extended exports" and "extended imports" into our analysis. That is, analyses of markets needs to recognize both the two modes of supply and the two-way trade in each mode. In other words, we need a set of concepts appropriate to markets that reflect the complex patterns of supply in an age when sales and capital investments have become global.

Hitherto the conception and measurement of intra-industry trade has been confined solely to direct exports and imports of goods. Using the notion of two modes of supply, one can examine and measure the intra-industry international production between two countries in precisely the same way one examines and measures intra-industry (direct) trade. One can also sum the two sources of supply to a particular national market and the flows in the reverse direction. Such two-way flows might be called *extended intra-industry trade.*

This concept of extended intra-industry trade is closely related to several other areas of recent research. One is *intra-industry foreign direct investment.* (For early discussions of this phenomenon, see Grubel [1979], Dunning [1981], and Dunning and Norman [1986]). What the concept of international production does is to shift the focus to the output of foreign investors, rather than keeping it on the foreign investment itself. From the standpoint of evaluating adjustment to globalization and competition in markets, comparative advantage, and other market-related features, intra-industry production is clearly more useful than intra-industry investment.[5]

It is logical to begin the analysis of these phenomena by examining the supply to markets by both modes and the flows in the reverse direction. Extended intra-industry trade provides a general measure of the inter-penetration of national markets. Such features as intra-industry merchandise goods trade, intra-industry foreign direct investment, and intra-firm trade are an integral part of the complex patterns of supply in modern global markets.

Extended intra-industry trade can be measured in precisely the same way as traditional intra-industry trade. For example, one can use the Grubel-Lloyd measure (adjusted or unadjusted) (see Grubel-Lloyd 1975) or any other measure of the extent of this trade. All that is required is the redefinition of exports and imports to include both modes of supply to foreign markets.

There are two choices. One can consider intra-industry trade by mode 1 and mode 2 separately. Intra-industry trade by mode 1 is the conventional type of intra-industry trade. Intra-industry trade by mode 2 is the parallel two-way flows of international production. The measurement of these flows and the indices is straightforward in principle. The Grubel-Lloyd indices for both modes separately is

$$A_i^j = \left\{\left(X_i^j + M_i^j\right) - \left|X_i^j - M_i^j\right|\right\} / \left(X_i^j + M_i^j\right) \quad j = 1 \text{ or } 2. \tag{5}$$

For $j = 1$, this gives the standard Grubel-Lloyd index of intra-industry merchandise trade. For $j = 2$, this gives the Grubel-Lloyd index of intra-industry international production. Summing across industries, one obtains a Grubel-Lloyd index of intra-industry specialization for international production in the economy. It shows the proportion of the total international production flows in both directions, which is intra-industry.

The second alternative is to aggregate trade flows by both modes and measure extended intra-industry trade as

$$\text{EIIT}_i = (X_i + M_i) - |X_i - M_i|$$
$$= \left(\sum_{j=1,2} X_i^j + \sum_{j=1,2} M_i^j \right) - \left| \sum_{j=1,2} X_i^j - \sum_{j=1,2} M_i^j \right|. \quad (6)$$

The Grubel-Lloyd measure of intra-industry "trade" in the industry is

$$A_i = \{(X_i + M_i) - |X_i - M_i|\} / (X_i + M_i). \quad (7)$$

This index may measure the extent of extended intra-industry trade in bilateral flows or in the total trade of a country, depending on the definition of "exports" and "imports." Clearly, by summing across industries, an index of the economywide level of extended intra-industry trade can be generated. Like the Grubel-Lloyd index for goods trade alone, this is a weighted average of the extended intra-industry trade of the individual industries, and it lies in the closed interval [0,1].

The two modes of supply may have quite different implications in terms of their derived demand for labor and capital across countries. For an "importing" country, one possible objection to the use of these summations is that supply to the country by the mode of a foreign affiliate takes place in the home or importing country and, therefore, employs domestic labor and other factors, whereas goods imports do not. For the "exporting" country, the opposite is true. These two sources of supply may, therefore, have different implications for national income and for employment of factors on both the source country and the country of sale. One solution to these differences is to separate that component in the index that in effect reflects two-way direct trade from that part that reflects "two-way production."

This separation can easily be done with a decomposition. Given the definitions above, extended intra-industry "trade" in an industry can be decomposed into three components:

1. Two-way exchange of international trade in goods (mode 1)
2. Two-way exchange of international production (mode 2)
3. Two-way exchange of international trade for international production (modes 1 and 2)

Algebraically, the decomposition of the term in numerator of equation 7 is given by

$$\left(\sum_{j=1,2} X_i^j + \sum_{j=1,2} M_i^j \right) - \left| \sum_{j=1,2} X_i^j - \sum_{j=1,2} M_i^j \right| \equiv 2 \min(X_i, M_i)$$
$$= \sum_{j=1,2} 2 \min\left(X_i^j, M_i^j\right) + 2 \min\left(X_i', M_i'\right). \quad (8)$$

where $X_i' = X_i - \sum_{j=1,2} \min\left(X_i^j, M_i^j\right)$ and $M_i' = M_i - \sum_{j=1,2} \min\left(X_i^j, M_i^j\right)$.

Hence the index is an exactly decomposable index.

The three terms in equation 8 are the three components of extended intra-industry trade. This decomposition essentially looks at international trade in goods and in international production first and measures the overlap of exports and imports in these. It then measures the overlap in total exports and imports that is not due to overlap within the two modes. The first term is clearly standard intra-industry trade. The second term is the analogous flows for international production by MNCs. The third term is the mixture of the flows for the two modes, one flow in one direction and the other flow in the reverse direction; for example, a parent company might establish a foreign affiliate and then import the output of that affiliate to be used as an intermediate input in its home production.

Extended intra-industry trade and its decomposition therefore offers a potentially powerful tool for disentangling two-way flows of trade and international production and for gaining greater insight into the relationship between them.

4. APPLICATION TO U.S. BILATERAL INTRA-INDUSTRY SUPPLY

Clearly the data requirements for constructing the extended intra-industry trade index are more demanding than those for a standard intra-industry trade index. We still of course need data for the latter as an input into the index. In addition, however, we require data on affiliate production, and therein lies the key difficulty. Affiliate production data is neither as readily nor as comprehensively available as trade data, nor can it be obtained on a multilateral basis at suitable levels of country and industry disaggregation.

It can for some countries at least be obtained, however, on a bilateral basis. The most comprehensive data available are for the United States (U.S. Bureau of Economic Analysis [BEA]). This agency not only reports information on U.S. overseas affiliate production but also reports data on overseas affiliate production in the United States. These data are collected on a consistent basis and disaggregated by country and industry. Since the United States is the largest source of and host to FDI in the world, it is an especially good case to take from the standpoint of applying this new measure. In the next section we report the results of calculating extended bilateral intra-industry trade and its components for the United States.

Data on the activities of U.S. MNCs were obtained from the BEA, which provide annual data for the period 1989–94 on sales of affiliates of U.S. firms in alternative national markets and on sales of foreign affiliates in the U.S. market. We consider bilateral supply between the United States and five of its major industrial trading partners: Canada, Germany, France, the United Kingdom, and the Netherlands. For each country (host market for U.S. FDI or origin for inward U.S. FDI) under consideration, the disclosed sales figures for non-service-sector industries (International Standard Industry classification) were matched to the U.S. merchandise trade data,[6] classified according to the 1987 standard industry code (SIC). The resulting "industries" correspond therefore to between the two- and three-digit level of the SIC: somewhat, but not excessively, more aggregated than the level used in conventional intra-industry analysis. The sample size is considerably constrained by nondisclosure where bilateral annual data is concerned. In order to restrict the loss of industry observations, averages were taken for two periods, 1989–91 and 1992–94.

First, the Grubel-Lloyd measure (equation 5) of intra-industry exchanges were calculated separately for trade and international production. These measures are reported in summary fashion in Table 1 for U.S. bilateral exchanges with the five industrial trading partners. The specific industries and the number of industries covered in each bilateral exchange vary, and comparisons across countries need to be undertaken with caution. The number of industry observations for each period ranges from 12 to 21. For the share of arms-length (mode 1) intra-industry trade in gross trade, there are, rather surprisingly, low values in the case of Canada and France in the period 1989–91. The indices for the other countries are more in line with values from earlier studies with more extensive industry coverage; the indices range from 0.50 to 0.74. For intra-industry production (mode 2), the values are in the 0.35–0.65 range over the two periods. This shows that

Table 1: *Bilateral Intra-Industry Indices for Trade and International Production Separately*[(1)]

With	Trade	(paired)[(4)]	Foreign Production	(paired)[(4)]	Number of Observations	(paired)[(4)]
(1989–91 average[(2)])						
Canada	0.00	(0.00)	0.35	(0.42)	14	(11)
United Kingdom	0.60	(0.63)	0.58	(0.56)	21	(18)
Germany	0.74	(0.73)	0.61	(0.64)	12	(11)
France	0.00	(0.00)	0.46	(0.44)	12	(11)
Netherlands	0.50	(0.53)	0.42	(0.41)	13	(11)
(1992–92 average)						
	Trade		**Foreign Production**			
Canada	0.37	(0.40)	0.45	(0.49)	13	(11)
United Kingdom	0.66	(0.66)	0.58	(0.57)	19	(18)
Germany	0.67	(0.72)	0.64	(0.70)	15	(11)
France	0.50	(0.46)	0.44	(0.46)	13	(11)
Netherlands	0.44	(0.49)	0.48	(0.48)	15	(11)

(1) Grubel-Lloyd index.

(2) Arithmetic average.

(3) Sales by relevant foreign affiliates.

(4) Industries for which measures available in both periods.

substantial intra-industry exchange of supplies is a feature of international production, just as it is of traditional intra-industry trade.

Interestingly, the highest values for the shares of two-way exchange via mode 2 are for the two countries (Germany and the United Kingdom), for which the Grubel-Lloyd indices for mode 1 exchange are also highest. Although the country coverage is small, this is in line with a complementarity thesis.

In Table 1 we also identify averages for observations in which data for the same industries are available for both periods, called "paired" (both paired between periods) and those which are not, called "unpaired." Thus for the 11 common industries in the two periods for U.S.-Canada cross supply, the Grubel-Lloyd index for trade rises from 0.00 to 0.40 and for production from 0.42 to 0.49. Although the paired indices rise or fall for particular countries in the case of two-way trade, the paired indices for two-way production consistently increase. This is indicative at least of the growing importance of two-way exchange by this mode of supply.

If we cease to treat the two modes of intra-industry supply separately, we can estimate the Grubel-Lloyd index for extended intra-industry trade as described in equation 7. We do this for the same data set summarized in Table 1. The results of this estimation exercise are reported in summary fashion in Table 2. Crude averages for each period (and paired observations) are reported, as well as the ranges of values of the index across industrial sectors. It is clear from the averages that, at this particular level of aggregation, intra-industry supply is important, with intra-industry supply accounting on average for at least 50% of

Table 2: *Bilateral Intra-Industry*[(1)] *Indices For Extended Trade*[(2)]

With	Average[(3)]	(paired)[(4)]	Range[(3)]
(1981–91)			
Canada	0.31	(0.39)	0.00–0.97
United Kingdom	0.70	(0.71)	0.17–0.99
Germany	0.66	(0.69)	0.10–0.98
France	0.51	(0.49)	0.11–0.95
Netherlands	0.52	(0.55)	0.18–0.96
(1992–94)			
Canada	0.52	(0.58)	0.07–0.92
United Kingdom	0.70	(0.69)	0.14–0.98
Germany	0.66	(0.71)	0.16–0.95
France	0.51	(0.53)	0.12–0.95
Netherlands	0.59	(0.57)	0.18–1.00

(1) Extended intra-industry exchanges by means of arms-length trade and sales by foreign affiliates.

(2) Grubel-Lloyd index.

(3) For same number of observations reported on in Table 1.

(4) For same number of observations common to both periods as identified in Table 2.

total supply for all countries in the second period and all but Canada in the first period. There is, however, considerable variation across industrial sectors. The lower values in the industry range are consistently less than 0.20 and the upper values consistently greater than 0.90.

The potential importance of cross hauling of supply by both mode 1 and mode 2 and of distinguishing between intra-industry and extended intra-industry trade indices is forcibly illustrated by the difference in these two indices (in the first column of Tables 1 and 2). In most observations the proportion of intra-industry extended trade is higher than the proportion of intra-industry trade as conventionally measured for direct exports and imports alone.

In Table 3 we explore the relative importance (on average) of modes 1 and 2 forms of intra-industry supply, using the decomposition in equation 8. Each component is now expressed as a proportion of the total supply (i.e., by both mode 1 and mode 2). Presented in this form, it is evident that international production is the dominant mode of intra-industry supply for U.S. exchanges with these industrial countries, the international production component being well in excess of half of the total index in all but one case (Canada in 1992–94), when international production is still the dominant mode of intra-industry supply. Interestingly, this pattern of dominance is the same for both U.S.-Canada exchanges and for U.S. exchanges with the more distant European countries. It should be emphasized that lower values for intra-industry supply for United States–Canada than for United States–United Kingdom exchange imply higher values of inter-industry supply for the former than the latter country pairing. Supply of foreign markets by either mode 1 or mode 2 can be important in the case of both intra- and inter-industry specialization.

Table 3: *Decomposition of U.S. Bilateral Intra-Industry Supply*

		Components[2]		
With	**Total[1]**	**Trade**	**Foreign Production**	**Mixed Trade Production**
(1989–91 averages)				
Canada	0.31	0.00	0.22	0.09
United Kingdom	0.70	0.14	0.44	0.11
Germany	0.66	0.17	0.46	0.02
France	0.51	0.00	0.38	0.13
Netherlands	0.52	0.18	0.29	0.06
(1992–94 averages)				
Canada	0.52	0.17	0.24	0.11
United Kingdom	0.70	0.16	0.44	0.11
Germany	0.66	0.18	0.48	0.01
France	0.51	0.11	0.32	0.08
Netherlands	0.59	0.18	0.30	0.11

(1) *Indices of intra-industry supply given in Table 2 for maximum number of observation available for each time period.*

(2) *Decomposition of total index given by equation 7 in text.*

The residual component of mixed trade and international production in Table 3 is small relative to the other two components. The mixing of supply in one direction by mode 1 only with mode 2 only in the other direction is ≤ 0.13 for all observations, but it is not negligible. As shown above, we can formally regard "exports" and "imports" and the two-way exchange of "exports" and "imports" as the aggregation of the two modes of supply. Because of the mixed component, the index of extended intra-industry supply must exceed the average of the indices of intra-industry supply by mode 1 and mode 2 (see equation 8 and the component columns of Table 3). This is true of all observations.

In Table 4 we provide industry results for U.S.-U.K. intra-industry supply for the 1992–94 period. The general pattern of the results described above still holds when we examine these disaggregated results. Two-way foreign production (mode 2) remains the dominant component of intra-industry

Table 4: *Sector Breakdown of Total Intra-Industry Supply and Its Components for U.S.–U.K. Exchanges: 1992–2*

		Components		
Sector	**Total**	**Trade**	**Foreign Production**	**Mixed Trade Production**
Agriculture, forestry, and fisheries	0.63	0.13	0.36	0.13
Audio, video, and communications equipment	0.91	0.27	0.36	0.28
Construction	0.90	0.00	0.90	0.00
Electrical machinery	0.80	0.32	0.44	0.04
Electronic components	0.63	0.23	0.39	0.00
Fabricated metal products	0.95	0.14	0.78	0.04
Ferrous metal industries	0.60	0.15	0.44	0.00
Instruments and related products	0.90	0.15	0.70	0.06
Lumber and furniture	0.70	0.12	0.30	0.28
Mining and other extraction	0.14	0.00	0.00	0.13
Misc. plastic products	0.80	0.17	0.63	0.00
Motor vehicles and equipment	0.21	0.05	0.11	0.05
Nonferrous metal industries	0.96	0.11	0.17	0.68
Other transport equipment	0.85	0.57	0.05	0.23
Paper and allied products	0.98	0.09	0.82	0.07
Printing and publishing	0.65	0.11	0.54	0.00
Rubber products	0.53	0.16	0.34	0.03
Stone, clay, and glass products	0.42	0.06	0.36	0.00
Textile products and apparel	0.75	0.18	0.57	0.00
Average	0.70	0.16	0.44	0.11

supply, being the largest component of the total index for 16 of the 19 industries covered by the empirical analysis. There are some exceptions: "other transport equipment" where arms-length trade and "nonferrous metals" dominate, and "mining" where mixing of modes dominates. Further empirical work is clearly required to test whether aggregation effects and trade in components is affecting the results for specific sectors. But the consistency and pattern of the results across industrial sectors gives some encouragement to the idea that the extended intra-industry trade concept offers important insights. One would anticipate tradability issues, transport costs, and trade policy barriers to produce the type of cross-industry variations in the levels and components of intra-industry supply that are shown in Table 4, with foreign production being the particular dominant mode of two-way supply in construction, fabricated metal products, and paper and allied products, for example, and arms-length trade being more important in sectors that are not subject to high natural trade barriers (e.g., electrical products) or to very high fixed costs of production (e.g., other transport equipment). Similarly, the mixing of trade and production modes is likely to be fashioned by natural resource endowments and is therefore likely to be important in mining and nonferrous metal industries.

Finally in Table 5 we report on detailed results for four specific industries for which results were available for U.S. exchanges with all five countries. Again the pattern of results across bilateral exchanges for each industry shows some consistency and is in line with expectations. Foreign production is generally the dominant mode of intra-industry supply in these industries; arms-length trade is, for instance, of greater relative importance where tradability is greater (e.g., in electronic components, textiles, apparel). There are, however, country-specific effects (e.g., distance between trading patterns, natural resources endowments, policy incentives) that mean that the level and composition of intra-industry supply does vary for the same industry in these U.S. bilateral exchanges. Of course it has to be recognized that there may also be some third-country effects that distort the measures, with sales, for instance, from U.S. production in one EU country going to another EU country. Again this is more likely to occur in industries whose products are more tradeable within the EU.

The data requirements of implementing these measures are exacting. Our calculations have not taken into account two features of international intra-industry exchanges.

The first complication is that we were forced by data limitations to aggregate statistics of goods trade based on the country of production with statistics of international production that were based on the nationality of the company. We did not make adjustments to the statistics of both direct trade and international production for that part of international production that is

Table 5: *Decomposition of U.S. Bilateral Intra-Industry Supply: Selected Industries*

	Total	Trade	Foreign Production	Mixed Trade Production
	Components			
(1992–94 averages)				
a) Electronic components				
Canada	0.53	0.42	0.11	0.00
United Kingdom	0.63	0.23	0.39	0.00
Germany	0.89	0.25	0.61	0.04
France	0.62	0.30	0.33	0.00
Netherlands	0.88	0.10	0.71	0.07
b) Lumber and furniture				
Canada	0.19	0.00	0.19	0.00
United Kingdom	0.70	0.12	0.30	0.28
Germany	0.87	0.19	0.68	0.00
France	0.95	0.31	0.50	0.14
Netherlands	0.36	0.25	0.11	0.00
c) Plastic products				
Canada	0.91	0.21	0.35	0.35
United Kingdom	0.80	0.17	0.63	0.00
Germany	0.95	0.18	0.78	0.00
France	0.59	0.11	0.42	0.06
Netherlands	0.18	0.03	0.15	0.00
d) Textiles and apparel				
Canada	0.92	0.31	0.39	0.22
United Kingdom	0.75	0.18	0.57	0.00
Germany	0.94	0.25	0.64	0.06
France	0.91	0.08	0.75	0.08
Netherlands	0.94	0.23	0.58	0.13

returned to the home or parent country. These might be called *boomerang Seffects*. This lack of adjustment gives a hybrid measure of extended trade rather than one based purely on the nationality of the company.

The omission of these adjustments means that our calculations are an approximation of the concept of extended supply. However, it is unlikely that they have resulted in a substantial bias in the case of bilateral U.S. trade, given the size of that market and the fact that most overseas affiliates locate there to penetrate that market alone. As a high-wage country, the United States is not a major export platform for MNCs of other countries, and, conversely, shipments from U.S. affiliates overseas back to the United States are small relative to total U.S. imports. Moreover, in the context of intra-industry trade analysis, part of these flows offset each other. For example, exports by U.K. affiliates in the United States back to the United Kingdom should be excluded from U.S. exports to the United Kingdom and from U.K. international production in the United States. (If one considers the multilateral statistics of trade between the United States and all countries, they would cancel exactly.) When calculating measures of intra-industry trade, these nonexclusions cancel out in the numerator of the indices of extended intra-industry trade, but their nonexclusion would still bias the proportions of intra-industry trade by each mode and the decomposition of total extended intra-industry trade.

The second complication is what might be called *third-country effects.* When applying these concepts to bilateral exchanges, one needs to recognize that production of foreign affiliates of country A located in country B may export to country C rather than selling their output locally in country B. In a study of bilateral extended trade these flows of exports to C should be excluded from country A's extended supply to country B. Conversely, firms in country A may supply firms in country B by establishing affiliates in a third country C and exporting from there to B. To capture these flows would require a trilateral or multilateral approach, but, unfortunately, statistics of international production are not available for countries other than the United States and Japan.

5. EXTENSIONS

In this paper we set out the rationale for a family of new measures that make for a more direct "like for like" comparison of arms-length trade and international production. The measures explicitly recognize that two-way exchanges of goods, two-way exchanges of international production, and two-way exchanges of one for the other all simultaneously occur, and offer a way of disentangling them. We focus on U.S. bilateral exchanges of trade and production.

The results of calculating our new measure of extended intra-industry trade and its components are very revealing. In particular they demonstrate just how important two-way international production is relative to two-way trade. It accounts for most two-way extended trade. Given the growth of FDI relative to arms-length trade over the last decade, many analysts and policymakers may have had suspicions that this was so; as far as we are aware, this is the first time it has been demonstrated in this way. Moreover, when the mode of supply of foreign markets by international production is recognized, these new measures show that levels of two-way interpenetration of markets is even higher than the levels of two (intra-industry) direct merchandise trade.

The results could be extended in a number of ways. Estimates could be made of the boomerang effects that were not excluded from the flows of international trade and international production. The calculations of net extended trade ($X_i - M_i$), rather than the usual variable for direct trade only ($X_i^1 - M_i^1$), could be used for tests of comparative advantage.

In the context of this book these results raise the question of the relationship between measures of extended trade on the one hand and fragmentation on the other. Fragmentation is by its nature intra-industry trade. A substantial part of the two-way exchange of goods may indeed be vertical specialization among countries of goods in process, that is, fragmentation. A number of models now exist that yield fragmentation in the international trade of both goods and services (see the papers in Cheng and Kierzkowski, forthcoming). Using a unit value approach, Hine, Greenaway, and Milner (1994) have attempted to separate "vertical" and "horizontal" intra-industry trade in final products, but there has been no empirical separation of intra-industry trade in intermediates (i.e., fragmentation) from intra-industry trade in final goods.

There is a second connection between two-way extended trade and fragmentation of goods trade. Much outsourcing and trade in intermediate inputs is associated with FDI. Recently Cheng, Qiu, and Tan (forthcoming) have explored the effects of FDI on fragmentation in goods trade. As we have shown in this paper, the international production resulting from FDI can itself be regarded as a component of the extended supply from one country to another. Part of the production of goods at some stages in a chain of production within one country may be produced by companies owned by the residents of this country, but part may be produced by companies owned by residents of other countries (i.e., foreign direct investors). If these two sets of companies have distinct products and technologies, one can regard production *within* the country as fragmented. This is an extension of the standard concept of fragmentation, which considers the location of production but ignores the nationality of the companies producing the goods.

References

Agmon, T. (1979), "Direct investment and intra-industry trade: substitutes or complements?" in Giersch, H., ed., *On the Economics of Intra-Industry Trade* (Tübingen: JCB Mohr).

Armington, P. (1969), "A Theory of Demand for Products Differentiated by Place of Production," *IMF Staff Papers*, 16, 159–78 (Washington, D.C.).

Baldwin, R. E. and Kimura, F. (1998a), "Measuring U.S. International goods and Services Transactions," in R. E. Baldwin, R. E. Lipsey and J. D. Richardson, eds., *Geography and Ownership as Bases for Economic Accounting* (Chicago and London: The University of Chicago Press).

Baldwin, R. E. and Kimura, F. (1998b), "Application of a Nationality-Adjusted net Sales and Value Added Framework: The Case of Japan," in R. E. Baldwin, R. E. Lipsey and J. D. Richardson, eds., *Geography and Ownership as Bases for Economic Accounting* (Chicago and London: The University of Chicago Press).

Cantwell, J. (1994), "The Relationship Between International Trade and International Production," in Greenaway, D. and Winters, L. A., eds., *Surveys in International Trade* (Oxford: Basil Blackwell).

Carr, D. L. Markusen, J. R. and Maskus, K. (1998), "Testing the Knowledge-capital Model of the Multinational Enterprise," paper presented to The First Annual Workshop on Empirical Research on International Trade and Investment, Fredensborg, Denmark, 19–21 June, 1998.

Cheng, L. K. and H. Kierzkowski, eds. (1999), this volume.

Cheng, L. K., Qiu, L. D. and Tan, G. (1999), "Foreign Direct Investment and International Fragmentation of Production," this volume.

Dunning, J. (1981), "A Note on Intra-industry Foreign Direct Investment," *Banca Nazionale del Lavoro Quarterly Review*, 139, 427–37.

Dunning, J. H. and Norman, G. (1986), "Intra-industry investment," in H. P. Gray, ed., *Research in International Business and Finance*, 5 (Greenwich, Conn.: JAI Press).

Ethier, W. J. (1994), "Multinational Firms in the Theory of International Trade," in Bacha, E., ed., *Economics in a Changing World* (London: Macmillan).

Ethier, W. J. (1996), "Theories about Trade Liberalisation and Migration: Substitutes or Complements," in Lloyd, P. J. and L. Williams, eds., *International Trade and Migration in the APEC Region* (Oxford: Oxford University Press).

Feenstra, R. C. (1996), "US Imports, 1972–94: Data and Concordances," NBER Working Paper, No 5515.

Feenstra, R. C. (1997), "US Exports, 1972–94: with State Exports and Other US Data," NBER Working Papers, No. 5990.

Greenaway, D., Hine, R. C. and Milner, C. R. (1995), "Vertical and Horizontal Intra-industry Trade," *Economic Journal*, 95, 1505–1514.

Greenaway, D., Hine, R. C. and Milner, C. R. (1994), "Vertical and Horizontal Intra-industry Trade: A Cross-country Analysis for the United Kingdom," *Economic Journal*, 105, 1505–1519.

Grubel, H. G. (1979), "Towards a Theory of Two-way Trade in Capital Assets," in H. Giersch (ed), *On the Economics of Intra-industry Trade* (Tübingen: J. C. B. Mohr).

Grubel, H. G. and Lloyd, P. J. (1995), *Intra-industry Trade* (London: Macmillan).

Harris, R. G. (2001), "A Communication Based Model of Global Production Fragmentation," in S. Arndt and H. Kierzkowski, eds, *Fragmentation: New Production Patterns in the World Economy* (Oxford: Oxford University Press).

Jones, R. W. and Kierzkowski, H. (1990), "The role of Services in Production and International Trade: A Theoretical Framework," in R. W. Jones and A. Krueger, eds., *The Political Economy of International Trade* (Oxford: Basil Blackwell).

Julius, D. (1990), *Global Companies and Public Policy* (London: R11A).

Kimura, F. (1998), "Japanese Multinationsl and Regional Integration in Asia," in K. Fukasaku, F. Kimura and S. Urata, eds., *Asia and Europe: Beyond Competing Regionalism* (Brighton, UK and Portland: Oregon Sussex Academic Press).

Markusen, J. R. (1995), "The Boundaries of Multinational Enterprises and the Theory of International Trade," *Journal of Economic Perspectives*, 9, Spring, 169–189.

Markusen, J. R. (1998), "Multinational Firms, Location and Trade," *The World Economy*, 21, 733–756.

Petri, P. (1997), "Foreign Direct Investment in a Computable General Equilibrium Framework," paper presented to the Conference on Making APEC Work: Economic Challenges and Policy Alternatives, Keio University, March 13–14, 197.

United Nations Conference on Trade and Development UNCTAD) (1995), *World Investment Report: Transnational Corporations and Competitiveness* (New York: UNCTAD).

Endnotes

*We would like to acknowledge the comments of Bob Baldwin, Fukunari Kimura, and others at the Workshop in Hong Kong.

[1]For the owner of a technology that may be sold in a foreign country, there is a third alternative of licensing another agent of production. This alternative could be defined as the third mode and analyzed in the same way, but it does not involve production and supply by the owner of the technology and will not therefore figure in our analysis.

[2]To compare the shares of the markets located in Western Europe, one needs to exclude the part of the sales of affiliates of MNEs that are exported to third countries. No data is available for individual countries in Western Europe.

[3]Thus in calculating direct cross-border sales from the United States to foreigners they excluded not only U.S. exports shipped by U.S. affiliates of foreign firms but also U.S. exports to U.S. foreign affiliates of U.S. firms (Baldwin and Kimura, 1998a, p. 14 and Table 1.1). The second exclusion is generally inappropriate when examining supplies to particular geographic markets. Parallel adjustments were made to imports and to international production. Moreover, they were primarily interested in the balance of sales between U.S. firms and foreigners ("net sales").

[4]For the purpose of examining the effects of extended supply on factor markets and domestic incomes, it may be preferable to define trade on the basis of value added by factors owned by residents of the country. This approach has been suggested by Baldwin and Kimura (1998a) and applied by Baldwin and Kimura (1998b) and Kimura (1998) to Japanese trade. In both

modes the value added by factors residing in the country that supplies the goods or services is less than the gross value of output. Foreign affiliates import management and other labor and, conversely, the affiliates in other countries may import labor from the home country, and both import intermediate and fixed capital inputs. What contributes to national incomes in both modes is the value added by residents of the home country. This definition would not involve any double counting or inconsistency. The supply by both modes would be part of national income. From the point of view of international trade statistics, this aggregation has the necessary property that the "exports" of one country are equal to the "imports" of the other, and world "exports" equal world "imports."

[5]In principle this concept of extended intra-industry trade, like the standard concept, can be applied to trade in services as well as to goods. Indeed, it is particularly important in the analysis of trade in services. Since the conclusion of the General Agreement on Trade in Services (GATS), it has been recognized that the cross-border delivery of services can be accomplished through the establishment of foreign affiliates in the country in which the service is delivered, the so-called "commercial presence" mode of supply. The four modes of supply recognized in GATS are a further breakdown of the two modes listed above, which is useful for analysing the particular characteristics of production and trade of services. In this service sector, supply by the mode of international production is called "trade" in services. Jones and Kierzkowski (1990) and Harris (2000) explore the relationship between trade in services and fragmentation.

[6]Obtained from the National Bureau of Economic Research on disk with Feenstra (1996, 1997).

Chapter 7

FRAGMENTATION, INTERNALIZATION, AND INTERFIRM LINKAGES: EVIDENCE FROM THE MICRO DATA OF JAPANESE MANUFACTURING FIRMS*

Fukunari Kimura

1. FRAGMENTATION AND INTERNALIZATION IN THE EMPIRICAL CONTEXT

The concept of fragmentation has been developed in the context of international trade theory and thus has naturally inherited a strong flavor of location theory. Deardorff (1998, p. 1) defines fragmentation as "the splitting of a product process into two to more steps that can be undertaken in different locations but that lead to the same final product." Two key ideas relevant to the concept of fragmentation are production blocks and service links (Jones and Kierzkowski 1990). Owing to the drastic cost reduction of service links such as transportation, telecommunication, and coordination, a set of production processes can be split into multiple production blocks in different locations. The concept of fragmentation has been developed primarily in conjunction with international trade theory; it relates to local production conditions and the costs of connecting different locations.

To apply the concept of fragmentation to an empirical situation, however, we must consider another important concept, that of the firm. Whether split production blocks are under a single firm (or firm group) or not is crucial when we empirically investigate fragmentation patterns. A firm has firm-specific assets, such as production technology and managerial ability, which determine how effectively it uses local conditions by setting up a production block. The implications of intrafirm fragmentation of the production process are quite different from those of arm's-length division of labor.

The theoretical treatment of the concept of the firm is not always explicit in the fragmentation literature. When Helpman (1984) formalizes the activities of multinational enterprises (MNEs) in an analytical model he discusses the geographical fragmentation of headquarter activities and production activities within a firm. Jones and Kierzkowski (1990) also seem to take intrafirm fragmentation for granted. When we write theoretical models for fragmentation, however, the base model is often along the traditional international trade model, and thus we do not necessarily distinguish intrafirm fragmentation from arm's-length fragmentation (see e.g., Jones and Kierzkowski, 2001 and Deardorff 2001). Actually, we do not have to limit our application of the concept of fragmentation to intrafirm fragmentation. Various patterns of arm's-length fragmentation, such as cross-border original equipment manufacturing (OEM) and subcontracting, also exist.

However, we still have to stress that intrafirm fragmentation is an important fragmentation mode and that this mode requires special attention. Cheng, Qiu, and Tan (2001) analyze foreign direct investment (FDI) as a vehicle for technological transfer. FDI is not a simple international capital flow but is accompanied by the flow of firm-specific assets such as technology and managerial ability. The source of location advantage for foreign affiliates comes from the combination of local conditions and firm-specific assets. In addition, FDI generates a special kind of service link that result in huge reductions in the cost of transactions between different locations. Jones and Kierzkowski (2001) and Harris (2001) emphasize strong increasing returns in the production of linked services. FDI is actually intended to overcome the disadvantage of international service links by controlling all the service links within a firm.

Furthermore, a firm's decisions about fragmentation are closely related to its decisions about internalization. Although the international business literature has long been claiming the importance of internalization decisions (see Dunning 1993), international trade theory has directed less attention to such decisions. Actually, the decisions a firm makes about what to do internally and how to fragment activities geographically are closely linked. How a firm chooses to proceed depends on its specific assets as well as on the economic environment, which may include the possibility of establishing effective interfirm linkages. The analysis of intrafirm fragmentation therefore requires an integrated framework for fragmentation and internalization decisions.

There are various types or modes of internalization in the real world, which results in a variety of different kinds of fragmentation. Although in theory we often treat a firm as an individual economic agent, a firm's boundaries are not always clear-cut; they may have quite a bit of gradation. A firm may have multiple establishments and domestic/foreign affiliates. It may have a long-term outsourcing/subcontracting arrangement or an alliance with other firms, and thus we must take the issue of interfirm relationships into consideration. The integrated approach to fragmentation and internalization may allow us to investigate, for example, economic elements unique to international fragmentation vis-à-vis domestic fragmentation or no fragmentation. Fragmentation within a cone and fragmentation across cones (see Deardorff, forthcoming) may reveal different patterns in location and internalization.

Although we desperately need to accumulate empirical knowledge, statistical data for fragmentation/internalization patterns are very limited. One available data set is on Japanese corporate firms; it was compiled by the Ministry of International Trade and Industry (MITI), Government of Japan. Although it is difficult to capture the details of activities and the nature of transactions, the MITI database provides precious information on the structure of firms, including establishment holdings, domestic and foreign affiliate holdings, commissioned work and subcontracting arrangements, and others. This paper analyzes the questionnaire-level micro data of Japanese machinery industry in fiscal year (FY) 1994.

The existing literature suggests that Japanese MNEs are different from other MNEs. A number of articles including Yamawaki (1991) and Kimura and Baldwin (1998) reveal the fact that the internalization of sales activities is important feature for Japanese firms when they export their products to developed countries. Belderbos and Sleuwaegen (1996) claim that interfirm linkages (or *keiretsu*), which include both horizontal and vertical linkages, attracts Japanese FDI to other Asian countries. Markusen and Maskus (1999a, 1999b) claim that horizontal MNEs are overwhelmingly dominant in U.S. cases. However, the case of Japanese multinationals seems to show both horizontal and vertical patterns. As discussed by Kimura (1998, 2000a), the foreign affiliates' choice of location is closely associated with their choice of activities. All of this suggests that the Japanese machinery industry is an important case for the fragmentation and internalization literature.

Furthermore, as Kimura (1999) points out, large, research and development (R&D)-intensive, export-oriented, and foreign-exposed firms do not necessarily perform better in the Japanese manufacturing sector as compared to the U.S. manufacturing sector (see Doms and Jensen 1998). This suggests that interfirm linkages make small firms' efficiency-enhancing deinternalization strategy effective in Japan. The industrial organization literature finds considerable heterogeneity of industrial organization across countries in terms of the size distribution of firms and outsourcing/subcontracting patterns.[1] Owing to its technological nature, fragmentation and internalization are particularly important in the machinery industry.

The paper is organized as follows. Section 2 discusses the stratified structure of fragmentation and internalization as well as its statistical tractability. Section 3 provides an overview of internalization patterns in the Japanese machinery industry with the micro data set compiled by MITI. Section 4 presents findings on the relationship between interfirm linkages and internalization by examining subcontracting arrangements in Japan. Section 5 statistically analyzes corporate firms' choices about establishment/affiliate holding patterns by using subcontracting patterns as an anchor. Section 6 summarizes the results and presents a future research agenda.

2. STRATIFIED PATTERNS OF FRAGMENTATION AND INTERNALIZATION

In theory, we usually treat a firm as a profit-maximizing individual economic entity that conducts a well-defined set of economic activities. Within this context, the fact that a firm has a unique decision-making body is taken for granted, and a clearly defined headquarters dictates all the fragmented production blocks and the service links connecting them.

In reality, however, the structure of a corporate firm is much more complicated. First, it is not easy to identify a standard set of activities that all firms conduct. Even the most precise definition of activities does not necessarily describe what any given firm does. Production activities often have a complex, firm-specific structure. In addition, a firm has a number of activities or functions other than production activities. It may have a headquarters function that includes overall planning, financial management, personnel management, and legal services. It may involve widely defined production activities including R&D, technology management, production control, quality control, purchases, and inventory management of parts and components. In addition, it may conduct marketing activities including marketing surveys and planning, inventory control of products, logistics arrangements, and advertising. Considering firm-specific assets and the saving of transaction costs, a firm decides which activities and functions are internalized and which are left for other firms, and at the same time geographically locates the internalized activities and functions. Not all firms have a full set of the above-listed functions and activities within their boundaries.

Second, it is not easy to identify the boundaries of a firm in the empirical context. From the theoretical point of view a firm's managerial control is crucial to identifying its boundaries, but such control is not directly observable in statistics. Statistically available is information on ownership (e.g., stockholding shares), not information on the control itself. In addition, there is no guarantee that all headquarter functions reside in one geographical location. As a result of these uncertainties the economic meaning of the term "firm" is unclear. Although cross-country differences in the statistical treatment of the term are large, a firm is usually defined as an individual legal entity endorsed by official registration. It may have a single establishment or multiple establishments, and establishments are likely to be under the control of the firm. Difficulties arise when we try to consider affiliates. By definition, a parent firm holds all or part of its affiliate's capital stocks, but to what extent a parent firm controls its affiliate depends on the individual case. An affiliate, whether domestic or foreign, is a legal entity independent of its parent firm. Some affiliates are joint ventures with other firms, which serves to blur the boundaries even more.[2] In addition, there exists a large gray area. Even if there is no capital-holding relationship, some firms accept key executives from other firms. Some firms have explicit or implicit long-term outsourcing/subcontracting arrangements with other firms. Such relationships are looser than complete vertical integration but tighter than spot-market bidding. Last but not the least, firms can also form strategic alliances. Thus the boundaries of a firm are not at all clear-cut in reality.

Third, service links that connect domestically and internationally fragmented plants are not empirically observable in most cases. We have relatively abundant information on what each firm or each establishment is doing, but it is usually difficult to determine what sort of service links it has, and with whom. The concept of service links includes various types of devices that connect production blocks (e.g., transportation, communication, and coordination). Some intangible types of service links (e.g., coordination) are no doubt important, particularly when we analyze FDI, but they are hard to capture statistically. Even if we limit our discussion to commodity transactions, various forms of these transactions still exist. There are pure arm's-length transactions and intrafirm/cross-establishment transactions at the two extremes. In between there are intrafirm-group transactions with domestic/foreign affiliates and transactions with *keiretsu,* commissioned work contracts, and subcontracting arrangements. Transactions within the same firm nationality may also be qualitatively different from pure international arm's-length trade (Baldwin and Kimura 1998). These links in various modes are again very important for characterizing fragmentation, but data are limited.

Nevertheless, there is some statistical information on the pattern of fragmentation and internalization in the case of Japanese firms. *The Basic Survey of Business Structure and Activity,* produced by MITI, provides detailed data about firms in Japan and their foreign affiliates. This survey was conducted first in FY 1991, then in FY 1994, and annually afterwards. Its main purpose is to capture the overall picture of Japanese corporate firms in light of their activity diversification, globalization, and strategy for R&D and information technology. Its strength is its census coverage of samples and the reliability of its figures. We must, however, be careful to remember that it only covers large domestic firms in specific industries. The domestic firms covered have more than 50 workers; more than 30 million yen in capital; and establishments in mining, manufacturing, wholesale and retail trade, or the restaurant industry. Domestic and foreign affiliates are defined as those in which Japanese domestic firms hold more than 50% of the shares. We use the questionnaire-level FY 1994 data (hereinafter referred to as the MITI database).

The MITI database provides the number of establishments and domestic/foreign affiliates each firm has in various locations. The activities of establishments are not reported in detail. The activities of domestic/foreign affiliates are classified by industry code, although the classification is too rough to identify the precise form of fragmentation. Data about transactions between establishments or between a parent firm and each affiliate are not available. The MITI database also includes questions on using subcontractor(s) (using *shitauke*) or not, and working as a subcontractor (doing *shitauke*) or not (only for FY 1991 and 1994). The definition of subcontracting (*shitauke*) is not often explicitly specified, even though we casually use the term quite often. Statistically, MITI defines subcontracting as follows:[3] "A contractual arrangement between a firm and a "parent" firm with more capital or a larger number of employees. The former firm is commissioned to produce products, parts, attachments, materials, or components used as inputs in the parent firm's production, or to produce or repair facilities, equipment, tools, and other things the parent firm uses in production. When a firm purchases noncustomized parts, components, and other such items through a usual marketing channel, this is not regarded as subcontracting. In subcontracting, a parent firm assigns a job directly to a subcontractor, assigning a plan, quality, specs, form, design, and other elements."[4]

Although this definition is still fuzzy, the questionnaire for the statistics includes questions related to subcontracting, which seem relatively easy to fill out. In any case, we must be careful when we make an international comparison about subcontracting.

Figure 1 illustrates what we can capture from the MITI database. We can see how many establishments and domestic/foreign affiliates each firm has. In addition, we can determine whether a firm uses subcontractors or works as a subcontractor for other firms. Thus we can analyze the relationship between establishment/affiliate holding patterns and other characteristics of firms. Of course, this does not reveal the whole picture of internalization and fragmentation. However, we can at least analyze some important features of a firm's choice on scope of activities, modes of internalization, and interfirm relationships.

3. FRAGMENTATION AND INTERNALIZATION PATTERNS: AN OVERVIEW

Machinery industry, which includes general machinery, electric machinery, transport equipment, and precision machinery, has a sophisticated structure of fragmentation and internalization. The production process is typically multilayered, and each part requires different levels of technology and diversified factor inputs. Such technological characteristics mean that delicate decisions must be made as to choices about fragmentation and internalization. Advertising and sales activities are also important in the downstream, particularly for home electric appliances and office machines. The Japanese machinery industry is a virtual showcase of various types of fragmentation and internalization.

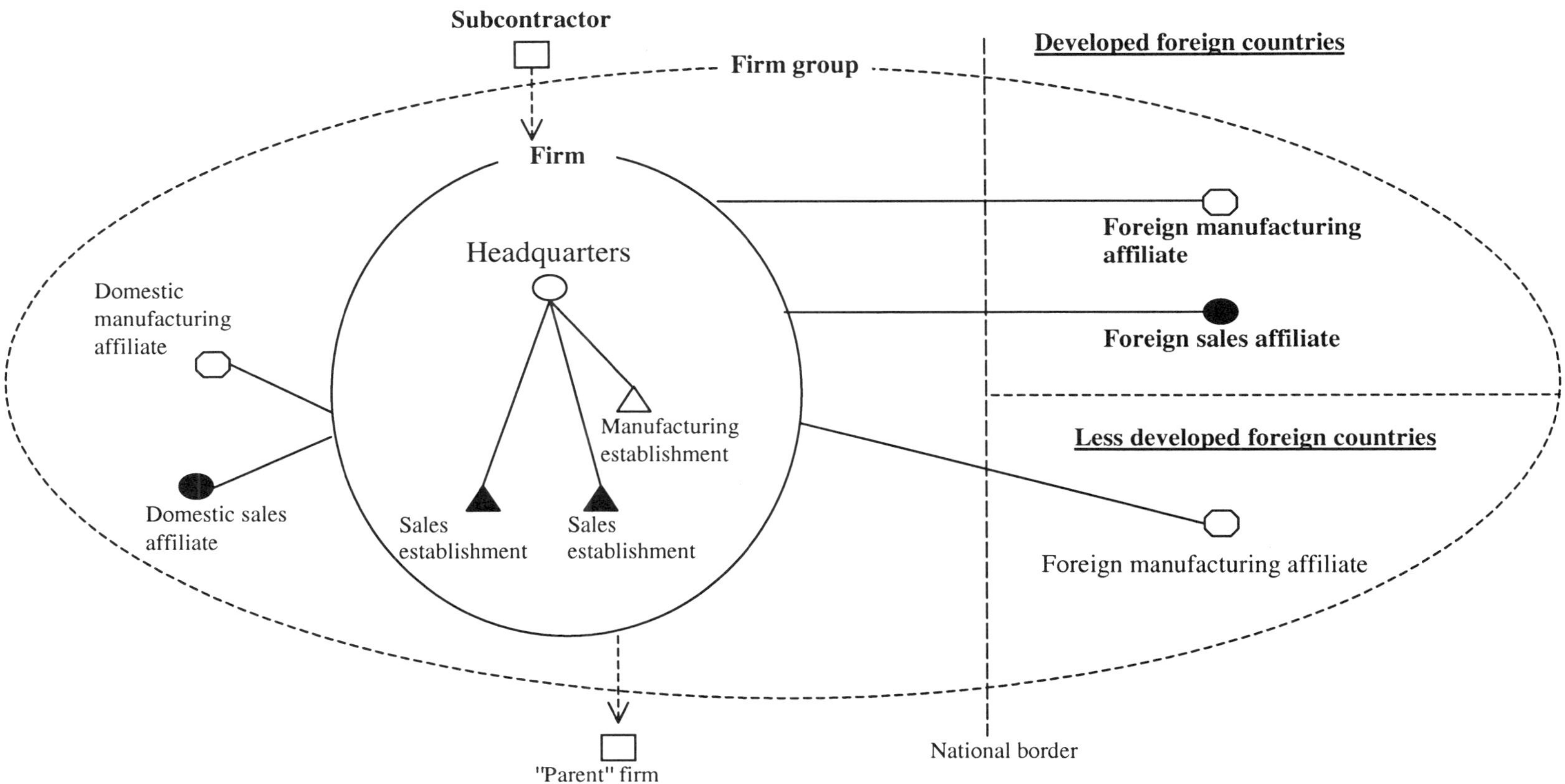

Figure 1: *The Internalization/Fragmentation Pattern: An Illustration.*

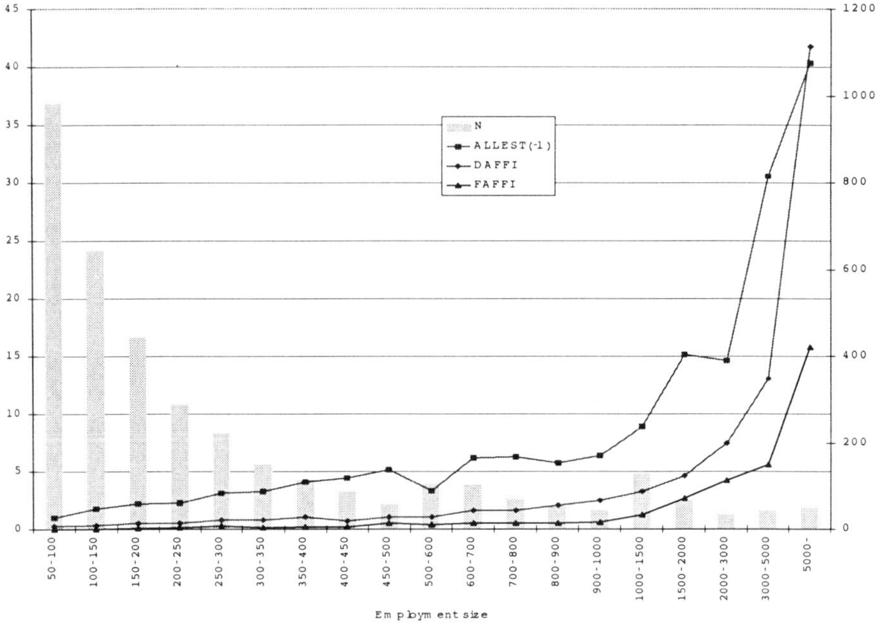

Figure 2: *Establishment/Affiliate Holding Patterns by Firm Size.*

Figure 2 presents the size distribution of Japanese machinery manu-facturing firms as well as their average number of establishment/affiliate holdings. Firms are classified by employment size, and N stands for the number of firms shown by the scale on the right-hand-side axis.[5] Because the MITI database covers only firms with 50 or more workers, the data are truncated on the left. However, the size distribution shows a normal decreasing pattern. ALLEST(-1) is the average number of establishments held (minus one for headquarters), DAFFI is the average number of domestic affiliates held, and FAFFI is the average number of foreign affiliates held. The scale for these three is shown on the left-hand-side axis. The domestic and foreign affiliates include majority-owned affiliates only. The numbers of establishments and domestic/foreign affiliates naturally increase as firm size grows. We, however, would like to claim below that this is not due to a simple size effect. Different establishment/affiliate holdings are strongly associated with different patterns of fragmentation and internalization. Firms vary not only in terms of size but also in terms of activity. Perhaps firm size is not a predetermined exogenous variable but an endogenously determined choice variable that comes together with internalization/fragmentation decisions.

We also claim that patterns of fragmentation and internalization are closely related to interfirm relationships. Figure 3 presents subcontracting patterns by firm size. PROBUSUB denotes the proportion of firms that

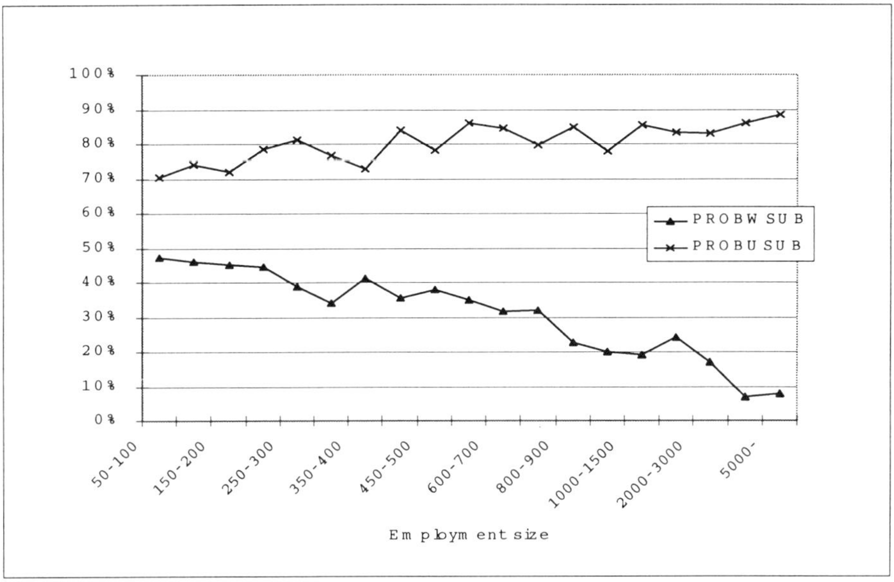

Figure 3: *Subcontracting Patterns by Firm Size.*

use subcontractor(s), and PROBWSUB stands for the proportion of firms that work as subcontractors. PROBUSUB starts at the 70% level and gradually increases as firm size gets larger. On the other hand, PROBWSUB steadily decreases from 47% to less than 10% as firm size expands. Obviously a large number of firms are both using subcontractor(s) and working as subcontractors, which means that the subcontracting system is multilayered in the machinery industry. As we discuss later, subcontracting arrangements strongly affect what to internalize and how to fragment activities.

Although it is difficult to identify the activities of parent firms and affiliates in detail, the MITI database provides industrial codes that they belong to. Tables 1 and 2 present the number of domestic affiliates and foreign affiliates that Japanese parent firms have. Each row presents the industry of parent firms, while each column shows the industry of domestic/ foreign affiliates. Industries 290, 300, 310, and 320 are general machinery, electric machinery, transport equipment, and precision machinery. Industries 120 to 340 are manufacturing industries, and industries 481 and 540 are wholesale trade and retail trade.[6] Because many domestic/foreign affiliates belong to the same industry as their parents,

Table 1: *Industries of Japanese Parent Firms and Domestic Affiliates, 1994*

(Number of Domestic Affiliates)

		Industry of Domestic Affiliates																										
		120	**130**	**140**	**150**	**160**	**170**	**180**	**190**	**200**	**210**	**220**	**230**	**240**	**250**	**260**	**270**	**280**	**290**	**300**	**310**	**320**	**330**	**340**	**050**	**481**	**540**	**Others**	**Subtotal**
Industry of	290	1	4	1	2	3	1	3	8	26		31	6		15	31	17	64	1093	72	53	41		22	5	992	105	849	3445
Parent firms	300	5	1	4	5	1	4	1	12	36	1	47	1	2	18	14	23	57	148	2113	34	56		157	3	749	98	1346	4936
	310	1	2	1	1	2	2	2	9	3	3	10	5		5	28	13	49	147	33	1380	9		18		636	336	767	3462
	320									3		2			3			4	16	31	6	250		1		112	13	101	542
	Subtotal	7	7	6	8	6	7	6	29	68	4	90	12	2	41	73	53	174	1404	2249	1473	356	0	198	8	2489	552	3063	12385

For the industrial codes, please refer to the appendix.

Data source: *MITI database.*

Table 2: *Industries of Japanese Parent Firms and Foreign Affiliates of Japanese Firms, 1994*

(Number of Foreign Affiliates)

		Industry of Foreign Affiliates																										
		120	**130**	**140**	**150**	**160**	**170**	**180**	**190**	**200**	**210**	**220**	**230**	**240**	**250**	**260**	**270**	**280**	**290**	**300**	**310**	**320**	**330**	**340**	**050**	**481**	**540**	**Others**	**Subtotal**
Industry of	290	1			4		2	1	2	2		6			4	8	2	13	561	23	20	13		6		491	12	143	1314
Parent firms	300			1						8		13			5	1	3	3	54	1129	9	25		42		723	30	238	2284
	310						2						2			4	1	1	30	7	602			2		188	22	161	1022
	320									4					1				1	8		151		1		99	2	11	278
	Subtotal	1	0	1	4	0	4	1	2	14	0	19	2	0	10	13	6	17	646	1167	631	189	0	51	0	1501	66	553	4898

For the industrial codes, please refer to the appendix.

Data source: *MITI database.*

large numbers are found in the cells of (290, 290), (300, 300), (310, 310), and (320, 320). However, a substantial number of domestic/foreign affiliates are classified in industries different from their parents'. In particular, there are many domestic/foreign affiliates that conduct wholesale activities.

Domestic affiliates are different from foreign affiliates in that they cover a wide variety of manufacturing industries. This reflects the fact that Japanese firms tend to establish domestic affiliates when they enter industries that are new to them and try to diversify the scope of their activities. Foreign affiliates tend to stick to the same production activities that parent firms conduct or to internalize wholesale/retail trade activities abroad. As is shown in Kimura (2000a), there are very few wholesale affiliates in developing Asian countries, whereas manufacturing affiliates are found in both developed and developing countries. These differences mean that the form of fragmentation differs according to destination.

4. INTERFIRM LINKAGES AND INTERNALIZATION

In Figures 2 and 3 we place the employment size of firms on the horizontal axis and treat that as an anchor of discussion. However, when we consider the internalization issue, the size of firms should be treated not as exogenous but as an endogenously determined varisable considering internalization and interfirm relationships. Because the empirical literature on this issue is thin, it is at this moment difficult to express the causal relationship of firm size, internalization, and interfirm relationships (i.e., it is not easy to say which of these determines the others). In this section, by using subcontracting patterns as a temporary anchor, we investigate the pattern of internalization and fragmentation. We empirically show that fragmentation, interfirm linkages, and internalization are closely related to each other.

Table 3 classifies the samples into four groups as follows:

Group 1: a firm not using subcontractor(s), not working as a subcontractor.
Group 2: a firm not using subcontractor(s), working as a subcontractor.
Group 3: a firm using subcontractor(s), working as a subcontractor.
Group 4: a firm using subcontractor(s), not working as a subcontractor.

The table presents the means and standard deviations of a number of variables by four firm groups. Although the micro data show fat tails, as usual, we observe clear differences in firm size and internalization patterns across the groups. Figures are for machinery industry firms located in Japan, not adding figures of domestic/foreign affiliates.

Table 3: *Characteristics of Firms in Japanese Machinery Industry, 1994, by Subcontracting Pattern*

Using Subcontractor(s):		Group 1	Group 2	Group 3	Group 4
Working as a		No	No	Yes	Yes
Subcontractor:		No	Yes	Yes	No
Firm Size					
	Number of regular workers (L)	477	212	282	925
	[number of persons]	(2992)	(272)	(560)	(3835)
	Total sales (SALES)	17790	4063	8253	43401
	[millions of yen]	(142895)	(5645)	(24497)	(288333)
Economic performance					
	Ratio of tangible assets to	7.21	5.71	5.89	7.77
	regular workers (KLRATIO)	(7.87)	(6.19)	(5.02)	(7.47)
	[millions of yen per person]				
	Ratio of operating surplus to	0.0561	0.0441	0.0481	0.0501
	total sales (PAIRATIO)	(0.0802)	(0.0529)	(0.2126)	(0.0648)
	Ratio of gross value added to	0.5418	0.5728	0.5368	0.4929
	total sales (VARATIO)	(0.2311)	(0.2574)	(0.2397)	(0.2112)
	Foreign sales (1) (PROBEX):	0.3236	0.1461	0.1944	0.4810
	1 positive; 0 zero	(0.4682)	(0.3542)	(0.3959)	(0.4998)
	Foreign sales (2) (REX):	0.0460	0.0156	0.0174	0.0745
	ratio to total sales	(0.1207)	(0.0660)	(0.0682)	(0.1440)
Product differentiation					
	R&D expenditure (1) (PROBRD):	0.4979	0.2809	0.4356	0.7009
	1 positive; 0 zero	(0.5003)	(0.4507)	(0.4960)	(0.4580)
	R&D expenditure (2) (RRD):	0.0117	0.0057	0.0068	0.0173
	ratio to total sales	(0.0261)	(0.0197)	(0.0208)	(0.0268)
	Ratio of advertisement	0.0032	0.0010	0.0015	0.0035
	Expenditure to total sales (RADV)	(0.0120)	(0.0024)	(0.0118)	(0.0064)
Firm structure:					
	Ratio of foreign stock	0.0146	0.0033	0.0027	0.0340
	Ownership (FRATIO)	(0.1067)	(0.0412)	(0.0412)	(0.1510)
	Number of establishments	3.88	2.49	2.94	7.22
	(ALLEST)	(5.77)	(3.25)	(4.13)	(16.95)
	Number of domestic affiliates	1.04	0.36	0.65	2.68
	(DAFFI)	(6.26)	(1.05)	(2.56)	(11.44)
	Number of foreign affiliates	0.32	0.03	0.14	1.03
	(FAFFI)	(1.90)	(0.21)	(1.46)	(4.61)
Number of firms in the sample		717	178	1327	1501

Machinery industry includes general machinery, electrical machinery, transportation equipment, and Figures are unweighted averages. Standard deviation is in parenthesis.

Data source: *MITI database.*
This table originally appeared in Kimura (1999)

The differences across groups 2, 3, and 4 reveal how subcontracting arrangements are associated with the characteristics of firms. On average, firms in group 2 are the smallest, and those in group 4 are the largest. (See L and SALES.) Small firms tend to work as subcontractors, while large firms are likely to use subcontractors. Midsize firms often place themselves in the middle of a multilayered subcontracting system. Firms in group 1, independent of subcontracting arrangements, fall between group 3 and group 4 in terms of average firm size, although the large standard deviation indicates that group 1 includes a wide variety of firms.

Among groups 2, 3, and 4, group 2 has the largest ratio of gross value added to sales (VARATIO), while group 4 has the smallest.[7] Although VARATIO may be influenced by a number of variables, this probably reflects the fact that group 2 firms are located in the upstream of the production value chain and group 4 firms are in the downstream. Large firms tend to commission subcontractors to conduct a part of their production process and concentrate themselves on downstream assembling and sales activities. This is consistent with the fact that subcontractors tend to be located in the upstream in the Japanese machinery industry.

The contrast across three groups in the areas of foreign sales (PROBEX, REX), R&D expenditure (PROBRD, RRD), and advertising expenditure (RADV) is also striking. Group 4 firms internalize foreign sales activities, R&D activities, and advertising, while group 2 firms tend to "deinternalize" these activities. Firms are different not only in size but also in the scope of their activities or in the pattern of internalization. And the internalization pattern is related to the positioning in subcontracting system.

The number of establishments and domestic/foreign affiliates also presents a systemic pattern. Group 2 firms have a small number of establish-ments and affiliates, while group 4 firms have many. This suggests that the number of establishments and affiliates is not a simple reflection of firm size but is systematically associated with the internalization pattern and interfirm linkages.

5. THE PATTERN OF ESTABLISHMENTS AND AFFILIATES HOLDINGS

Let us investigate the pattern of establishments and affiliates holdings in detail. Table 4 presents establishment/affiliate holding patterns. Again, samples are classified into four groups by subcontracting patterns. As for the establishment holdings, the average number of manufacturing establishments is around one, regardless of the groups. That is, including headquarters, firms have two production plants on average. This means that the number of sales establishments largely explains the difference in the number of total establishment holdings. The intensity of sales activities, particularly the geographical coverage, seems to determine the number of establishments.

Group 4 firms are the most sales oriented, while group 2 firms are the least. Group 4 firms also have a substantial number of foreign branches. We cannot directly observe what these foreign branches do, but they are likely to conduct sales activities. The share of R&D establishments for group 4 is far larger than those for other groups, which reflects both the firm size and the R&D intensity of group 4 firms.

Table 4: *Establishments/Affiliates Holding Pattern in Japanese Machinery Industry, 1994: By Subcontracting Pattern*

	Group (1)	Group (2)	Group (3)	Group (4)
Using Subcontractor(s):	**No**	**No**	**Yes**	**Yes**
Working as a Subcontractor:	**No**	**Yes**	**Yes**	**No**
Establishments:				
Number of establishments (ALLEST)	3.88	2.49	2.94	7.22
	(5.77)	(3.25)	(4.13)	(16.95)
Share of headquarters (HQEST)	26.53%	40.39%	34.53%	14.68%
Share of manufacturing establishments (MEST)	21.28%	35.65%	31.07%	16.64%
Share of sales establishments (SEST)	42.67%	20.08%	26.61%	53.08%
Share of R&D establishments (LEST)	2.08%	0.68%	0.82%	19.19%
Share of Foreign Branches (FEST)	2.16%	0.45%	1.85%	5.78%
Domestic affiliates:				
Number of domestic affiliates (DAFFI)	1.04	0.36	0.65	2.68
	(6.26)	(1.05)	(2.56)	(11.44)
Share of domestic manufacturing affiliates (DMAFFI)	75.95%	78.13%	75.93%	64.65%
Share of domestic sales affiliates (DSAFFI)	22.78%	18.75%	22.22%	34.34%
Share of domestic R&D affiliates (DLAFFI)	1.27%	3.13%	1.85%	1.01%
Foreign affiliates:				
Number of foreign affiliates (FAFFI)	0.32	0.03	0.14	1.03
	(1.90)	(0.21)	(1.46)	(4.61)
Share of foreign manufacturing affiliates (FMAFFI)	70.00%	100.00%	58.33%	57.30%

Table 4: (*Continued*)

	Group (1)	Group (2)	Group (3)	Group (4)
Using Subcontractor(s):	**No**	**No**	**Yes**	**Yes**
Working as a Subcontractor:	**No**	**Yes**	**Yes**	**No**
	Share of foreign sales affiliates (FSAFFI)			
	30.00%	0.00%	41.67%	41.57%
	Share of foreign R&D affiliates (FLAFFI)			
	0.00%	0.00%	0.00%	1.12%
Number of firms in the sample:	717	178	1327	1501

Machinery industry includes general machinery, electrical machinery, transportation equipment, and precision machinery.

Figures in the rows of numbers are unweighted averages, and standard deviation is in parenthesis.

**shows the share to total establishments, domestic affiliates, or foreign affiliates*

The shares for establishments do not sum up to 100% because there are categories not shown in the table.

For domestic/foreign affiliates, the number of affiliates and shares of them are from different data tables, which means that they may not be perfectly consistent.

Data source: *MITI database.*

Affiliates, both domestic and foreign, are on average more manufacturing-activities oriented than establishments. This tendency is particularly pronounced for firms in group 2. Sales activities are also conducted by domestic/foreign affiliates, but the number of sales affiliates is much smaller than the number of sales establishments. There is a very small number of R&D affiliates, both domestic and foreign. Affiliates tend to have a standard set of activities themselves as an independent economic entity. Because the cost of service links between a parent firm and an affiliate must be larger than the cost of service links between establishments, an affiliate is perhaps larger in size and has a wider scope of activities than an establishment. Affiliates need a stronger incentive to engage in fragmentation.

To analyze the establishment/affiliate holding patterns more formally, we conduct some regression analysis. We do not claim any causality among variables at this moment. The following regression equations are ad hoc ones in this sense; we are simply trying to find the conditional relationship among variables. In particular, we would like to analyze the establishment/affiliate holding patterns after controlling for the size of firm. Tables 5, 6, and 7 present regression results. Dependent variables are the number of establishments, the number of domestic affiliates, and the number of foreign affiliates, respectively. Each table reports coefficient estimates for four separate samples, groups 1 through 4. Because a large number of samples have zeros on the left-hand side, as shown at the bottom of each table, we use the Tobit estimation method.[8]

Table 5: *The Results of Tobit Estimation on Establishment Holdings: Firms in Japanese Machinery Industry, 1994*

Dependent Variable:	ALLEST Group (1)	Group (2)	Group (3)	Group (4)
Constant	−1.331111	−3.077472**	−1.664241**	−6.329005**
	(−1.513)	(−2.804)	(−3.837)	(−4.407)
L	0.0005846**	0.0095656**	0.0042580**	0.0015370**
	(5.781)	(6.439)	(16.203)	(11.970)
KLRATIO	0.0402294	−0.0258684	0.0478794	0.0820442
	(0.996)	(−0.285)	(1.634)	(1.214)
PAIRATIO	−3.615793	−11.20113	−1.203400	−9.144419
	(−0.901)	(−1.238)	(−0.771)	(−1.134)
VARATIO	0.6016679	2.057271	0.3085039	2.599959
	(0.440)	(1.412)	(0.489)	(1.057)
REX	8.255813**	8.298177	3.116157	17.51314**
	(3.272)	(1.107)	(1.481)	(4.883)
RRD	43.04992**	−39.84979	19.86132**	106.4363**
	(3.619)	(−1.514)	(2.842)	(5.472)
RADV	42.53543	347.3921*	18.05582	491.0863**
	(1.657)	(2.150)	(1.576)	(6.156)
FRATIO	0.0030865	−15.30876**	0.0154816**	0.0133351**
	(1.077)	(−2.980)	(4.509)	(4.150)
Log likelihood	−1674.0633	−307.7284	−2645.997	−4918.607
ALLEST = 0	287	88	572	426
ALLEST > 0	430	90	755	1075

Note: *t-statistics are in parentheses.*
***Significant at the 1% level.*
**Significant at the 5% level.*

Table 6: *The Results of Tobit Estimation on Domestic Affiliate Holdings: Firms in Japanese Machinery Industry, 1994*

Dependent Variable:	DAFFI Group (1)	Group (2)	Group (3)	Group (4)
Constant	−5.536768**	−3.707194**	−5.248998**	−7.04673**
	(−5.648)	(−3.134)	(−9.365)	(−8.059)
L	0.0018935**	0.0035140**	0.0026016**	0.0026070**
	(19.937)	(3.287)	(9.274)	(37.344)

Table 6: (*Continued*)

Dependent Variable:	DAFFI Group (1)	Group (2)	Group (3)	Group (4)
KLRATIO	0.0859555	0.0069139	0.1518184**	0.1468371**
	(1.902)	(0.074)	(4.432)	(3.692)
PAIRATIO	−24.112100**	−3.454752	−3.145610	−18.67215**
	(−2.969)	(−0.382)	(−0.832)	(−3.187)
VARATIO	−0.5595976	−0.5881295	−0.5297675	−2.41465
	(−0.372)	(−0.417)	(−0.669)	(−1.598)
REX	8.693556**	−2.045043	12.30221**	10.56460**
	(3.406)	(−0.274)	(5.420)	(5.108)
RRD	35.29872**	−17.82503	25.83491**	46.89560**
	(2.805)	(−0.721)	(3.466)	(4.065)
RADV	4.629549	429.3427**	1.833164	224.4725**
	(0.175)	(3.081)	(0.120)	(4.876)
FRATIO	−0.0020144	−0.2394897	−0.0040715	−0.0058316**
	(−0.555)	(−0.246)	(−1.123)	(−2.758)
Log likelihood	−839.99142	−120.0924	−1461.208	−2705.015
DAFFI = 0	524	148	969	878
DAFFI > 0	193	30	358	623

Note: *t-statistics are in parentheses.*
***Significant at the 1% level.*
**Significant at the 5% level.*

Table 7: *The Results of Tobit (Logit) Estimation on Foreign Affiliate Holdings: Firm in Japanese Machinery Industry, 1994*

Dependent Variables:	FAFFI Group (1) Tobit est.	PROBFAFFI Group (2) Logit est.	PROBFAFFI Group (3) Logit est.	FAFFI Group (4) Tobit est.
Constant	−9.160999**	−6.737357**	−4.381864**	−14.30956**
	(−5.499)	(−2.860)	(−8.955)	(−10.933)
L	0.0005388**	0.0040663**	0.0008739**	0.0009002**
	(5.236)	(2.717)	(3.556)	(13.375)
KLRATIO	0.0640074	0.2588995	0.0575223*	0.1260374**
	(1.019)	(1.796)	(2.515)	(2.598)

(*Continued*)

Table 7: (*Continued*)

Dependent Variables:	FAFFI Group (1) Tobit est.	PROBFAFFI Group (2) Logit est.	PROBFAFFI Group (3) Logit est.	FAFFI Group (4) Tobit est.
PAIRATIO	−27.00263*	−53.03841	−1.301052	−13.49021
	(−2.128)	(−1.517)	(−0.344)	(−1.858)
VARATIO	−3.341641	1.985271	0.3264255	−4.421982*
	(−1.470)	(0.692)	(0.451)	(−2.125)
REX	16.90321**	19.61765**	5.118746**	26.71565**
	(5.504)	(2.709)	(4.424)	(11.384)
RRD	49.65165**	10.00244	6.228899	75.11832**
	(3.226)	(0.359)	(1.671)	(5.687)
RADV	−5.494827	−444.7103	1.576559	136.6971*
	(−0.116)	(−1.712)	(0.191)	(2.365)
FRATIO	−0.001665	−0.1366981	0.0024584	0.0021308
	(−0.323)	(−0.369)	(0.967)	(0.993)
Log likelihood	−274.8302	−12.858442	−179.5171	−1087.549
FAFFI = 0 or PROBFAFFI = 0	663	173	1275	1274
FAFFI > 0 or PROBFAFFI = 1	54	5	52	227

The tobit estimations for Group (2) and (3) were not converged. Hence, the results of logit estimation are shown.

Note: *t-statistics are in parentheses.*

***Significant at the 1% level.*

**Significant at the 5% level.*

As already seen in Figure 2, the number of establishments and domestic/foreign affiliates is naturally correlated with the size of firm. In all regressions the coefficients for the employment size (L) are positive and significantly different from zero. However, the coefficients of some other explanatory variables also have significant signs, which means that firm size does not fully explain the number of establishments and affiliates. In particular, we would like to investigate variables, the coefficients of which have different signs across four groups.

First, let us look at the coefficients for the ratio of foreign sales to total sales (REX). Firms with large REX tend to have more establishments and more domestic affiliates in groups 1 and 4. However, in the case of firms in group 2, the coefficients for REX are not significantly different from zero for establishment and domestic affiliate holdings. Taking into consideration the fact that most of the firms in group 2 do not have foreign affiliates, we can say that these firms are working as subcontractors and are largely domestic market–oriented, whether or not they have establishments or domestic affiliates. On the other hand, the coefficients for REX are all positive and significant in the regressions for foreign affiliate holdings, which is partly due to parent companies' exports to foreign affiliates.

Second, the coefficients for the ratio of R&D expenditure to total sales (RRD) are positive and mostly significant except, again, for firms in group 2. This is consistent with the hypothesis that firms with technological advantage tend to have establishments and domestic/foreign affiliates, rather than going with arm's-length trade, to secure their firm-specific assets. The coefficients for the ratio of advertising expenditure to total sales (RADV) are also positive and significant for firms in group 4 in all the regressions, probably for the same reason. The patterns for RADV are less clear for firms in groups 1, 2, and 3.

Third, the coefficients for the ratio of gross value added to total sales (VARATIO) are negative for firms in group 4 in the affiliates holding regressions, and such a pattern looks stronger for foreign affiliates than for domestic affiliates. This may suggest that a part of manufacturing activities is fragmented to affiliates and that such fragmentation is more drastic in the case of foreign affiliates than in the case of domestic affiliates.

Fourth, the negative signs of coefficients for the ratio of operating surplus to total sales (PAIRATIO) are of interest, particularly for firms in groups 1 and 4 in the regression for domestic-affiliate holdings. Why is the poor performance of a firm associated with larger holdings of domestic affiliates? This question may be answered in several ways. Because we use single-year data, there may be some bias in the sample; for example, if the particular year has many new entries, firms that recently established affiliates may perform poorly in the short run. Or they may be earning profits through their affiliates rather than through parent firms. Or parent firms may perform poorly from the beginning and may use affiliates, particularly domestic ones, as a buffer for restructuring.

Fifth, the coefficients for the ratio of foreign stock ownership (FRATIO) are positive for establishment holdings but negative for domestic affiliate holdings for firms in group 4. Firms with large foreign ownership ratios may behave differently in internalization.

Overall, we found that there seems to be a strong association among firm size, internalization/fragmentation patterns, and interfirm relationship.

6. SUMMARY AND FUTURE RESEARCH AGENDA

In this paper we use firm-level micro data and investigate the pattern of fragmentation and internalization of Japanese machinery manufacturing firms. We focus on the relationship between the internalization and establishment/affiliate holding pattern. Although we faced difficulties in obtaining statistical information, we were able to confirm that fragmentation, internalization, and interfirm linkages are systematically related. Developing an integrated analytical framework is essential to applying the concept of fragmentation in an empirical context.

Even with a serious lack of statistical data, additional work can be done. First, we can go back to the traditional location analysis and integrate the concept of internalization with it. Differences between domestic affiliates and foreign affiliates may indicate the patterns of within-cone and across-cone fragmentation. A careful analysis of foreign affiliates in developed and developing countries using the concept of internalization may deepen our understanding of horizontal and vertical multinationals. Second, we can statistically capture service links connecting fragmented blocks to some extent. Because it is difficult to pinpoint the activities of parent firms and affiliates, service link data, including intrafirm trade, are important in capturing the nature of fragmentation. We need to think seriously about the differences between intrafirm trade and arm's-length trade.

Although this paper makes only a modest contribution to the theoretical literature by accumulating some empirical knowledge related to fragmentation, it makes a few noteworthy points. First, although the fragmentation literature has inherited a strong flavor of location analysis from the tradition of international trade theory, we may need to pay more attention to the issue of internalization and the boundaries of a firm. Second, interfirm linkages such as subcontracting clearly influence decisions about fragmentation and internalization, and thus we may need to incorporate such linkages somehow with theoretical models, allowing for various types of internalization. Third, a more fundamental challenge for theory is the question of what can be an anchor of discussion. This paper claims that firm size, fragmentation and internalization patterns, and interfirm linkages are all related and are perhaps jointly determined. However, we cannot make everything endogenous at the same time in theory. A reasonable description of causal relationships between a corporate firm's structure, behavior, and performance is needed.

References

Baldwin, Robert E. and Fukunari Kimura (1998), "Measuring U.S. International Goods and Services Transactions," in Baldwin, R. E., R. E. Lipsey, and J. D. Richardson (eds.), *Geography and Ownership as Bases for Economic Accounting* (Chicago: The University of Chicago Press), 9–36.

Belderbos, Rene and, Leo Sleuwaegen (1996), "Japanese Firms and the Decision to Invest Abroad: Business Groups and Regional Core Networks." *The Review of Economics and Statistics*, 78/2, 214–220.

Cheng, Leonard K.; Larry D. Qiu, and Guofu Tan (2001), "Foreign Direct Investment and International Fragmentation of Production," in Arndt, Sven and Henryk Kierzkowski, eds., *Fragmentation: New Production Patterns in the World Economy,* Oxford University Press.

Deardorff, Alan V. (1998), "Fragmentation in Simple Trade Models," Research Seminar in International Economics, School of Public Policy, The University of Michigan, Discussion Paper No. 422 (January).

Deardorff, Alan V. (2001), "Fragmentation across Cones," in Arndt, Sven and Henryk Kierzkowski, eds., *Fragmentation: New Production Patterns in the World Economy,* Oxford University Press.

Doms, Mark E. and J. Bradford Jensen (1998), "Comparing Wages, Skills, and Productivity between Domestically and Foreign-Owned Manufacturing Establishments in the United States," in Baldwin, Robert E., Robert E. Lipsey, and J. David Richardson (eds.), *Geography and Ownership as Bases for Economic Accounting* (Chicago: University of Chicago Press).

Dunning, John H. (1993), *Multinational Enterprises and the Global Economy* (Wokingham: Addison-Wesley).

Harris, Richard G. (2001), "A Communication Based Model of Global Production Fragmentation," in Arndt, Sven and Henryk Kierzkowski, eds., *Fragmentation: New Production Patterns in the World Economy*, Oxford University Press.

Helpman, Elhanan (1984), "A Simple Theory of International Trade with Multinational Corporations." *Journal of Political Economy* 92 (June), 451–471.

Jones, Ronald W. and Henryk Kierzkowski (1990), "The Role of Services in Production and International Trade: A Theoretical Framework," in Jones, Ronald W. and Anne O. Krueger (eds.), *The Political Economy of International Trade: Essays in Honor of Robert E. Baldwin* (Oxford: Basil Blackwell).

Jones, Ronald W. and Henryk Kierzkowski (2001), "A Framework for Fragmentation," in Arndt, Sven and Henryk Kierzkowski, eds., *Fragmentation: New Production Patterns in the World Economy,* Oxford University Press.

Kimura, Fukunari (1998), "Japanese Multinationals and Regional Integration in Asia," in Fukasaku, Kiichiro, Fukunari Kimura, and Shujiro Urata (eds.), *Asia & Europe: Beyond Competing Regionalism* (Brighton: Sussex Academic Press), 111–133.

Kimura, Fukunari (1999), "Small and Medium Enterprises in Japan: Subcontracting," prepared for the World Bank Workshop on "The Role of Small and Medium Enterprises in Development," Chiang Mai, Thailand on August 13–14, 1999, forthcoming in the conference volume.

Kimura, Fukunari. (2000a), "Location and Internalization Decisions: Sector Switching in Japanese Outward Foreign Direct Investment," in Ito T. and A. O. Krueger (eds.), *The Role of Foreign Direct Investment in Economic Development* NBER-EASE volume 9 (Chicago: The University of Chicago Press), 79–108.

Kimura, Fukunari (2000b), "Higashi Ajia Shokoku ni okeru Seizougyou Jigyousho no Kibo Bunpu to Keizai Hatten (Size Distribution of Manufacturing Establishments and Economic Development in East Asian Countries)," *Ajia* Keizai 41, No. 5 (May), 7–36 (in Japanese).

Kimura, Fukunari and Robert E. Baldwin (1998), "Application of a Nationality-Adjusted Net Sales and Value Added Framework: The Case of Japan," in Baldwin, R. E., R. E. Lipsey, and J. D. Richardson (eds.), *Geography and Ownership as Bases for Economic Accounting* (Chicago: The University of Chicago Press), 49–80.

Markusen, James R. and Keith E. Maskus (1999a), "Multinational Firms: Reconciling Theory and Evidence," NBER Working Paper No. 7163 (June).

Markusen, James R. and Keith E. Maskus (1999b), "Discriminating among Alternative Theories of the Multinational Enterprise," NBER Working Paper No. 7164 (June).

Sengenberger, W., G. Loveman and M. J. Piore (eds.), (1990), *The Re-emergence of Small Enterprises: Industrial Restructuring in Industrialised Countries* (Geneva: International Institute for Labour Studies).

Van Ark, Bart and Erik Monnikhof (1996), "Size Distribution of Output and Employment: A Data Set for Manufacturing Industries in Five OECD Countries, 1960s-1990," Economics Department Working Papers No. 166, OECD.

Yamawaki, Hideki (1991), "Exports and Foreign Distributional Activities: Evidence on Japanese Firms in the United States," *Review of Economics and Statistics* 73 (May), 294–300.

APPENDIX: INDUSTRY CLASSIFICATION OF THE MITI DATABASE

Manufacturing sector

120	Food processing
130	Beverages, tobacco, and animal feed
140	Textiles
150	Apparel
160	Wood and wood products
170	Furniture and fixtures
180	Pulp, paper, and paper products
190	Publishing and printing
200	Chemicals
210	Petroleum and coal products

220	Plastic products
230	Rubber products
240	Leather and leather products
250	Ceramics, clay, and stone products
260	Iron and steel
270	Nonferrous metal
280	Metal products
290	General machinery
300	Electric machinery
310	Transport equipment
320	Precision machinery
330	Arms
340	Other manufacturing
Nonmanufacturing sector	
050	Mining
481	Wholesale trade
540	Retail trade
Others	Services and others

Endnotes

[*]The author thanks the conference participants for comments and suggestion. Comments by Leonard Cheng and Henryk Kierzkowski were particularly useful. The research assistance provided by Takamune Fujii is gratefully acknowledged. The MITI database was prepared and analyzed in cooperation with the Research and Statistics Department, Minister's Secretariat, Ministry of International Trade and Industry, Government of Japan. However, opinions expressed in this paper are the author's.

[1]Sengenberger, Loveman, and Piore (1990) and van Ark and Monnikhof (1996) conduct comparison studies of industrial organization in Organization of Economic Cooperation and Development countries. Kimura (2000b) provides an overview of the size distribution of manufacturing firms in East Asian countries.

[2]Until January 1998 pure stockholding companies were banned in Japan under the antitrust law. The removal of the ban is expected to provide more flexibility in firms' choices over fragmentation/internalization patterns. In addition, Japan is currently in the process of a so-called "accounting big ban," which includes the introduction of a consolidated accounting system for parent companies and their affiliates. The affiliates include those for which a parent company has more than a 50% stock share and those for which a parent company has more than a 40% stock share and has effective managerial control.

[3]The definition is based on *Kougyou Jittai Kihon Chousa (Basic Survey of the Industrial Sector)* (1976) by the Small and Medium Enterprise Agency (SMEA) and MITI.

[4]Translated from the Japanese by the author. The MITI statistics also use the concept of outsourcing (*gaichu*). The difference between outsourcing and subcontracting is again unclear, but it seems that subcontracting is a subset of outsourcing.

[5]"Employment" includes all the workers of a firm as a whole, including workers in all establishments that the firm has, but not including workers in domestic/foreign affiliates.

[6]For the detailed industry code, please refer to the appendix.

[7]Gross value added is calculated simply by subtracting the value of purchases from the value of sales.

[8]Since the second and third Tobit regressions in Table 7 do not converge, because of the small number of non-zero observations on the dependent variable, the results of Logit estimation are presented. The results, however, are not very reliable.

Chapter 8

SANYAL AND JONES ON FRAGMENTATION AND TRADE: EMPIRICAL EVIDENCE FOR SOUTH KOREA

Ulrich Kohli[*]

1. INTRODUCTION

The main purpose of this paper is to put to the test of the data what is arguably the first model of fragmentation and trade, the model of Sanyal and Jones (S&J) (1982). Models of trade in intermediate goods have received a great deal of attention in the literature since it became widely recognized that most international trade is in raw materials and nonfinished goods and that even most so-called "finished" products must still be processed in a number of ways (e.g., unloading, transportation, insurance, repackaging, advertising, retailing) at home (imports) or abroad (exports) before they are ready to meet final demand. This means that a substantial proportion of their final price tag is actually accounted for by local value added. It also means that by the time these goods are absorbed by domestic households, firms, or governments, they have essentially become nontradables. The model of S&J incorporates this feature, and it is particularly intriguing in that, unlike earlier models, it postulates a very detailed and rather complex production structure in which the fragmentation of production is made explicit.

Much of the empirical work on import determination is consistent with the view that imports are an input to the technology.[1] Burgess (1974) has formally introduced what is now called the production theory approach to import determination. This work has been extended by Kohli (1978, 1991) and Woodland (1982) into the GNP/GDP function approach to modeling the demand for imports and the supply of exports. According to this approach all imports are treated as inputs to the technology. They are combined with domestic factor services (e.g., labor and capital) to produce goods intended for domestic use (e.g., consumption goods and investment goods), as well as goods destined for foreign markets. These exports, in turn, are used as inputs by the foreign technology.

Most of the previous work on the GDP function approach has allowed for joint production, and it has placed very few restrictions on the form of the technology, except for constant returns to scale, convexity, and free disposals. Assuming two domestic factors (labor, L, and capital, K), two middle products (imports, M, and exports, X), and two final goods (nontradables, Y and Z), the black box of technology is depicted in Figure 1. Nothing filters out about what goes on inside the box.

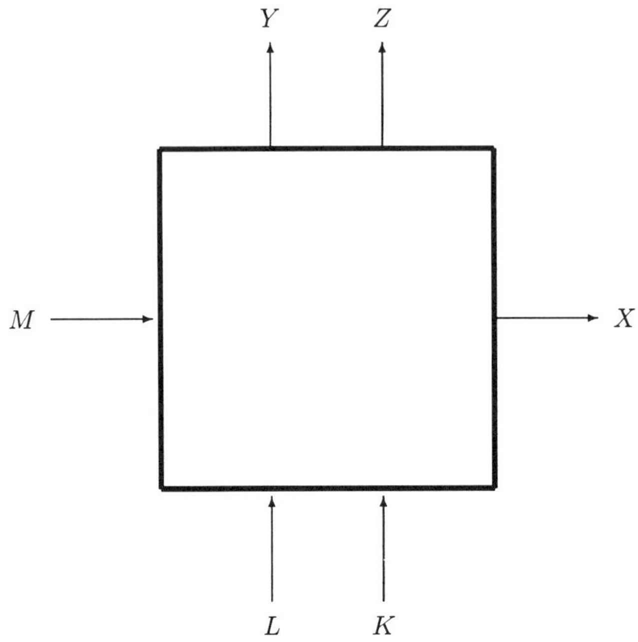

Figure 1: *The Black Box of Technology and Trade in Middle Products.*

S&J, on the other hand, make some very specific assumptions about the way inputs are transformed into outputs. Thus they assume that production is fragmented and that it takes place in several steps which they identify. More specifically, they assume that there are two production tiers. The two middle products are produced in the "input tier," and the two final goods are produced in the "output tier." Production is assumed to be nonjoint in input quantities at each level of production. There are thus four distinct sectors. Labor is mobile between sectors and tiers, whereas capital is only used in the input tier, and, moreover, it is sector specific. While labor and capital are used in the input tier, the production of final goods involves labor and one middle product. Trade takes place exclusively in middle products. Net exports of each middle product are equal to the difference between domestic production and domestic utilization. All these assumptions are captured by Figure 2, which provides us with an X-ray of the guts of the technology box.[2] Fragmentation thus makes it possible to exploit the gains from trade at the intermediate good level. A country can partly specialize in the production of those middle products in which it enjoys a comparative advantage.

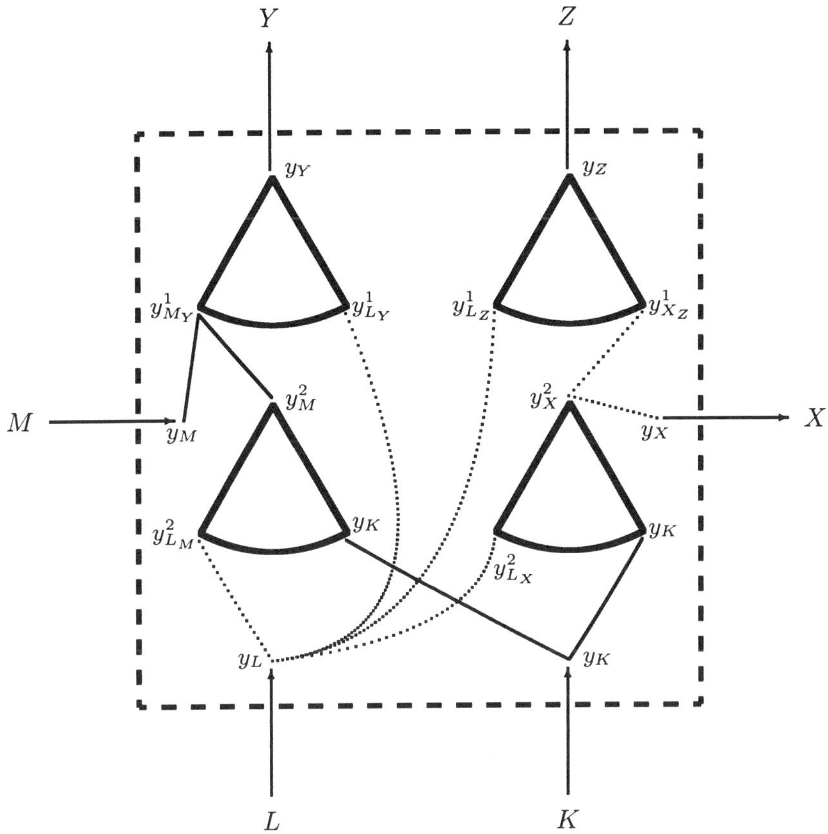

Figure 2: *The Sanyal & Jones Structure of Production: An X-Ray of the Black Box of Technology.*

In view of the originality and the sophistication of the S&J model, it seems worthwhile to confront it with the data and to attempt to quantify some of its predictions. More generally, considering the wide gap that exists between theory and empirical work in the international trade literature, we believe that it is important to attempt to estimate and test existing trade models. Furthermore, as discussed by Kohli (2001), there is an additional challenge here for the empirical economist, since the data that the model requires—specific capital in the input tier, production of middle products in the input tier, use of middle products in the output tier, and allocation of labor between the two tiers and within the two tiers—are generally not available.[3] However, there are ways to get around these difficulties under some circumstances. This paper thus illustrates how one can model some rather fancy production structures with aggregate data if one carefully selects the appropriate representation of the technology as a starting point.

Earlier empirical work conducted for the United States has actually cast doubts on the validity of some of the restrictions implied by the S&J production structure (see Kohli 2001). This might be due to the highly heterogeneous nature of the U.S. economy and to the fact that imports and exports only account for a relatively small share of GDP. Given the importance of foreign trade in newly industrialized countries, and the fact that East Asian economies are often cited as examples in the debate about fragmentation, it is tempting to estimate the S&J model for a country from that region. Data availability considerations led us to select South Korea for this exercise.[4] This paper also innovates in that we report for the first time estimates of the sectoral technologies and details about the flows of intermediate goods and services between tiers and between industries.

The remainder of this paper is organized as follows. We begin by discussing a general representation of the aggregate technology that is fit to describe a multiple-input multiple-output technology, and that in principle contains any production structure as a special case. In Section 3 we review the specific production structure postulated by S&J, and we examine the restrictions that follow. We next turn to the empirical implementation of the model; thus in Section 4 we discuss the functional form, the data, the estimation technique, and related issues. Empirical results are discussed in Sections 5 and 6, and Section 7 concludes.

2. GENERAL DESCRIPTION OF THE AGGREGATE TECHNOLOGY

We begin with a general description of the aggregate technology. We assume that there are M inputs and outputs. We denote the vector of input and output quantities by $y \equiv [y_m]$; an element of y is positive if the corresponding quantity is an output and negative if it is an input. The vector of input and output prices is denoted by $p \equiv [p_m]$; all elements of p are assumed to be strictly positive. Let T be the production possibilities set, that is, the set of all feasible input and output combinations.[5]

Let us partition the set of inputs and outputs into two subsets, A and B, that contain, respectively, I and $J \equiv M - I$ elements. We denote the vector of the I input and output quantities in A by $y_A \equiv [y_i]$, and the vector of the $J \equiv M - I$ input and output quantities in B by $y_B \equiv [y_j]$; the corresponding price vectors are denoted by $p_A \equiv [p_i]$ and $p_B \equiv [p_j]$. As we shall see later on, the partitioning of the inputs and outputs may be dictated by economic theory, by statistical considerations, or by technical requirements. Without loss of generality (by appropriately reordering the inputs and outputs), we may assume that A contains the first I input and outputs, and that B contains the last J ones: thus $y \equiv (y'_A, y'_B)'$ and $p \equiv (p'_A, p'_B)'$.

The aggregate technology can now be described by a restricted profit function, which is defined as follows:

$$\pi(p_A, y_B) \equiv \max_{y_A}\{p'_A y_A : (y_A, y_B) \in T\}. \tag{1}$$

The function $\pi(\cdot)$ is well defined for all quantity vectors y_B and for all strictly positive price vectors p_A. It is linearly homogeneous and convex in the elements of p_A; it is increasing (decreasing) in p_i if the corresponding element is an output (input); it is linearly homogeneous, concave, and decreasing in the elements of y_B. The restricted profit function treats the first I quantities as variable and the last J quantities as fixed. The restricted profit function is identical to the variable profit function, except that, although one generally allows for A to contain outputs as well as (variable) inputs, one usually assumes that all elements of B are inputs;[6] this is not the case here. Indeed, in what follows both A and B will generally be assumed to contain inputs as well as outputs.

The profit-maximizing supply of variable outputs and demand for variable inputs can be obtained by differentiation of the restricted profit function with respect to prices, a result implied by Hotelling's (1932) Lemma:

$$y_A(p_A, y_B) = \nabla_{p_A}\pi(p_A, y_B), \tag{2}$$

where $\nabla_{p_A}\pi(\cdot)$ is the vector of the partial derivatives of $\pi(\cdot)$ with respect to $p_A : \nabla_{p_A} \equiv [\partial\pi(p_A, y_B)/\partial p_i]$, $i = 1, \ldots, I$.

Under competitive conditions the marginal cost of each fixed output and the marginal product of each fixed input is equal to its price. Thus

$$p_B(p_A, y_B) = -\nabla_{y_B}\pi(p_A, y_B), \tag{3}$$

where $\nabla_{y_B}\pi(\cdot)$ is the vector of the partial derivatives of $\pi(\cdot)$ with respect to $y_B : \nabla_{y_B}\pi(\cdot) \equiv [\partial\pi(p_A, y_B)/\partial y_j]$, $j = I+1, \ldots, M$; the negative sign on the right-hand side of equation 3 stems from the facts that the marginal cost of a fixed output is equal to the marginal reduction in profits that its production entails, and inputs are measured negatively. These equations are inverse output (and input) supply (demand) functions: they express fixed output (input) prices in terms of variable element prices and fixed element quantities. Note that it follows from equations 2 and 3 and from the linear homogeneity of $\pi(\cdot)$ that

$$\pi(\cdot) = \sum_{i=1}^{I} p_i y_i = -\sum_{j=I+1}^{I+J} p_j y_j.$$

The substitution possibilities allowed for by the technology can be described by a set of substitution elasticities. The substitution matrix $\Sigma \equiv [\sigma_{mn}]$ is defined as follows:

$$\Sigma \equiv \begin{bmatrix} \Sigma_{pp} & \Sigma_{py} \\ \Sigma_{yp} & \Sigma_{yy} \end{bmatrix} \equiv \pi \begin{bmatrix} \pi_p^{-1}\pi_{pp}\pi_p^{-1} & \pi_p^{-1}\pi_{py}\pi_y^{-1} \\ \pi_y^{-1}\pi_{yp}\pi_p^{-1} & \pi_y^{-1}\pi_{yy}\pi_y^{-1} \end{bmatrix}, \quad (4)$$

where $\pi_p \equiv diag[\nabla_{p_A}\pi(\cdot)], \pi_y \equiv diag[\nabla_{y_B}\pi(\cdot)]$, $\pi_{pp} \equiv \nabla^2_{p_A p_A}\pi(\cdot) \equiv [\partial^2\pi(\cdot)/(\partial p_i \partial p_h)]$, $\pi_{py} \equiv \nabla^2_{p_A y_B}\pi(\cdot) \equiv [\partial^2\pi(\cdot)/(\partial p_i \partial y_j)]$, and so on; and $\pi \equiv \pi(\cdot)$ for short. The elements of Σ_{pp} are the elasticities of transformation and substitution between variable inputs and outputs.[7] The elements of Σ_{yy} are sometimes called inverse elasticities of transformation,[8] while the elements of Σ_{py} (or Σ_{yp}, since Σ is symmetric) are the elasticities of intensity. Even though these three sets of elasticities have quite different interpretations, we will use the same symbol in all cases to simplify the notation. The subscripts will indicate whether a particular elasticity is an element of Σ_{pp}, Σ_{yy}, or Σ_{py}.

The substitution matrix can be used to derive price and quantity elasticities, which are easier to interpret. Let $H \equiv [\eta_{mn}]$ be the matrix of price and quantity elasticities. H is defined as follows:

$$H \equiv \begin{bmatrix} H_{pp} & H_{py} \\ H_{yp} & H_{yy} \end{bmatrix} \equiv \begin{bmatrix} \partial \ln|y_i|/\partial \ln p_h & \partial \ln|y_i|/\partial \ln|y_j| \\ \partial \ln p_j/\partial \ln p_i & \partial \ln p_j/\partial \ln|y_k| \end{bmatrix}. \quad (5)$$

Thus the elements of $H_{pp} \equiv [\eta_{ih}]$ are the price elasticities of the supply of variable outputs and of the demand for variable inputs; note that these elasticities generally are functions of p_A as well as of y_B. The components of $H_{yy} \equiv [\eta_{jk}]$ are the quantity elasticities of the inverse supply (demand) functions of (for) the fixed outputs (inputs). We could also call them the inverse price elasticities of fixed output supply (input demand); they indicate the effect of a change in the composition of y_B on the prices of the fixed components. The elasticities in $H_{py} \equiv [\eta_{ij}]$ indicate the effect of a change in the fixed quantities (y_B) on the variable quantities (y_A),[9] while the elasticities in $H_{yp} \equiv [\eta_{ji}]$ indicate the effects of changes in the prices of the variable elements (p_A) on the prices of the fixed elements (p_B).

It can easily be seen that H can be calculated directly from Σ. Indeed, the definition of H together with equations (2)–(4) imply:

$$H = \pi^{-1} \begin{bmatrix} \Sigma_{pp}\pi_y P & \Sigma_{py}\pi_w Y \\ \Sigma_{yp}\pi_y P & \Sigma_{yy}\pi_w Y \end{bmatrix} = \begin{bmatrix} \Sigma_{pp} S_I & \Sigma_{py} S_J \\ \Sigma_{yp} S_I & \Sigma_{yy} S_J \end{bmatrix} = \Sigma \cdot S, \tag{6}$$

where $P \equiv diag(p_A)$ and $Y \equiv diag(y_B)$; $S_I \equiv diag\{[s_i]\}$, $S_J \equiv diag\{[s_j]\}$, $S \equiv diag\{[s_i],[s_j]\}$; $s_i \equiv p_i y_i / \pi$ and $s_j \equiv p_j y_j / \pi$ are the profit shares of the I variable elements and the J fixed elements, respectively, and S_I, S_J, and S are profit-share diagonal matrices of dimensions $I \times I$, $J \times J$, and $(I+J)\times(I+J)$, respectively. Hence, once a restricted profit function has been estimated, estimates of H can be obtained without difficulty.

3. THE SANYAL AND JONES STRUCTURE OF PRODUCTION

The description of the technology reviewed in the previous section is very general, and it can be used to describe a wide range of multiple-input, multiple-output technologies. In particular, it is consistent with joint production, as well as with cases where production is nonjoint or disjoint.[10] Under joint production, the choice of the elements of y_A and y_B is relatively unimportant, and it may be affected by convenience or by statistical considerations. Under less general production structures, those involving some form of nonjointness or disjointness for instance, this choice may be crucial. Thus, in many cases, particular production structures imply additional restrictions on the form of the restricted profit function, but these can only be imposed for particular partitions of the inputs and outputs.

We now turn to the specific production structure implied by S&J and investigate the nature of the restrictions that their model imposes on the form of the aggregate technology. Here we are faced with two difficulties. First, there is the question of the sector-specific capital in the input tier. Sector-specific capital is generally not observable; this is particularly true here, since not even the industries producing the middle products are observable. However, as shown by Kohli (1993a), if we are willing to assume that the relative supplies of the two sector-specific types of capital do not change much over time, we may aggregate them into a composite factor by invoking Leontief aggregation. This composite capital good then has all the properties of a public input that enters all industries (of the input tier) at the same time. The return to this composite factor is the sum of the returns of the two types of sector-specific forms of capital.

Second, there is S&J's assumption, reflected by Figure 2, that each middle product is used as an input by only one output-tier industry. In our empirical work we consider two types of middle products (importables and exportables) and two final goods (investment and consumption goods),[11] and it would be arbitrary to attribute either middle product to one or the other final good. In their paper, S&J do, however, provide a generalization of their model, one where both middle products are used by both output-tier industries. In what follows we mostly consider this generalized structure. Nevertheless, in the last part of our paper, we briefly investigate the original S&J structure by considering in turn the cases in which importables and exportables are attributed exclusively to one or the other final good industry.

In what follows we therefore consider six inputs and outputs: labor (L), capital (K), imports (M), exports (X), investment goods (I), and consumption goods (C). The first three elements are viewed as inputs to the technology ($y_L \leq 0$, $y_K \leq 0$, $y_M \leq 0$), while the last three elements are outputs ($y_X \geq 0$, $y_I \geq 0$, $y_C \geq 0$).

The (generalized) S&J production structure can be described as follows:

Tier 1 (output tier)

$$y_i = f^i\left(y^1_{M_i}, y^1_{X_i}, y^1_{L_i}\right),\quad y^1_{M_i}, y^1_{X_i}, y^1_{L_i} \leq 0;\ y_i \geq 0;\quad i = I, C \tag{7}$$

Tier 2 (input tier)

$$y^2_h = f^h\left(y^2_{L_h}, y_K\right),\quad y^2_{L_h} \leq 0, y^2_h \geq 0;\ y_K \leq 0;\quad h = M, X \tag{8}$$

Full employment of labor

$$y_L = \sum_h y^2_{L_h} + \sum_i y^1_{L_i}. \tag{9}$$

Net exports

$$y_h = y^2_h + y^1_{h_I} + y^1_{h_C},\quad h = M, X. \tag{10}$$

Under profit maximization, the following relationships hold in the price space:

Rental price of the composite capital good

$$p_K = -\sum_h p_h \frac{\partial f^h(\cdot)}{\partial y_K} \tag{11}$$

Wage rate

$$p_L = -p_h \frac{\partial f^h(\cdot)}{\partial y^2_{Lh}},\quad h = M, X \tag{12}$$

Output prices

$$\frac{p_h}{p_i} = -\frac{\partial f^i(\cdot)}{\partial y^1_{hi}},\quad i = I, C;\quad h = M, X \tag{13}$$

$$\frac{p_L}{p_i} = -\frac{\partial f^i(\cdot)}{\partial y^1_{Li}}, \quad i = I, C, \tag{14}$$

where the minus signs result from the convention that inputs are measured negatively. Following S&J, we assume that all production functions are well behaved (i.e., monotonically increasing, linearly homogeneous, and strictly quasi-concave).

For empirical implementation we now treat the quantities of the final goods and of the composite factor as given and the quantities of labor and traded goods as variable; that is, $y_A \equiv (y_M, y_X, y_L)'$ and $y_B = (y_I, y_C, y_K)'$. The motivation behind this particular partition will become clear later on. We can now describe the aggregate technology by the following restricted profit function:

$$\pi(p_M, p_X, p_L, y_I, y_C, y_K) \equiv \max\{p_M y_M + p_X y_X + p_L y_L : (y_A, y_B) \in T\}. \tag{15}$$

Function $\pi(\cdot)$ has all the properties of the restricted profit function as described in Section 2. Note that the value of $\pi(\cdot)$ is the balance of trade minus the wage bill. In absolute value this is also equal to the value of final output (domestic sales) minus the total return to the composite capital good. The fact that $\pi(\cdot)$ treats p_L as given and y_L as variable, rather than the reverse, does not match the usual assumption in international trade theory that the endowments of all factors are given. However, this treatment of labor as a variable input is necessary at this stage to enable us to impose the S&J production structure onto the aggregate technology; we will revert to the alternative treatment of labor later on when assessing our results.

As shown by Kohli (2001), the S&J production structure is disjoint. Furthermore, the output tier is nonjoint in input quantities, while the input tier is almost nonjoint in input prices and quantities. This implies that restricted profit function 15 has the following additive form:

$$\begin{aligned}\pi(p_M, p_X, p_L, y_I, y_C, y_K) = {} & \pi^{1_I}(p_M, p_X, p_L, y_I) + \pi^{1_C}(p_M, p_X, p_L, y_C) \\ & + \pi^{2_M}(p_M, p_L, y_K) + \pi^{2_X}(p_X, p_L, y_K),\end{aligned} \tag{16}$$

where $\pi^{1_i}(\cdot)$ ($i = I, C$) is linearly homogeneous, decreasing and convex in prices, and linearly homogeneous in y_i, and where $\pi^{2_h}(\cdot)$ ($h = M, X$) is linearly homogeneous and convex in prices, increasing in p_h and decreasing in p_L, and linearly homogeneous in y_K.

The $\pi^{1_i}(\cdot)$s can be interpreted as the restricted profit (or revenue) functions of the two industries in the output tier. They are defined as follows:

$$\pi^{1_i}(p_M, p_X, p_L, y_i) \equiv \max\{p_M y^1_{M_i} + p_X y^1_{X_i} + p_L y^1_{L_i} : y_i = f^i(y^1_{M_i}, y^1_{X_i}, y^1_{L_i})\},$$
$$i = I, C. \tag{17}$$

These revenue functions can also be interpreted as the negative of the industry cost functions. The $\pi^{2_h}(\cdot)$s, on the other hand, are the restricted profit functions of the two industries in the input tier. They yield the return to the specific capital, and they are defined as follows:

$$\pi^{2_h}(p_h, p_L, y_K) \equiv \max\{p_h y_h^2 + p_L y_{L_h}^2 : y_h^2 = f^h(y_{L_h}^2, y_K)\} \quad h = M, X. \quad (18)$$

Compared to the general specification given by function 15, the S&J production structure implies a number of important restrictions. Indeed, it is visible from expression 16 that

$$\frac{\partial^2 \pi(\cdot)}{\partial y_i^2} = 0, \quad \frac{\partial^2 \pi(\cdot)}{\partial y_I \partial y_C} = 0, \quad \frac{\partial^2 \pi(\cdot)}{\partial y_i \partial y_K} = 0, \quad \frac{\partial^3 \pi(\cdot)}{\partial p_M \partial p_X \partial y_K} = 0, \quad i = I, C.$$

We now examine how these restrictions can be imposed and/or tested in empirical work.

4. EMPIRICAL IMPLEMENTATION

4.1. Functional Form

We require a functional form that is flexible enough to represent an arbitrary technology, that is, a functional form that does not restrict a priori the size or the sign of the various elasticities of substitution. Moreover, we require a functional form that allows the restrictions implied by the S&J production structure to be imposed. Finally, given that curvature conditions often tend to be violated in empirical work based on flexible functional forms, we require a function that allows curvature conditions to be imposed globally without interfering with flexibility. For this purpose we use the flexible functional form introduced by Kohli (2001):

$$\pi = \frac{1}{2}\frac{\sum\sum\sum a_{ihj} p_i p_h | y_j |}{\sum a_i p_i} + \frac{1}{2}\frac{\sum\sum\sum b_{jki} | y_j || y_k | p_i}{\sum b_j | y_j |} + \sum\sum c_{ij} p_i | y_j |, \quad (19)$$

$$i, h = 1, \ldots, I; \quad j, k = I+1, \ldots, I+J,$$

where $a_{ihj} = a_{hij}$, $\sum_i a_{ihj} = 0$, $b_{jki} = b_{kji}$, $\sum_j b_{jki} = 0$, and where the a_is and b_js can be set subject to the conditions $a_i \geq 0$, $b_j \geq 0$, $\sum a_i = 1$, and $\sum b_j = 1$.[12]

This function is linearly homogeneous in prices and in quantities. Moreover, it is convex in prices as long as matrices $A^j \equiv [a_{ihj}]$ are positive semidefinite, and it is concave in quantities as long as matrices $B^i \equiv [b_{jki}]$ are negative semidefinite. If needed, these restrictions can be imposed.

Functional form (19) is a member of the family of Normalized Quadratic flexible functional forms introduced by Diewert and Wales (1987, 1988a). Two other examples are the Normalized Quadratic (NQ) variable profit function introduced by Diewert and Ostensoe (1988), and the Symmetric Normalized Quadratic (SNQ) variable profit function proposed by Kohli (1993b). However, expression (19) is more general than either one of these two functions. Indeed, it contains them both as special cases. Thus, functional form (19) is identical to the SNQ if $\beta_j A^j = A$ and $\alpha_i B^i = B$, where $\beta_j \geq 0$, $\alpha_i \geq 0$. It is identical to the NQ if, moreover, $\beta_j = 0 (j = I+1,\ldots, I+J-1)$ and $\alpha_i = 0 (i = 1,\ldots, I-1)$.

Expression (19) has $IJ(I + J)/2$ parameters, which is more than what is required by a flexible functional form. This larger number of parameters adds somewhat to the complexity of this function, and it may lead to problems of multicollinearity when the model is estimated. However, this function will be extremely useful for our purposes since, as we shall see, it allows for nonjointness and disjointness to be imposed and/or tested.

Compared to other functional forms available in the literature, function (19) is quite unique in that it allows nonjointness restrictions to be incorporated and global curvature conditions to be imposed globally without interfering with flexibility. Neither the NQ nor the SNQ allow for imposition of nonjointness and disjointness, whereas the Generalized-Linear Generalized-Leontief introduced by Hall (1973) cannot be forced to satisfy curvature conditions globally without wrecking its flexibility if the number of either the fixed or the variable components exceeds two. The same is true for the Mean-of-Order-Two Mean-of-Order-Zero functional form proposed by Diewert (1973) and applied by Kohli (1991). In the case of the Translog variable profit function (Diewert 1974), finally, convexity can only be imposed locally, whereas concavity can be imposed locally or globally, but, in the latter case, only at the cost of destroying the flexibility of the function; in any case the Translog is not well suited to modeling the restrictions implied by nonjointness and disjointness.

The derived variable output supply and input demand equations are obtained by differentiation of function (19) with respect to the prices of the variable components:

$$y_i = \frac{\sum\sum a_{ihj} p_h \mid y_j \mid}{\sum a_i p_i} - \frac{1}{2}\frac{a_i \sum\sum\sum a_{ihj} p_i p_h \mid y_j \mid}{(\sum a_i p_i)^2} + \frac{1}{2}\frac{\sum\sum b_{jki} \mid y_j \mid\mid y_k \mid}{\sum b_j \mid y_j \mid} + \sum c_{ij} \mid y_j \mid, \quad (20)$$

$$i = 1,\ldots, I.$$

Similarly, the fixed component inverse demand and supply functions are as follows:

$$\pm p_j = \frac{1}{2}\frac{\Sigma\Sigma\Sigma a_{ihj} p_i p_h}{\Sigma a_i p_i} + \frac{\Sigma\Sigma b_{jki} \mid y_k \mid p_i}{\Sigma b_j \mid y_j \mid}$$

$$-\frac{1}{2}\frac{b_j \Sigma\Sigma\Sigma b_{jki} \mid y_j \mid\mid y_k \mid}{(\Sigma b_j \mid y_j \mid)^2} + \sum c_{ij} p_i, \tag{21}$$

$$j = I+1, \ldots, I+J,$$

the minus sign holding if fixed component *j* is an output.

In order to derive the substitution elasticities, we first note that the elements of the Hessian of $\pi(\cdot)$ are as follows:

$$\pi_{ii} = \frac{\Sigma a_{iij} \mid y_j \mid}{\Sigma a_i p_i} - 2a_i \frac{\Sigma\Sigma a_{ihj} p_h \mid y_j \mid}{(\Sigma a_i p_i)^2} + a_i^2 \frac{\Sigma\Sigma\Sigma a_{ihj} p_i p_h \mid y_j \mid}{(\Sigma a_i p_i)^3} \tag{22}$$

$$\pi_{ih} = \frac{\Sigma a_{ihj} \mid y_j \mid}{\Sigma a_i p_i} - a_i \frac{\Sigma\Sigma a_{ihj} p_h \mid y_j \mid}{(\Sigma a_i p_i)^2} - a_h \frac{\Sigma\Sigma a_{ihj} p_i \mid y_j \mid}{(\Sigma a_i p_i)^2}$$

$$+ a_i a_h \frac{\Sigma\Sigma\Sigma a_{ihj} p_i p_h \mid y_j \mid}{(\Sigma a_i p_i)^3} \tag{23}$$

$$\pi_{ij} = \frac{\Sigma a_{ihj} p_h}{\Sigma a_i p_i} - \frac{1}{2} a_i \frac{\Sigma\Sigma a_{ihj} p_i p_h}{(\Sigma a_i p_i)^2} + \frac{\Sigma b_{jki} \mid y_k \mid}{\Sigma b_j \mid y_j \mid}$$

$$-\frac{1}{2} b_j \frac{\Sigma\Sigma b_{jki} \mid y_j \mid\mid y_k \mid}{(\Sigma b_j \mid y_j \mid)^2} + c_{ij} \tag{24}$$

$$\pi_{jj} = \frac{\Sigma b_{jji} p_i}{\Sigma b_j \mid y_j \mid} - 2b_j \frac{\Sigma\Sigma b_{jki} \mid y_k \mid p_i}{(\Sigma b_j \mid y_j \mid)^2}$$

$$+ b_j^2 \frac{\Sigma\Sigma\Sigma b_{jki} \mid y_j \mid\mid y_k \mid p_i}{(\Sigma b_j \mid y_j \mid)^3} \tag{25}$$

$$\pi_{jk} = \frac{\Sigma b_{jki} p_i}{\Sigma b_j \mid y_j \mid} - b_j \frac{\Sigma\Sigma b_{jki} \mid y_k \mid p_i}{(\Sigma b_j \mid y_j \mid)^2} - b_k \frac{\Sigma\Sigma b_{jki} \mid y_j \mid p_i}{(\Sigma b_j \mid y_j \mid)^2}$$

$$+ b_j b_k \frac{\Sigma\Sigma\Sigma b_{jki} \mid y_j \mid\mid y_k \mid p_i}{(\Sigma b_j \mid y_j \mid)^3}, \tag{26}$$

$$i, h = 1, \ldots, I; \quad j = I+1, \ldots, I+J.$$

The elements of Σ can then easily be calculated with the help of equation (4).

4.2. Imposing the Sanyal and Jones Production Structure

In our empirical application we consider three variable components (M, X, L) and three fixed components (I, C, K). In order to be able to impose the S&J production structure we begin by setting $a_M = a_X = 0$ and $a_L = 1$. Thus we introduce here the same type of asymmetry as is present, for instance, in the NQ restricted profit function proposed by Diewert and Ostensoe (1988). While asymmetrical treatment of variable components is generally not desirable, it is necessary here if we want to replicate the S&J production structure given that their model does treat labor, imports, and exports in a very asymmetrical way. To restore some symmetry, however, we apply the same treatment to the fixed components and thus set $b_I = b_C = 0$ and $b_K = 1$.

In view of Theorem 3 and of equations (22)–(26), it is apparent that the generalized S&J production structure requires the following restrictions:

- disjointness of the input and output tiers

$$b_{IKM} = b_{IKX} = b_{IKL} = b_{CKM} = b_{CKX} = b_{CKL} = 0; \quad (27)$$

- nonjointness in input quantities (output tier)

$$b_{ICM} = b_{ICX} = b_{ICL} = 0; \quad (28)$$

- almost nonjointness in input prices and quantities (input tier)

$$a_{MXK} = 0. \quad (29)$$

Furthermore, it is possible to impose the original S&J production structure that is somewhat more restrictive. According to the original model, each industry in the output tier uses only one traded input. However, there is no indication as to which one, of importables and exportables, is used in the investment-good industry as opposed to the consumption-good industry. We will therefore let the data speak and will consider both hypotheses in turn.

- original S&J production structure, version $MI : XC$ (importables used in the investment-good industry and exportables used in the consumption-good industry)

$$a_{MMC} = a_{MXC} = a_{MXI} = a_{XXI} = c_{MC} = c_{XI} = 0; \quad (30)$$

- original S&J production structure, version $MC : XI$ (importables used in the consumption-good industry and exportables used in the investment-good industry)

$$a_{MMI} = a_{MXI} = a_{MXC} = a_{XXC} = c_{MI} = c_{XC} = 0. \quad (31)$$

4.3. Curvature Conditions

As anybody who has worked with flexible functional forms is well aware, it is common in empirical work that the required curvature conditions are not met. Yet these curvature conditions are an important part of the theory, and it makes little economic sense to consider a model that does not satisfy them. As noted above, restricted profit function (19) is globally convex in prices if and only if matrices A^j are positive semidefinite, and it is globally concave in fixed quantities if and only if matrices B^i are negative semidefinite. To impose these restrictions we use the technique of Wiley, Schmidt, and Bramble (1973). As shown by these authors, a sufficient condition for a matrix A^j to be positive semidefinite is that it can be written as

$$A^j = T^j T^{j'},$$

where $T^j \equiv [\tau_{ihj}]$ is a lower triangular matrix. Diewert and Wales (1987), exploiting a result by Lau (1978), have shown that the above condition is also necessary for A to be positive semidefinite. Thus, with I = 3, and recalling that $\sum_i a_{ihj} = 0$, we define

$$T^j \equiv \begin{bmatrix} T_{11j} & 0 \\ \tau_{21j} & \tau_{22j} \end{bmatrix}.$$

We then have

$$A^j \equiv \begin{bmatrix} a_{11j} & a_{12j} \\ a_{21j} & a_{22j} \end{bmatrix} = \begin{bmatrix} \tau_{11j}^2 & \tau_{11j}\tau_{21j} \\ \tau_{11j}\tau_{21j} & \tau_{21j}^2 + \tau_{22j}^2 \end{bmatrix}.$$

Applying this reparameterization to all matrices $A^j(j = I + 1, \ldots, I + J)$ ensures that the resulting restricted profit function is convex in output prices.

Similarly, a matrix B^i is negative semidefinite if it can be written

$$B^i = -M^i M^{i'},$$

where $M^i \equiv [\mu_{jki}]$ is a lower triangular matrix. This reparameterization can be applied to all matrices $B^i (i = 1, \ldots, I)$ to ensure that function (19) is globally concave in quantities.

4.4. Technological Change

It is well known that growth in inputs alone is not sufficient to explain all output growth. Since our data is uncorrected for the effects of technological change, it is important to allow for the technology to shift over time. This is achieved by adding a time index (t) as an additional variable and by letting this index interact with all prices and quantities. We therefore rewrite the variable profit function as follows:

$$\pi = \frac{1}{2}\frac{\sum\sum\sum a_{ihj} p_i p_h \mid y_j \mid}{\sum a_i p_i} + \frac{1}{2}\frac{\sum\sum\sum b_{jki} \mid y_j \mid\mid y_k \mid p_i}{\sum b_j \mid y_j \mid}$$

$$+\sum\sum c_{ij} p_i \mid y_j \mid + \sum\sum d_{ij} p_i \mid y_j \mid t \qquad (32)$$

$$i, h = M, X, L; \quad j = I, C, K.$$

This extension adds nine unknown parameters to the model, and it implies that the time index appears as an additional variable in each inverse output supply and direct input demand equation. Naturally equations (22)–(26) and restrictions (30) and (31) must be modified accordingly.

4.5. Data

Our empirical application is for South Korea. We require price and quantity series for all six inputs and outputs. The necessary data are taken from Kohli and Werner (1998). They are annual and for the period 1960–1991. All prices are normalized to unity for 1985, and current values are expressed in billion wons; t is defined as a time trend with unit annual increments and normalized to zero for 1985. Additional details can be found in Werner (1994).

4.6. Stochastic Specification and Estimation Technique

We assume that the demand and supply functions (20) and (21) are exact, except for errors in optimization. All demand and supply equations are converted into current value terms by multiplying by the corresponding exogenous price or quantity. Furthermore, to avoid likely problems of heteroskedasticity, we divide both sides of all equations by p_L and $-y_K$. The model is estimated using the algorithm of Berndt, Hall, Hall, and Hausman (1974); this is essentially a nonlinear version of Zellner's (1962) method for seemingly unrelated regressions. It follows from the GDP identity that equations (20) and (21), once that they are expressed in value terms, add up to zero. The system of demand and supply functions is therefore singular, and one equation must be omitted for estimation purposes. Fortunately the estimation results do not depend on which equation is left out. All in all, we have 160 observations (five equations times 32 annual observations) to estimate up to 36 unknown parameters. Hypothesis testing is carried out using the likelihood ratio test statistic, which is χ^2 distributed.

5. EMPIRICAL RESULTS

We began by estimating the unrestricted version of the model. It turned out that several of the τ_{ihj} and μ_{jki} estimates were essentially zero. This is not unusual in a model of this size, and it suggests that a semiflexible functional form would have been sufficient.[13] We purposely did not restrict the flexibility of the function, however, since this might have interfered with the testing for disjointness and nonjointness. Estimates of the remaining (nonzero) parameters, together with their asymptotic *t*-values,[14] are reported in Table 1. Curvature conditions were imposed, and thus they are met globally. We also verified that all monotonicity conditions are satisfied for all observations.

Table 2 shows 1991 estimates of the price and quantity elasticities contained in definition (5). When interpreting these elasticities one must remember that they are defined for given prices of imports, exports, and labor services, for given outputs of investment goods and consumption

Table 1: *Parameter Estimates (asymptotic t-values in parentheses)*

	Unrestricted Model		Generalized S&J Model	
c_{MI}	–0.8089	(–4.53)	–0.5716	(–4.48)
c_{MC}	–1.3722	(–10.03)	–1.1548	(–6.88)
c_{MK}	1.9944	(6.10)	1.3087	(3.49)
c_{XI}	0.2860	(1.62)	–0.1045	(–0.66)
c_{XC}	1.2107	(5.94)	0.6507	(4.13)
c_{XK}	–1.1514	(–2.79)	0.2221	(0.61)
c_{LI}	–0.3208	(–7.05)	–0.3183	(–5.54)
c_{LC}	–0.5956	(–7.88)	–0.4917	(–8.81)
c_{LK}	–0.1609	(–1.11)	–0.3460	(–3.01)
d_{MI}	0.0062	(0.79)	0.0034	(0.55)
d_{MC}	–0.0685	(–9.03)	–0.0660	(–7.48)
d_{MK}	0.0980	(6.94)	0.0862	(5.00)
d_{XI}	0.0163	(2.05)	0.0025	(0.39)
d_{XC}	0.0733	(8.72)	0.0567	(6.90)
d_{XK}	–0.0850	(–5.27)	–0.0371	(–2.26)
d_{LI}	0.0011	(0.22)	0.0110	(3.18)
d_{LC}	0.0354	(7.06)	0.0395	(10.78)
d_{LK}	0.0087	(0.87)	0.0003	(0.04)
τ_{MMI}	0.2287	(2.39)	0.2288	(1.59)
τ_{MXI}	0.0281	(0.31)	–0.0761	(–0.59)
τ_{MMC}	–0.0089	(–0.16)	0.0296	(0.68)
τ_{MXC}	–0.0233	(–0.34)	0.0476	(1.03)
μ_{IIM}	–0.6117	(–6.82)	—	
μ_{ICM}	–0.1856	(–1.03)	—	
μ_{IIX}	0.1220	(0.51)	—	
μ_{ICX}	0.4165	(3.71)	—	

Table 2: *1991 Price and Quantity Elasticities*

$$\eta_{mn} \equiv \partial \ln z_m(p_M, p_x, p_L, y_I, y_C, y_K) / \partial \ln v_n,$$

$$z_n \in \{-y_M, y_X, -y_L, p_I, p_C, p_K\}, v_m \in \{p_M, p_X, p_L, y_I, y_C, -y_K\}$$

	$n = M$	$n = X$	$n = L$	$n = I$	$n = C$	$n = K$
η_{Mn}	–0.019	–0.003	0.022	0.837	2.534	–2.371
η_{Xn}	0.002	0.001	–0.002	0.274	1.827	–1.102
η_{Ln}	0.010	0.002	–0.011	0.297	0.597	0.106
η_{In}	0.738	–0.330	0.592	0.342	0.226	–0.568
η_{Cn}	1.827	–1.800	0.974	0.185	0.380	–0.565
η_{Kn}	3.791	–2.407	–0.384	1.031	1.253	–2.284

goods, and for a given endowment of capital. It is the demand for imports and labor, the supply of exports, and the prices of investment goods, consumption goods, and capital services that do the adjusting. This partition between exogenous and endogenous variables was dictated by the requirements of the S&J structure of production. Consider η_{MM}, for instance. Its 1991 value is –0.019, which indicates that a 10% increase in the price of imports would, ceteris paribus, lead to a 0.2% reduction in the demand for imports. The other things being held constant are not only the price of exports and the endowment of capital but also the wage rate and the output of investment goods and consumption goods. The impact of the import price increase is rather small, but this is not surprising, since the quantity of imports can only decrease to the extent that the demand for labor rises and/or the supply of exports falls.[15] This experiment might be relevant in a Keynesian context with wage rigidities and the output of nontraded goods being demand determined. In international trade theory, however, one is accustomed to treating the prices of goods and the endowments of factors as given. Fortunately, it is possible to express the comparative statistics of the model for changes in the quantities of goods and in the rental prices of domestic factors as functions of changes in goods prices and factor endowments. The resulting elasticities, which correspond to the traditional GDP setting,[16] are reported in Table 3.

We now find that the own price elasticity of the demand for imports is –0.877. This indicates that a 10% increase in the price of imports will reduce the demand for imports by about 9%, *for given domestic factor endowments*

Table 3: *1991 Price and Quantity Elasticities GDP Function Setting (indirect estimates)*

$$\varepsilon_{mn} \equiv \partial \ln z_m(p_M, p_X, p_I, p_C, y_L, y_K) / \partial \ln v_n,$$

$$z_m \in \{-y_M, y_X, y_I, y_C, p_L, p_K\},\ v_n \in \{p_M, p_X, p_I, p_C, -y_L, -y_K\}$$

	$n = M$	$n = X$	$n = I$	$n = C$	$n = L$	$n = K$
ε_{Mn}	–0.877	1.522	–1.832	1.187	4.185	–3.185
ε_{Xn}	–1.114	2.155	–2.744	1.703	3.019	–2.019
ε_{In}	1.614	–3.303	4.316	–2.627	0.036	0.964
ε_{Cn}	–0.856	1.678	–2.150	1.328	1.646	–0.646
ε_{Ln}	–1.849	1.823	0.018	1.009	–0.649	0.649
ε_{Kn}	5.092	–4.411	1.750	–1.432	2.348	–2.348

and for given prices of exports, investment goods, and consumptions goods. In this context the output mix is variable, and the increase in the price of imports has a strong negative effect on the supply of exports and of consumption goods. The output of investment goods, on the other hand, benefits from the relative price change. At the same time the return to labor falls, while the return to capital increases. An increase in the price of exports has the opposite effect: that is, it favors the supply of exports—the own price elasticity of the supply of exports is in excess of 2—and the supply of consumption goods: exports and consumption goods can thus be viewed as complements in production. Moreover, it pulls along the demand for imports but penalizes the supply of investment goods. The income distribution changes in favor of workers. An increase in the price of investment goods alters the output mix in favor of those goods, and it mostly benefits capital owners. The impact of a change in the price of consumption goods is qualitatively similar to the one resulting from a change in export prices. An increase in the endowment of labor favors foreign trade and the output of consumption goods, whereas an increase in the capital stock tends to increase the supply of investment goods.[17]

We next turn to the test of the restrictions implied by the generalized S&J production structure. We reestimated the model, successively imposing restrictions (27), (28), and (29). The parameter estimates corresponding to the generalized S&J model are shown in the right-hand side of Table 1; all regularity conditions are satisfied for all observations. The test results are reported in Table 4. We find that none of the implied hypotheses can be rejected. Thus it appears that the South Korean data is compatible with production disjointness, that is, the division of the production structure into two tiers. Moreover, production can be viewed as nonjoint in input quantities

Table 4: *Test Results*

Restriction(s)	Null Hypothesis	df	$\chi^2_{.95}$	Test Statistic
1. *Disjointness of input and output tiers (DJIO)*				
(28)	Unrestricted	6	12.59	8.40
2. *Nonjointness in input quantities, output tier (NJIQ)*				
(29)	DJIO	3	7.81	3.15
3. *Almost nonjointness in input prices and quantities, input tier (ANIPQ)*				
(30)	DJIO & NJIQ	1	3.84	0.00
4. *Original Sanyal & Jones structure (MI:XC)*				
(31)	Generalized S&J	8	15.51	81.86
5. *Original Sanyal & Jones structure (MC:XI)*				
(32)	Generalized S&J	8	15.51	83.79

Table 5: *Generalized S&J Production Structure 1991 Price and Quantity Elasticities*

$$\eta_{mn} \equiv \partial \ln z_m(p_M, p_X, p_L, y_I, y_C, y_K)/\partial \ln v_n,$$

$$z_n \in \{-y_M, y_X, -y_L, p_I, p_C, p_K\}, v_m \in \{p_M, p_X, p_L, y_I, y_C, -y_K\}$$

	$n = M$	$n = X$	$n = L$	$n = I$	$n = C$	$n = K$
η_{Mn}	−0.018	0.006	0.012	0.481	1.957	−1.437
η_{Xn}	−0.004	0.003	0.001	−0.058	1.059	−0.001
η_{Ln}	0.006	−0.001	−0.005	0.257	0.401	0.342
η_{In}	0.448	0.074	0.478	0	0	0
η_{Cn}	1.487	−1.096	0.609	0	0	0
η_{Kn}	1.906	−0.001	−0.906	0	0	0

in the output tier, and almost nonjoint in input prices and quantities in the input tier.[18]

Table 5 contains 1991 estimates of the price and quantity elasticities, conditional on the S&J production structure. Note that in this case all quantity elasticities of the endogenous prices (η_{jk}, j, k = I, C, K) are necessarily zero. We find that the remaining elasticities are roughly similar to the ones shown in Table 2.

We finally tested both versions of the original S&J specification. It is apparent from the last two lines of Table 4 that both restriction sets (30) and (31) are very severely rejected. It thus appears that importables as well as exportables are used in the production of both outputs.

6. PRODUCTION FRAGMENTATION: FURTHER EMPIRICAL RESULTS

Once the model has been estimated subject to the restrictions implied by S&J, it becomes possible to identify several fragmented technologies. Using equation (32) as a starting point and taking equations (27)–(29) into account, the aggregate restricted profit function can be rewritten as

$$\pi(\cdot) = r^2(p_M, p_X, p_L)y_K - c^I(p_M, p_X, p_L)y_I - c^C(p_M, p_X, p_L)y_C, \qquad (33)$$

where $r^2(\cdot)$ is the *unit* restricted profit function of the input tier,[19] and $c^I(\cdot)$ and $c^C(\cdot)$ are, respectively, the *unit* cost functions of the investment good and consumption good industries:[20]

$$r^2(\cdot) = \frac{1}{2}\frac{\sum\sum a_{ihK}\, p_i p_h}{\sum a_i p_i} + \sum c_{iK}\, p_i + \sum d_{iK}\, p_i t \qquad (34)$$

$$c^j(\cdot) = -\frac{1}{2}\frac{\sum\sum a_{ihj}\, p_i p_h}{\sum a_i p_i} - \sum c_{ij}\, p_i - \sum d_{ij}\, p_i t \qquad (35)$$

$$i, h = M, X, L; \quad j = I, C,$$

where $a_{MXK} = 0$.

As implied by Hotelling's (1932) lemma, the demand and supply of the various inputs and outputs at each production stage can be obtained by differentiation:

$$y_i^2 = \frac{\partial r^2(\cdot)}{\partial p_i} y_K \qquad (36)$$

$$y_i^1 = -\frac{\partial c^j(\cdot)}{\partial p_i} y_j \qquad (37)$$

$$i, h = M, X, L; \quad j = I, C.$$

Multiplying through by the corresponding prices yields nominal values. Moreover, second-order differentiation of the fragmented technologies makes it possible to compute the corresponding elasticities of substitution and transformation. For the output-tier industries, for instance, the Allen-Uzawa elasticities of substitution (σ_{ih}^j) can be obtained as

$$\sigma_{ih}^j = \frac{c^j(\cdot)c_{ih}^j(\cdot)}{c_i^j(\cdot)c_h^j(\cdot)}, \quad i, h = M, X, L; \quad j = I, C, \qquad (38)$$

where $c_i^j(\cdot) \equiv \partial c^j(\cdot)/\partial p_i$ and $c_{ih}^j(\cdot) \equiv \partial^2 c^j(\cdot)/(\partial p_i \partial p_h)$.

As already mentioned, we verified that the estimated aggregate restricted profit functions satisfy all regularity conditions for all observations. This does not necessarily mean, however, that the estimated tier and sectoral restricted profit functions will do likewise, although convexity was imposed and therefore holds globally. Monotonicity is a different matter, however, and it is necessary to check each observation individually. Thus the two output-tier unit cost functions must be increasing in all their arguments, whereas $r^K(\cdot)$ must be increasing in the prices of imports and exports and decreasing in the price of labor services. It turns out that the investment-good cost function satisfies all regularity conditions for all observations but that monotonicity is only satisfied from 1968 to 1973 in the case of consumption-good cost function, and from 1970 to 1990 in the input tier. This casts some doubt on the validity of our empirical estimates of the S&J production structure. It must be said, however, that the S&J production structure imposes some formidable restrictions on the data. Thus the aggregate restricted profit function must be decreasing in p_M and increasing in p_X, but two of its components—see expression (35)—are in fact decreasing in p_X, whereas the other one—see expression (34)—is increasing in p_M. This is because, while imports are a net input to the technology, importables are an output in tier 2. Similarly, while exports are a net output for the aggregate technology, exportables are an input in the production of investment and consumption goods. It is probably not very surprising therefore if some of the monotonicity conditions are violated at the sectoral level, even though they are satisfied overall. In what follows we only report results for 1971, a year when all regularity conditions were rigorously met.

We show in Figure 3 the 1971 estimated values of the input and middle product flows at each stage of production. The elasticities of substitution in the investment good and consumption good sectors are also reported.[21] Compared to Figure 2, Figure 3 is somewhat more complex, since it describes the *generalized* version of the S&J model. That is, both production functions in the output tier allow for three inputs. The figures shown are expressed in billion wons at 1971 prices. The values shown outside the box are aggregate data, and they are readily observable. The values inside the box were calculated once the S&J structure had been imposed. Thus we find, for instance, that labor performed most of its services in the output tier, particularly in the consumption-good industry where the wage bill amounted to about 1,329 billion wons. This compares to 182 billion wons in the investment-good industry and to 211 billion wons in the input tier. The input tier produced importable middle products for about 129 billion wons. Actual imports for 923 billion wons brought the total to 1,052 billion wons, of which just over half (604 billion wons) was used up by the consumption good industry, and the remainder by the investment good industry. The input tier

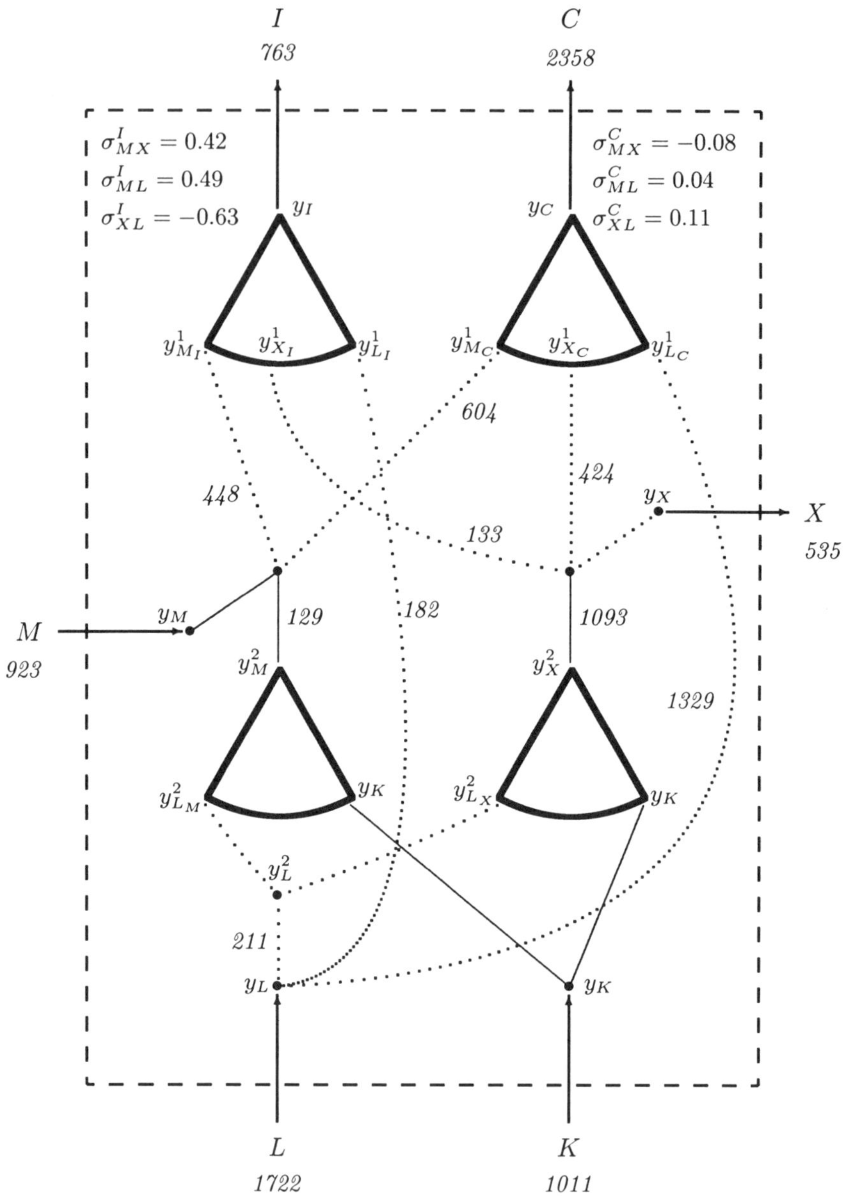

Figure 3: *The Generalized Sanyal & Jones Model: 1971 Estimates of Inter-Industry Flows of Resources and Middle Products (Billion Wons at 1971 Prices).*

also produced exportables for about 1,093 billion wons. Nearly half of those (535 billion wons) were sold to the rest of the world. About three-quarters of the remainder (424 billion wons) flowed to the consumption-good sector, and the rest went to the investment-good sector. Thus the investment-good sector seems to have used importables in relatively large amounts, while the consumption-good sector was relatively more intensive in labor services. The elasticity of substitution estimates show that the substitution possibilities are relatively limited in both sectors, particularly in the consumption good industry. There are even some hints of complementarity between exportables and labor in the investment-good industry, and between importables and exportables in the consumption-good industry.

7. CONCLUSIONS

The main purpose of this paper is to estimate and to test the model of S&J with data for a newly industrialized country, namely South Korea. We found that the data gave some support to several important features of the S&J model. In particular, the hypothesis that production is fragmented between tiers cannot be rejected. This lends credence to the belief that fragmentation enables countries to specialize in the production of those middle products in which they have a comparative advantage, a driving force in the globalization process.

Needless to say, the empirical estimates reported in the two previous sections must be interpreted with care, since they reflect the specific form of the S&J model. The S&J production structure, although very sophisticated compared to that of competing models, might still be too restrictive in some important ways to do full justice to the data. It thus would be worthwhile in future research to attempt to extend it. Generalization could be pursued in several directions. Thus one could allow for capital—either mobile or sector-specific—to be used in the output tier also. Another possible avenue would be to allow for the production of exportables to involve the use of importables. Although such extensions might alter the S&J model in significant ways, it is likely that the building blocks discussed in this paper would still have their place in a more realistic construct. Thus a country's aggregate technology probably contains numerous tiers and many occurences of disjointness and nonjointness of all kinds and at different levels. Whether such a complex web can be effectively modeled empirically remains an open question, but the notion that production at home and abroad is fragmented and that nearly all trade takes place in middle products is almost certain to survive.

References

Berndt, Ernst R., Bronwyn H. Hall, Robert E. Hall, and Jerry A. Hausman (1974) "Estimation and Inference in Nonlinear Structural Models," *Annals of Economic and Social Measurement* 3/4, 653–665.

Burgess, David F. (1974) "Production Theory and the Derived Demand for Imports," *Journal of International Economics* 4, 103–117.

Diewert, W. Erwin (1973) "Functional Forms for Profit and Transformation Functions," *Journal of Economic Theory* 6, 284–316.

Diewert, W. Erwin (1974) "Applications of Duality Theory," in M.D. Intriligator and D.A. Kendrick (eds.) *Frontiers of Quantitative Economics*, Vol. 2 (Amsterdam: North-Holland).

Diewert, W. Erwin and Lawrence A. Ostensoe (1988) "Flexible Functional Forms for Profit Functions and Global Curvature Conditions," in William A. Barnett, Ernst R. Berndt, and Halbert White (eds.) *Dynamic Econometric Modeling* (Cambridge, England: Cambridge University Press).

Diewert, W. Erwin and Terence J. Wales (1987) "Flexible Functional Forms and Global Curvature Conditions," *Econometrica* 55, 43–68.

Diewert, W. Erwin and Terence J. Wales (1988a) "Normalized Quadratic Systems of Consumer Demand Functions," *Journal of Business & Economic Statistics* 6, 303–312.

Diewert, W. Erwin and Terence J. Wales (1988b) "A Normalized Quadratic Semiflexible Functional Form," *Journal of Econometrics* 37, 327–342.

Gourieroux Christian, Alberto Holly and Alain Monfort (1982) "Likelihood Ratio Test, Wald Test and Kuhn-Tucker Test in Linear Models with Inequality Constraints on the Regression Parameters," *Econometrica* 50, 63–80.

Hall, Robert E. (1973) "The Specification of Technology with Several Kinds of Output," *Journal of Political Economy* 81, 878–892.

Hotelling, Harold (1932) "Edgeworth's Taxation Paradox and the Nature of Demand and Supply Functions," *Journal of Political Economy* 40, 577–616.

Kohli, Ulrich (1978) "A Gross National Product Function and the Derived Demand for Imports and Supply of Exports," *Canadian Journal of Economics* 11, 167–182.

Kohli, Ulrich (1983) "Nonjoint Technologies," *Review of Economic Studies* 50, 209–219.

Kohli, Ulrich (1985) "Technology and Public Goods," *Journal of Public Economics* 26, 379–400.

Kohli, Ulrich (1991) *Technology, Duality, and Foreign Trade: The GNP Function Approach to Modeling Imports and Exports* (Ann Arbor, MI: University of Michigan Press).

Kohli, Ulrich (1993a) "U.S. Technology and the Specific-Factors Model," *Journal of International Economics* 34, 115–136.

Kohli, Ulrich (1993b) "A Symmetric Normalized Quadratic GNP Function and the U.S. Demand for Imports and Supply of Exports," *International Economic Review* 34, 243–255.

Kohli, Ulrich (2001) "Sanyal and Jones on Trade in Middle Products: Estimates," *Journal of International Trade & Economic Development*, forthcoming.

Kohli, Ulrich and Augustin Werner (1998) "Accounting for South Korean GDP Growth: Index-Number and Econometric Estimates," *Pacific Economic Review* 3, 133–152.

Lau, Lawrence J. (1978) "Testing and Imposing Monotonicity, Convexity, and Quasiconvexity Constraints," in M. Fuss and D. McFadden (eds.) *Production Economics: A Dual Approach to Theory and Applications*, Vol. 1 (Amsterdam: North-Holland).

Leamer, Edward E. and Robert M. Stern (1970) *Quantitative International Economics* (Boston, MA: Allyn and Bacon).

McFadden, Daniel (1978) "Cost, Revenue and Profit Functions," in Melvyn Fuss and Daniel McFadden (eds.) *Production Economics: A Dual Approach to Theory and Applications*, Vol. 1 (Amsterdam: North-Holland).

Sanyal, Kalyan K. and Ronald W. Jones (1982) "The Theory of Trade in Middle Products," *American Economic Review* 72, 16–31.

Werner, Augustin (1994) "Analyse de la croissance du PNB de la Corée du Sud: Estimations paramétriques et non-paramétriques," mémoire de diplôme, Université de Genève.

Wiley, D. E., W. H. Schmidt, and W.J. Bramble (1973) "Studies of a Class of Covariance Structure Models," *Journal of the American Statistical Association* 68, 317–323.

Woodland, Alan D. (1977) "Joint Outputs, Intermediate Inputs and International Trade Theory," *International Economic Review* 18, 517–533.

Woodland, Alan D. (1982) *International Trade and Resource Allocation* (Amsterdam: North-Holland).

Zellner, A. (1962) "An Efficient Method for Estimating Seemingly Unrelated Regressions and Tests for Aggregation Bias," *Journal of the American Statistical Association* 57, 348–368.

Endnotes

[*]I am grateful to Ronald W. Jones for many useful comments. Any errors or omissions are my own. This research was partially financed by the Swiss National Science Foundation under grant #12-45777.95.

[1]See Leamer and Stern (1970) for a review of the early literature.

[2]The rounded cones represent one-output two-input neoclassical production functions allowing for substitution between inputs; the continuous lines indicate mandatory directions, whereas the dotted lines show the alternative paths that may be followed. See section 3 for a complete definition of the notation. This is actually the original version of the S&J model. In what follows we mostly consider a generalized version.

[3]While data on *direct* factor requirements in particular industries are widely available, what is needed here are data on *total* (i.e., direct and *indirect*) factor requirements. That is, the factor requirements of intermediate products—which are netted out at the aggregate level—used by any industry would have to be known as well and allocated to that industry. The car-manufacturing industry, for instance, uses goods (such as sheets of steel and plastics) produced by other industries, which themselves use inputs (such as iron ore or petroleum products) supplied by yet another set of manufacturing or mining industries. The amount of labor and capital inbedded in the final product is the sum of all direct and indirect factor requirements. While input-output tables might give us some indication as to what those might be, these tables do not allow for substitution between inputs, and even less for any flexibility in that process.
[4]In what follows we use the terms "Korea" or "South Korea" interchangeably to designate the Republic of Korea.
[5]We assume that T is a closed, non-empty, convex cone, that it is bounded from above, and that it allows for the free disposals of inputs.
[6]See Diewert (1974) for an analysis of the properties of the variable profit function. Note also that the restricted profit function defined by expression (1) is equal to minus a joint cost function, except that we allow for fixed inputs as well as variable ones, and for variable outputs as well as fixed ones; see Hall (1973) and McFadden (1978) for a discussion of the properties of the joint cost function.
[7]If elements i and h are both inputs, σ_{ih} is the Allen-Uzawa elasticity of substitution between them.
[8]If both elements j and k are inputs, σ_{jk} is the Hicksian elasticity of complementarity between them.
[9]If all elements of y_B were outputs, and if all elements of y_A were inputs, H_{py} would indicate the marginal input requirements of output components.
[10]See Kohli (2001) for a definition of disjointness.
[11]The disaggregation of final output between consumption goods and investment goods seems a natural choice from a national accounts, macroeconomic, or growth-theory perspective. Note that the distinction between tradables (middle products) and nontradables (end products) is already accounted for. Another possibility would be to disaggregate final output between goods and services, but there seems to be no compelling reason to do this, and it would raise additional data problems. Some people might argue that our disaggregation is somewhat arbitrary, since a particular good (e.g., a motor vehicle) may end up being classified as an export, a consumption good, or an investment good based on the accident of who purchases it. However, this misses the point: goods that are exported, consumed, or purchased by business, in spite of how it may appear, are fundamentally different products. A car that is exported does not go through the same chanels as a car that ends up in a dealer's showroom: the former is still a middle product at this stage (it may end up in a showroom in some foreign country), and as a result it typically contains less domestic value added than a car purchased by a domestic household. Similarly, a car-rental company replenishes its fleet by buying its vehicles wholesale rather than retail, which again suggests that these contain less value added than cars purchased by households.
[12]As long as all fixed quantities and all prices are nonzero, the numerators in expression (19) are necessarily strictly positive.
[13]See Diewert and Wales (1988b).
[14]These must be interpreted with care, however, since A and B do not have full rank; see Gourieroux, Holly, and Monfort (1982).
[15]The supply of exports is actually found to increase, which means that the adjustment rests on the demand for labor.

[16]See Kohli (1991).
[17]These results are qualitatively similar to the ones obtained by Kohli and Werner (1998), although they tend to be somewhat larger in absolute value. This probably has to do with the fact that the disaggregation, the time period, and the functional forms are not the same. Moreover, the estimates contained in Table 3 are obtained indirectly, using the values shown in Table 2 as a starting point.
[18]This is in contrast with the results of Kohli (2001) obtained for the United States for which the S&J production structure does not pass the test of the data, essentially because it seems that the technology cannot be broken into two tiers.
[19]It is not possible to separately identify the importable and exportable production technologies, since we do not know how the composite capital stock is actually divided up between the two industries within the input tier. It is therefore not possible to determine how the labor used in the input tier is split up between the two industries either.
[20]It is important to point out that the functional form is the Symmetric Normalized Quadratic in both cases. Hence, both expression (34) and expression (35) are flexible representations of the disaggregate technologies; see Diewert and Wales (1987).
[21]We do not show here the estimates of the elasticities of transformation of the input tier, since they are all very close to zero, revealing essentially a Leontief-like technology.

Chapter 9

TAIWAN'S HIGH-TECH INDUSTRIES

Chi Schive and Regina Yeu-Shyang Chyn

1. INTRODUCTION

In the decade following the mid-1980s Taiwan's small and medium-sized enterprises (SMEs), the major players in the country's export sector, increased in number (Hu and Schive 1998). They not only mastered the technology of their original activities but also, as outsourcing and the downsizing of large companies became prevalent, began undertaking original design manufacturing (ODM). Since the mid-1990s SMEs have advanced even further; where they once handled only manufacturing and product development, they now became involved in logistics operations, as well (Chart 1). As is illustrated by the rapid development of Taiwan's information technology (IT) industry, the country's comparative advantage shifted from traditional labor-intensive production to sophisticated high-tech operations.

Behind this megatrend has been the development of production fragmentation, which is driven by both inward and outward foreign direct investment (FDI) (Cheng, Qiu, and Tan, 2001). Outward FDI is attributable to the emergence of Taiwanese multinationals. Taiwanese companies' growing services links (i.e., coordinating activities related to the production block) have also contributed to the flourishing of fragmentation in the past decade (Jones and Kierzkowski 1990).

Two issues are related to production fragmentation in Taiwan. (1) Two seemingly competing forces—industrial clustering and globalized production—have worked and developed together in the country. Not only are these two forces the new sources of Taiwan's competitiveness, but they also act as a strong indication that production fragmentation will take place beyond Taiwan as well as within it. (2) The rapid development of air express delivery and e-commerce is helping forge a higher value-added operation (global logistics), which in turn has empowered Taiwan to take on more in global production fragmentation. The question now arises as to whether this operation will become a new trend and will enhance Taiwan's competitive advantage.

This paper first addresses how the international division of labor developed before the mid-1980s, particularly in conjunction with inward FDI and the core-satellite industrial organization.[1] It then discusses the development of Taiwan's high-tech industry. The role of SMEs, which provide evidence of domestic fragmentation, is covered as well.

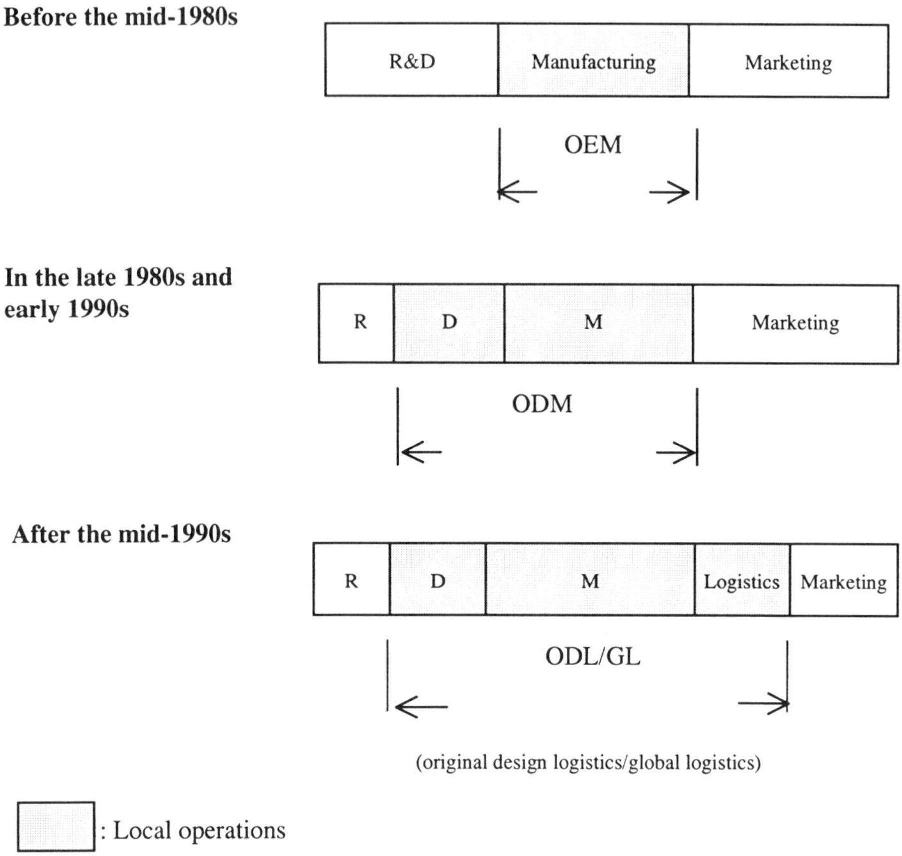

Chart 1: *Changes in Local Business Operations.*

2. DIVISION OF LABOR

Trade and Foreign Direct Investment

FDI has played an important role in integrating Taiwan in the global production network (i.e., international fragmentation). On the surface it would seem that FDI, being one of many types of capital, has never been the major source of Taiwan's capital formation. During the period 1965–86 FDI's share in the country's total capital formation never exceeded 4.32%, which it reached during the period between 1969 and 1972. Most of the time, the percentage remained around 2%. If we focus only on the manufacturing sector, the same measure never exceeded 7.33%, again peaking between 1969 and 1972. However, in the 1970s, when Taiwan's exports registered an average growth rate of around 25%, a stable 20% of its total exports was attributable to foreign business.[2] In the mid-1970s about 40% of foreign firms' production was for direct export, and 15% went to downstream production that also ended up in exports (the so-called indirect export eligible for tariff rebate). Foreign companies also purchased locally. The local content ratio of foreign-invested firms' total purchase of materials increased from 40.8% in 1972 to 50.95% and 52.4% in 1975 and 1978, respectively, although export-oriented foreign firms did buy less from the local market (Schive 1990, 1993). In short, FDI has not contributed much to Taiwan's capital formation, but it has contributed significantly to the economy's export capacity expansion and to the enlargement of the local market by gradually increasing local purchase, creating linkages ex post and ex ante.

On the one hand, FDI came to Taiwan to take advantage of inexpensive labor and to promote export. Over time foreign investors have revealed a strong industrial preference. For instance, the electrical and electronics industry attracted 36.6% of total FDI before the mid-1980s, while the chemicals industry received 17.49%, and the machinery and basic metals industries took 9.29% and 8.18%, respectively (Table 1).

The uneven distribution of arrived FDI among industries can also be detected from foreign firms' share of industry's total exports. The FDI concentrated in the electrical and electronics industry, which took the lead in total export, contributed an incredible 42.1% to that industry's exports in 1987. Next in line for total exports was the chemicals industry, in which FDI contributed 78.92% toward exports (Table 2). FDI has been instrumental in developing the export capacity of several important industries.

Trade-oriented foreign investors came to Taiwan, as might be expected, to exploit the comparative advantage that existed at the time. But doing this came at a price. In addition to contributing to Taiwan's poorly developed marketing skills, the flow of FDI was closely connected to technology

Table 1: *Taiwan's Inward FDI by Industry, 1985 (Percent)*

	Overseas Chinese		Non-Chinese		Total	
	Approval	Arrival	Approval	Arrival	Approval	Arrival
Primary	2.10	—	0.19	—	0.55	—
Manufacturing:	60.20	72.52	83.61	93.84	79.04	89.83
Textiles	6.20	15.47	0.81	1.22	1.86	3.90
Chemicals	4.07	17.44	17.35	17.51	14.76	17.49
Basic metals and products	2.66	7.18	8.09	8.41	7.03	8.18
Machinery	2.34	5.24	11.09	10.23	9.38	9.29
Electrical and electronics	3.40	5.77	35.19	43.74	28.98	36.60
Services:	37.70	27.48	16.20	6.16	20.41	10.17
All	100.00	100.00	100.00	100.00	100.00	100.00

Source: *Investment Commission, MOEA, Statistics on Overseas Chinese and Foreign Investment, Technical Cooperation, Outward Investment, Outward Technical Cooperation, 1987; and an Analysis of Operations and Economic Effects of Foreign Enterprises in Taiwan, 1985.*

Table 2: *Inward FDI share of Manufacturing Exports*

Unit NT$ million,

	1987		1997	
	Value	**Share**	**Value**	**Share**
Food & beverage processing	1,820.3	8.96	1,729.7	12.24
Textile	15,134.0	13.85	6,821.3	2.04
Garment & footwear	7,872.2	4.58	5,239.4	3.74
Lumber & bamboo products	1,501.8	3.81	755.0	4.06
Leather & fur products	62.2	0.13	252.0	0.61
Rubber & plastic products	17,068.0	18.06	13,250.9	6.03
Chemicals	18,451.5	78.92	17,719.2	18.98
Non-metallic minerals	1,393.6	2.12	8,346.4	6.00
Basic metals & metal products	10,810.4	21.65	22,776.2	12.02
Machinery equipment	15,627.2	11.42	151,570.7	42.72
Electronic & electric appliances	163,727.6	42.21	179,235.8	12.71

Sources: *1. Ministry of Finance,* Monthly Statistics of Exports and Imports, Taiwan Area, R.O.C. *July 1999.*

2. Investment Commission, Ministry Of Economic Affairs, The Economic Development Contribution of Foreign Direct Investment in ROC Survey on 1994, 1997, *July 1999.*

transfer. An earlier study of foreign firms' investment behavior indicates that the degree of foreign capital involvement in a local company is closely linked to the possibility that the company will use foreign technology. The same study also reveals that nearly all foreign-majority-owned companies in the electrical and electronics industry in Taiwan received foreign technology from their foreign partners. FDI does serve as an effective conduit of technology transfer among industries, particularly new ones (Schive 1990).

Core-Satellite Operations

SMEs have played an ongoing part in Taiwan's rapid economic and industrial development. The large number of SMEs in the country points to a well-developed vertical and horizontal division of labor in the domestic industry. The division, or fragmentation, has occurred with only small entry barriers and low transaction costs (Hu 1999; Chang 1991).

In the early 1950s and again in the 1980s, the government, witnessing SMEs' disadvantages in operating scope, marketing and distribution, management, financing, and research and development (R&D), tried to promote SME mergers but without success.[3] But in the 1960s a relationship developed naturally between SMEs and large enterprises, which together formed core-satellite operations. The first such operations formed in the machinery industry; next in line was the electrical and electronics industry. Singer, which invested in Taiwan in the mid-1960s, is known for providing generous technical assistance to its parts suppliers, thereby forging a tightly bounded core-satellite operation. Whenever and wherever outsourcing was profitable, relationships formed spontaneously. Moreover, a large number of competing and complementary SMEs in the same or related industries made up a strong industrial network or cluster in the late 1970s.

The globalization of production, logistics, distribution, and marketing provides opportunities for companies to expand business activities or extend along the value curve of the industry. In contrast, core-satellite relationships among domestic companies can be developed by backward and/or forward integration among related companies. Nevertheless, the internal integration of production, logistics, distribution, and marketing provides an alternative by which a company can expand its territory. Along the backward integration chain, core factories assemble parts and components provided by satellite factories into finished products such as automobiles, motorcycles, and televisions. In a forward integration relationship, manufacturers of raw materials such as steel sheets and petrochemicals serve as the core, providing intermediate inputs for downstream SMEs such as producers in metal processing and plastic products. In the third type of core-satellite relationship the core role is taken by trading firms or a turnkey plant exporter, both of which are non-participants in production but which contract out production to satellite factories.

The core-satellite system has prevailed in Taiwan ever since its first appearance there in the 1960s. The government established an agency to take full advantage of the trend. Three years after the introduction of such an office in 1990, there were 127 registered cases, with 1,963 formally participating firms from 23 industries, including traditional and emerging high-tech industries (Table 3). Members of the core-satellite system receive

Table 3: *Core-Satellite Industrial Groupings, 1993*

Type of Industry	No. of Groupings	No. of Participating Firms	Type of Industry	No. of Groupings	No. of Participating Firms
Autos	16	523	Footwear	3	20
Motorbikes	4	88	Furniture	2	14
Bicycles	4	58	Umbrella	1	11
Sewing machines	4	56	Metal processing	9	183
Machinery	19	203	Chemicals	6	90
Electric machinery, cables, and home appliances; electronics; information	38	555	Toys	1	8
			Food	4	38
			Ceramics	2	19
			Glass	1	10
			Textile	9	59
Photography equipment	1	6	Aerospace	2	14
Sporting	1	8	Total	127	1,963

Source: *Core-Satellite Development Center, Ministry of Economic Affairs.*

bank loans at preferential rates, and the core company helps the satellite companies obtain credit guarantees. By 1998 the number of firms registered and the number of firms involved in the activities of the registered firms had increased a great deal. The latter was roughly six times greater than the former.[4] Moreover, many firms formed similar groupings without reporting them to the government.

Before the launch of the government initiatives to promote the core-satellite system, there were already fairly well-developed streams of integration among SMEs and large companies. The role of the government has been to embrace market force and build on the existing structure. In addition to promoting the core company in its effort to provide technical assistance, information-sharing, and credit enhancement to satellite companies, the Core-Satellite Development Center has organized training programs; helped form technical, marketing, procurement, and investment partnerships; introduced standards; advanced management methods in production, finance, and information; and so forth. Though the exact outcome of the government's promotion measures has yet to be evaluated and should not be exaggerated, no negative consequences of the measures have been identified.

One might say that a kind of open grouping or natural clustering (Porter 1998) has permitted Taiwan to exercise its industrial organizational flexibility to the greatest extent possible. Moreover, through core-satellite practices, industries have acquired the ability to coordinate activities within the production block as well as by connecting foreign markets by themselves or via trading companies. More recently this ability has helped Taiwanese firms move into global fragmentation.

2. THE RISE OF HIGH-TECH INDUSTRIES IN TAIWAN

High-tech industries in Taiwan are industries that manufacture products with high value added, sophisticated technology, skilled manpower, and a high R&D ratio. Electronics, chemicals, transport equipment, machinery, and the like are among the largest high-tech industries in Taiwan. The share of these industries in manufacturing production rose from 27.4% in 1986 to 43.4% in 1997. Their share in exports expanded from 27.6% to 48.65% during the same period. The largest high-tech sector, electrical and electronic products (which largely comprise IT components) accounts for 23.3% of manufacturing production in 1997.

The Birth of the IT Industry

While its foundation had been in place for some time, Taiwan's IT industry took off in the early 1980s, when minicomputers and computer terminals began to be produced. In 1982 a local television manufacturer began large-scale production of terminals, which were being exported in increasingly large numbers. At the same time other key products such as minicomputers, disk drives, and printers began to be produced and exported as well.

The IT industry as a whole was highly export oriented; in 1982 the ratio of exports to total sales was 94%. At the same time the domestic informatics market was relying heavily on informatics imports; its import ratio was 97% during the same year. There was domestic demand for more sophisticated equipment, while export-oriented firms saw Taiwan as a supplier of low-cost components and peripherals.

Monitors enjoyed a dominant share of total informatics exports during the early 1980s, implying a vested comparative advantage for Taiwan. The industry was built on Taiwan's prosperous television industry, which comprised the bulk of the world supply of television sets. In many cases the monitor producers that made inroads in foreign markets were either current or former producers of television sets. As the market for black-and-white televisions shrank later, producers began to produce monitors instead, which was a readily available alternative (Schive and Simon 1986).

The government, the domestic private sector, and foreign-based firms have played different roles in the development of Taiwan's high-tech industries. Local firms have entered the industries in five different ways:

(1) Companies have been founded by individuals with firsthand experience in IT technologies (e.g., university professors) who were attracted to the growing demand for IT products, or by former video-games manufacturers who went out of business after the government closed all video-game parlors.

(2) Companies have been founded by former marketers who dealt with the local market in their capacity as employees of foreign multinationals. After these companies acquired technical know-how through their involvement in the industry, they moved to manufacturing themselves. In some cases the foreign partners would invite these companies to join in the manufacturing process. In most cases these companies maintained their links with foreign partners to gain access to technology and market information.

(3) Companies were organized by large multinationals through FDI.[5] Generally these foreign firms preferred to retain 100% ownership in their investments. They saw their business activities in Taiwan as a means to support overall marketing and development activities in response to changing global demands and new technological opportunities.

(4) Companies arose in response to government initiatives. In an effort to encourage entry into particular product segments and technology areas, the government helped absorb partial startup costs in terms of capital investment and, on occasion, by helping companies secure foreign technology and then diffusing it to domestic firms. In some cases the government contributed as much as 50% of start-ups paid-in capital while allowing them to remain

private. Such cases often involved large-scale efforts to develop new, specialized technologies.

(5) Companies were formed by large local firms eager to expand product lines. These firms acquire technology through licensing or other cooperative agreements with foreign companies. Some of them have also invested overseas in their dominant products.

The Development of the Semiconductor Industry

The roots of Taiwan's semiconductor industry can be traced to the 1966 establishment of the General Instrument subsidiary in the export process zone in south Taiwan, followed by the establishment of Texas Instruments, RCA, and Philips (1969–71). Although these firms' main activity was assembly, they helped develop a pool of skilled labor and well-trained technicians and managers in the early 1970s. The firms also brought foreign packaging and testing technology and quality control systems into Taiwan, hence laying the groundwork for integrated circuit (IC) development. Moreover, a semiconductor laboratory at Chiao Tung University, which specialized in science and engineering, had already developed the capacity to produce ICs. Later, similar labs were established within other educational and research institutions, thus nurturing a breed of specialized engineers.

The year 1974 was a watershed year for Taiwan's semiconductor industry. That year represents a leap from assembly to IC fabrication. In 1974 the Electronics Research and Service Organization (ERSO) was formed with a mission to establish a prototype IC fabrication plant and transfer relevant technology to commercial firms. ERSO initiated a cooperative agreement with RCA to acquire technology, training, and operation management. At the same time it recruited experts from abroad and from other electronics research institutions in Taiwan. Three years later ERSO came up with a technology that was well received by the market and not too costly to produce. In 1980 United Microelectronics Corporation (UMC), a joint venture between the government (15%), a public investment bank (25%), and domestic private enterprises (60%), was established. It began mass production in 1982 and by 1983 was the most profitable company in Taiwan. With such encouragement, IC-related companies began to flourish.

Since the 1980s the IT industry has seen a trend of shortening product life cycles, constantly changing production-process technology, and intensifying competition. As a result, profit margins have been narrowing, and the pressure to reduce cost has become tremendous. Organizational flexibility, market adaptability, and effective cost control may afford Taiwanese enterprises an advantage in surviving the new trends. A well-developed division of labor among large companies and SMEs, complemented by a comprehensive supply of peripheral industries, has provided Taiwan's IT industry with robust flexibility and speed in product supply. We elaborate on this point in the next section.

3. PRODUCTION FRAGMENTATION AND HIGH-TECH INDUSTRIES IN TAIWAN

Local Production

The rapid development of the semiconductor industry in Taiwan has helped boost the IT industry by providing a stable supply of upstream materials. By joining the IT industry, the semiconductor industry has also broadened the production scope of the IT industry and increased the complementarity of members of the emerging industry.

The semiconductor industry comprises many firms of different sizes. These firms produce a wide spectrum of products, from logic, analog, DRAM, and SRAM to input/output devices and microprocessors. The industry itself is very competitive. It has ever-more new entrants and also enjoys competitive prices on equipment and materials in the international marketplace. Unlike traditional industries, Taiwan's semiconductor industry is not dominated by export markets. Its exports, in turn, are not aimed at the U.S. market. While the semiconductor industry, as an upstream supplier, has injected extra strength into the IT industry, other information industries, such as electronic components in the midstream and personal and notebook computers, monitors, and the like in the downstream, have constituted a remarkable market for the semiconductor industry.

The semiconductor industry in Taiwan has been disintegrated into IC fabrication (63%), IC design (13%), IC packaging (17%), and IC testing (6%). The fabrication segment is engaged in custom ASIC (2.7%), foundry (35.3%), and standardized products (61.9%) for information, memory, communications, consumer electronics, and so forth. IC design includes system IC, network IC, memory IC, and multimedia IC. Since fabrication is highly capital intensive, the industry comprises mainly large enterprises. Because the minimum efficiency scale in the testing and designing industries is relatively lower, there are many SMEs in these two industries. (Chart 2).

Chart 2 presents the location of Taiwan's semiconductor industry by major operations and county. The well-known Hsin-Chu Science-Based Industrial Park hosts most of the production of almost all kinds of semiconductor products in Taiwan. All but 1 of the country's 20 IC fabrication companies are located in the park, which also hosts 19 of Taiwan's 36 IC packaging companies and 45 of its 76 IC design shops. The nearby Taoyuan and Taipei counties are two other major home bases for IC businesses, especially for IC design shops located on the outskirts of the Taipei metropolitan area. A newly established computer software industrial

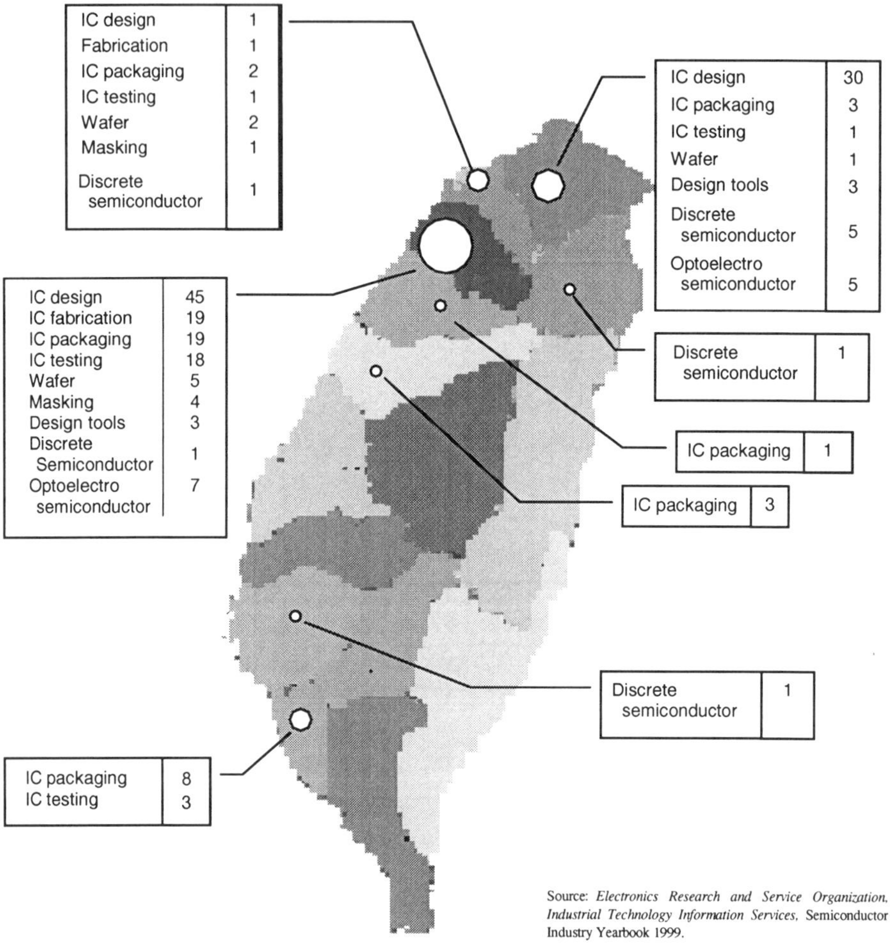

Source: *Electronics Research and Service Organization, Industrial Technology Information Services*, Semiconductor Industry Yearbook 1999.

Chart 2: *Distribution of Semiconductor Industries, 1999.*

spark is located in a suburb of Taipei. The Kao-Hsiung export processing zone in the south of Taiwan has attracted 8 of the country's 36 IC packaging houses. Only 6 such firms are scattered in other areas of Taiwan. Semiconductor companies in Taiwan are concentrated mainly in north Taiwan, with some in the south and very few elsewhere. Locating a new business near existing similar ones is common; it offers newcomers an opportunity to compete as well as to assimilate into the industry. Industrial clustering goes beyond the principle of "birds of a feather flock together." Seeing low-priced computers gaining popularity, major computer manufacturers on the world market have adapted to the new trend by outsourcing. An integrated service chain from designing, to packaging, and to testing will be the top choice of these multinationals, which must ensure short delivery times and high-quality products. This trend has generated an advantage for Taiwan, which has just such an integrated production chain in the semiconductor industry. To keep their competitive advantage and market share, complementary industries have formed strategic alliances among themselves. For example, Taiwan Semiconductor Manufacturing Company (TSMC) as a foundry maker is allied with a packaging enterprise and a testing enterprise. As the foundry's scale of operation continues to grow larger and more complicated, foundry makers may also enter the testing and packaging industries themselves and make it more difficult for SMEs to compete. A way out of the straightjacket for SMEs in the fields is to secure market orders by forming strategic alliances with upstream and midstream industries.

Changing Global Production Fragmentation

Before the mid-1980s Taiwan's trade-related activities consisted simply of manufacturing or assembling the base of imported materials or intermediates aimed mainly at the U.S. market. Without advanced technology and the ability to engage in self-design, Taiwan was only able to handle original equipment manufacturing (OEM). Technology and management skill were brought in through OEM orders and/or inward investments from Japan and the United States. During the late 1980s and early 1990s Taiwan became an active source country in FDI, first in Association of Southeast Asian Nations (ASEAN) countries, then in the People's Republic of China (PRC). Taiwan then provided intermediates, equipment, and technology plus management to the host economies and exported finished products from there to the United States. While still producing for exports, Taiwan's local companies moved from OEM to ODM. Industries at this stage could design products to satisfy buyers' requests, (i.e., could engage in product development) (Chart 3).

Since the mid-1990s the introduction of electronic networks, combined with the availability of powerful personal computers and dramatically lower-priced telecom services, has been enabling companies to communicate and

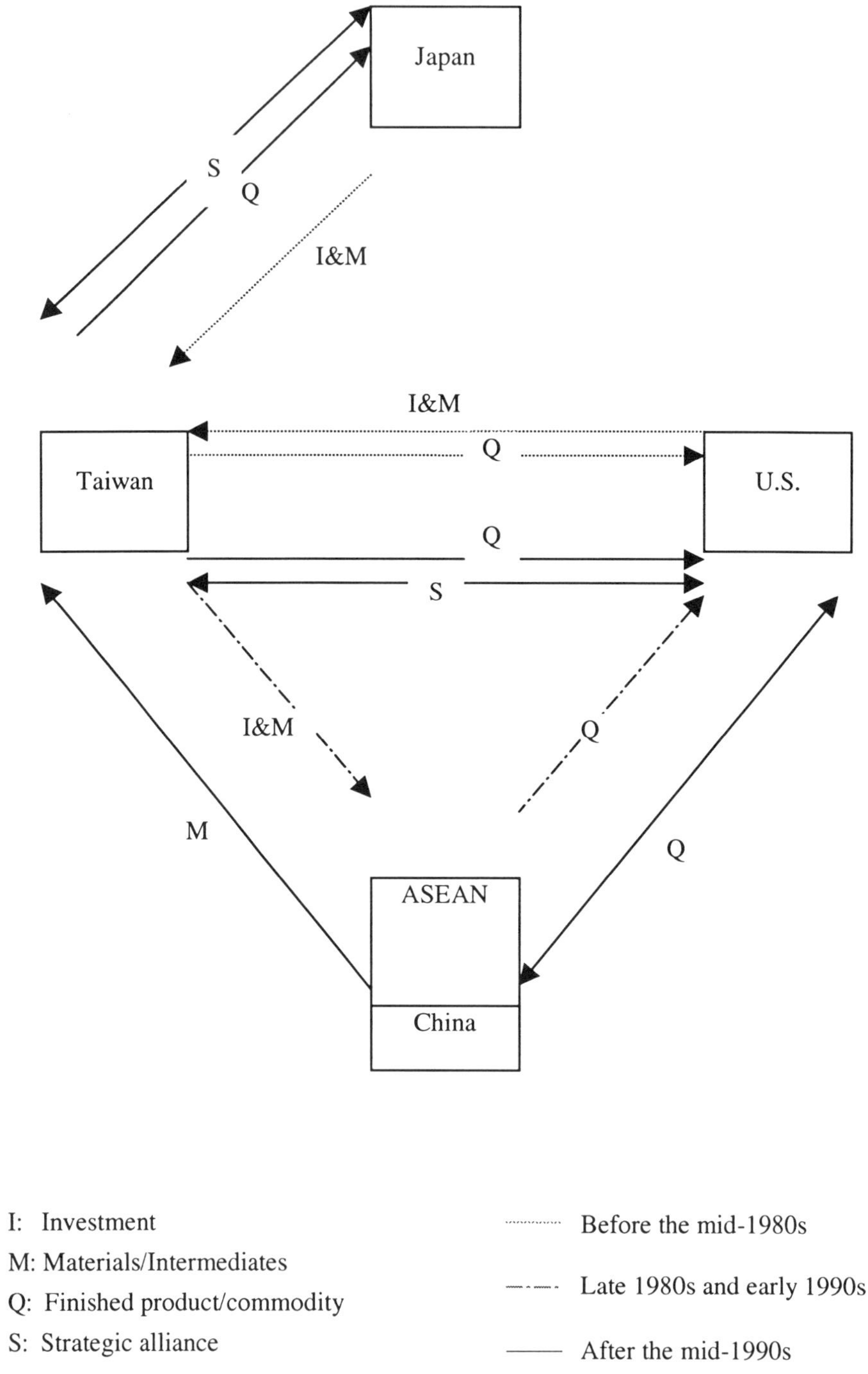

Chart 3: *Division Labor between Taiwan and Its Major Trade Partners.*

exchange data far more quickly and cheaply than ever before. At the same time consumers have become more sophisticated, and competition and the pursuit of growth keep companies constantly in search of better business practices that improve efficiency. As more business interactions move onto electronic networks like the Internet, basic organization of corporate operations has changed. Some companies find it is to their advantage to outsource non-core activities, while others may find that activities central to their business can be performed better, faster, and more efficiently by new, specialized competitors. Moreover, the combination of rising customer demand for technical support services and an increasing need to build customer loyalty and support to maintain market share has pushed large manufacturers to focus more on the customer end of the value chain. Some business-management consultants even believe the real money in manufacturing today lies downstream, not in the production function (Wise and Baumgartner 1999). In this case, multinationals would shed logistical operation to concentrate on building customer relationships. And manufacturing suppliers like Taiwan are offered the opportunity to extend the supply chain by carrying out global logistic management. (Chart 3)

Facing increasingly sophisticated demand from global business, how has Taiwan, previously an OEM/ODM player, responded? Although it has long had a robust manufacturing base, how has it moved into the global logistics arena?

Global logistics is built on three things: (1) a logistics infrastructure, including a strong logistic information system and framework of standardized operational procedures; (2) a production base that includes a frictionless production network and a comprehensive integration of production activities and supply raw materials, components, and other inputs, regionally or worldwide; and (3) management talents in international business interactions and finance. Since the early 1990s Taiwan has switched its development emphasis from industrial growth to macro reengineering of the economy mapped out by an initiative to strengthen Taiwan's international operations capability and develop it as a center for regional operations. Liberalization efforts and physical development have been devoted to better regional telecom operations and sea/air transportation, which has enhanced the country's logistics infrastructure. Transshipment management and cooperation with big international players has been promoted to generate a bigger market for the manufacturing sector, which over the past 30 years has already built a viable production-integration and technology foundation. Vast-scale deregulation in the financial sector since the early 1990s has created many new financial instruments and has enhanced the facilitation of cross-border financial flows.

For all domestic development, the manufacturing sector's overseas activities have been internalized into Taiwan's production network, which has helped reduce the cost of international transactions and has made the launch of global logistics management an even more natural choice for Taiwan.

Table 4: *Procurement Order Sharing in Taiwan's Manufacturing, 1998 (Percent of Total)*

	% of Mfg	Domestic Production	Foreign Production
Manufacturing	100.00	89.35	10.65
Information electronics	35.71	82.03	17.97
Electronics, electrical machinery	34.90	82.06	17.94
Precision instruments	0.81	80.92	19.08
Metal machinery	23.91	97.62	2.38
Chemicals	23.35	89.05	10.95
Consumer products	17.04	93.50	6.50

		Domestic Production		Foreign Production			
	Total Order	Own Production	Contract Out	Japan U.S. Europe	Mainland China	Southeast Asia	Others
Manufacturing	100.0	84.95	4.40	2.37	4.42	3.38	0.49
Information, electronics	100.0	76.97	5.06	4.28	8.83	4.59	0.28
Electronics, electrical machinery	100.0	76.94	5.12	4.37	8.64	4.66	0.28
Data storage processing	100.0	68.42	5.05	8.89	10.45	6.92	0.27
Telecom	100.0	73.86	3.81	0.09	7.14	15.10	0.00
Electronic components	100.0	82.06	4.93	1.95	8.15	2.68	0.23
Precision instruments	100.0	78.12	2.79	0.61	17.08	1.37	0.03
Metal machinery	100.0	92.61	5.01	0.71	1.29	0.26	0.12
Chemicals	100.0	86.95	2.10	1.44	2.46	6.00	1.05
Consumer products	100.0	88.20	5.30	1.95	2.25	1.63	0.67

Source: *Ministry of Economic Affairs, Statistics Department,* Survey on Manufacturing Activities, *July 1999. In Chinese.*

As overseas investment by Taiwanese companies expands, a purchase-order distribution trade between domestic and overseas production bases is emerging. As Taiwanese enterprises move forward along the supply chain to global logistics, this trend is becoming increasingly noticeable. In 1998 the domestic manufacturing industry allocated nearly 11% of its orders to overseas production and retained the remaining 89% (Table 4). Information electronics has a far higher overseas production ratio (18%) than the other three major manufacturing sectors—metal and machinery, chemicals, and consumer consumption. Moreover, information electronics' domestic production has the lowest ratio (77%) of in-house production; all other sectors have ratios over 86%. Obviously, production in high-tech industries is more fragmented, both domestically and overseas, by a significant degree. Since the high-tech industry is the mainstay of Taiwan's economy today, its increasing fragmentation presents the rationale for the country's development of global logistics.

This phenomenon is more noticeable when one examines the information electronics industry more closely. The domestic production ratios of data storage and processing (73.5%) and telecom devices (77.7%) are much lower than those of the industry's more traditional sectors, such as machinery and audio and video equipment. The same is true for own-production ratios. However, the ratio of domestic production and the ratio of in-house production of the electronic components industry are higher than those of machinery and audio and video equipment industries.

Taiwanese companies pass orders on to overseas manufacturers for five main reasons: (1) lower production costs; (2) customer requests; (3) availability of overseas subsidiaries; (4) inability of domestic production to absorb all orders; and (5) lack of domestic facilities availability for certain types of production. Among the domestic manufacturing industry's orders fulfilled by overseas production, about 22% go to Japan, the United States, and Europe; 41.5% go to the PRC; and 32% go to Southeast Asia. The information electronics industry has a higher than average ratio, with 23.8% going to Japan, the United States, and Europe; 49% to the PRC; and 25.5% to Southeast Asia. The ratios of the electronics and electric machinery industry are similar.

Since costs in the United States, Japan, and Europe are much higher than costs in Taiwan, the reasons for sending orders to be produced in these economies may be related to global logistics. However, the fact that a large proportion of all Taiwanese industries send a high proportion of orders to the PRC for production is undoubtedly explained by cost.

Because Taiwan lacks natural resources, Taiwanese industries depend heavily on imported raw materials. Their reliance on imported components changes with the evolution of the industry. As a result of backward integration in the 1970s, upstream industries were developed, and Taiwan was able to rely on its own supply of components except in the case of key parts. This reliance is known as the result of secondary import substitution. However, given the increasing need for more sophisticated products since the late 1980s, the country's reliance on imported components has risen slightly. In 1998 the manufacturing industry's spending on raw materials, components, and semi-finished products accounted for 57% of its total expenditure on intermediate inputs, with information electronics spending the largest proportion (64%) among manufacturing industries. Of the manufacturing sector's input of all types of intermediates, 57.8% were from domestic sources and the rest were from overseas sources. The information electronics industry was more dependent, on average, than the manufacturing industry, with a ratio of overseas supply to total supply of 52.3% (Table 5). While the manufacturing industry as a whole purchased 21.38% of its total intermediate input, or 50.7% of its imported intermediates, from Japan and the United States jointly, the information electronics industry purchased a larger proportion from these countries: 31.97% of total intermediates, or 61% of imported inputs.

Table 5: *Sources of Materials/Intermediate Goods of Firms in Taiwan, 1998 (Percent of Total)*

	Mfg	Information Electronics	Electronics Electrical Machinery	Data Storage Processing	Telecom Devices	Electronic Components	Precision Instruments
Total	100.00	100.00	100.00	100.00	100.00	100.00	100.00
Domestic	57.84	47.70	47.32	49.76	44.95	39.29	64.26
Foreign	42.16	52.30	52.68	50.33	55.05	60.71	35.74
Japan	12.19	17.91	17.93	12.85	9.00	23.15	16.97
U.S.	9.19	14.06	14.29	17.58	22.40	12.87	3.78
Europe	5.26	7.28	7.39	2.95	6.59	15.09	2.46
Mainland China	1.95	1.85	1.86	1.73	3.59	2.06	1.50
Southeast Asia	4.85	6.71	6.82	8.44	13.18	5.57	2.16
Others	8.73	4.49	4.39	6.79	0.28	1.97	8.87

Source: *Ministry of Economic Affairs, Statistics Department, Survey on Manufacturing Activities, July 1999. In Chinese.*

Table 6: *Market Destination of Taiwan's High-Tech Production, 1998 (Percent of Total)*

	Mfg	Information Electronics	Electronics Electrical Machinery	Data Storage Processing	Telecom Devices	Electronic Components	Precision Instruments
Total	100.00	100.00	100.00	100.00	100.00	100.00	100.00
Domestic	59.21	31.97	32.03	13.94	36.01	33.49	29.58
Foreign	40.79	68.03	67.97	86.06	63.99	66.51	70.42
Japan	3.33	5.45	2.09	4.22	4.14	5.08	2.42
U.S.	15.31	28.93	29.18	43.77	36.19	23.62	18.09
Europe	7.85	15.41	15.48	25.42	11.28	11.73	12.27
Mainland China	3.98	2.10	4.95	1.08	4.36	2.30	27.41
Southeast Asia	6.52	10.76	10.84	6.15	4.17	17.50	7.16
Others	3.80	5.37	5.42	5.43	3.85	6.26	3.08

Source: *Ministry of Economic Affairs, Statistics Department, Survey on Manufacturing Activities, July 1999. In Chinese.*

The manufacturing industry in general has an export ratio of 40%. In the information electronics industries, however, the ratio is above 60%, and in the data-storage and -processing sector it exceeds 86% (Table 6). This implies that the high-tech industry is much more outward looking than other manufacturing industries in Taiwan, which reflects its strong link to global logistics.

The deepening international specialization of the semiconductor industry, if not enlarging the spectrum of IT products manufactured in Taiwan, has at minimum provided a niche for Taiwan in the world market.

5. GLOBAL LOGISTICS AND THE HIGH-TECH INDUSTRY

In the past, big domestic computer producers such as ACER, Tatung, and Mitac would forecast the forthcoming year's OEM orders and plan and produce ahead. The products would then go to global buyers in the United States or Japan, which would take care of distribution, delivery, inventory, and after-sale services. Thanks to technological breakthroughs, global leaders such as Compaq and Dell have adopted a build-to-order (BTO) strategy to keep costs low and the market secure. The trend toward a BTO mode implies a change in the logistics of Taiwanese firms. Since U.S. and Japanese companies will not give Taiwan a production order until they receive a purchase order, these companies shorten the supply chain and reduce their inventory costs, and pass the job on to OEM/ODM producers. If a Taiwanese firm wants to win OEM orders from these big players, it must be strong in global logistics management. As a result, OEM orders mainly go to large companies such as Acer and Mitac. SMEs have to work with these core firms for orders, hence forging a close link.

In a BTO mode it is to the advantage of an OEM firm to have a warehouse located where the client chooses. The distributor, upon receiving the order, goes directly to the production point or shipment warehouse to obtain the ordered products. With sound global logistic management empowered by logistics businesses such as Federal Express and United Parcel Services, firms gain a significant competitive edge. Therefore, increasing overseas investment to build a strong global network based on further horizontal and vertical division of labor is a must for most of Taiwan's high-tech industries.

The lengthening value chain implies increased business risks as well as increased opportunities. Taiwan seems ready to take advantage of the opportunities. It has a robust manufacturing base and is building its capacity as a regional operations center, and it has set up a large number of overseas production bases by increasing outward investment since the mid-1980s. Although outward investment's main purpose was to take advantage of the low cost of labor and land in the host country, the availability of support to global operations has been increasingly important.

In 1998 more than 71% of Taiwanese companies that had invested overseas had investments in the PRC, and about 17% had investments in the United

States. Investor companies usually prefer to invest alone. Investments are usually made in cash. However, most SMEs have made investments in the form of machinery, equipment, raw materials, components, and semi-finished or finished products. While most parent companies are equipped for production, R&D, design, marketing and distribution, finance, planning, procurement, and services, the subsidiaries may comprise only a production department and a finance and marketing and distribution department. It is worth noting that 30% of overseas subsidiaries already have R&D departments. Currently subsidiaries' technology is predominantly supplied by parent companies (although this trend is declining). Whether the increasing R&D capability of overseas subsidiaries has had any impact on Taiwan's domestic industries remains to be seen (Ministry of Economic Affairs 1999).

Since the mid-1980s Taiwan has seen increasing exports induced by outward investment. Intermediates from Taiwan accounted for 33.4% of its total exports in 1986. That figure has risen steadily; it reached 61.1% in 1999. However, looking at the company level, the supply of intermediates has been more localized to reduce distant supplies from Taiwan and thus to lower costs, as contacts with the host economy have increased and linkage have expanded (Table 7). Among all Taiwan-invested companies, regardless of size, the supply of raw materials, components, and semifinished goods received from Taiwan shrank from 1996 to 1997. Although there is substitution between supplies from Taiwan and from the host economy, the host economy has increased the supply

Table 7: *Sources of Raw Materials, Components and Semi-finished Products of Overseas Subsidiaries (Percent, by Size of Companies)*

	All		Small Enterprises		Medium Enterprises		Large Enterprises	
Supply by	**1996**	**1997**	**1996**	**1997**	**1996**	**1997**	**1996**	**1997**
Raw materials	100	100	100	100	100	100	100	100
Taiwan	50	46	52	47	49	46	45	43
Taiwan subsidiaries in host economy	15	18	18	21	14	15	11	12
Non-Taiwanese firms in host economy	22	23	20	21	24	25	24	25
Imports from elsewhere	13	14	11	11	13	14	20	20
Components, semi-finished products	100	100	100	100	100	100	100	100
Taiwan	54	50	55	50	57	52	49	47
Taiwan subsidiaries in host economy	16	18	19	22	10	12	10	12
Non-Taiwanese firms in host economy	22	23	19	20	25	27	27	27
Imports from elsewhere	9	10	7	8	8	9	14	15

Source: *Ministry of Economic Affairs,* Survey on Foreign Investment by the Manufacturing Industry, *1999.*

more frequently on the part of other Taiwanese firms in the host economy. This economy. This could also imply that Taiwanese firms have developed products that meet the demand of other Taiwanese firms in terms of either quality or quantity. It is also likely that linkages took place more easily within groups of similar language, ethnicity and culture.

FDI tends to maintain a close tie with the motherland, and in this Taiwan is no exception. It registered a lion's share of raw materials supply to Taiwanese subsidiaries abroad, and among these the highest percentage (above 55%) went to the United States, followed closely by PRC and the Philippines (Table 8).[6] As far as the supply of components and semifinished products is concerned, the relationship between Taiwan and its foreign operations in the above three countries appeared to be even stronger.

Table 8: *Sources of Intermediate Inputs of Taiwan's Overseas Subsidiaries (Percent of Total)*

Supply by		All	U.S.	Japan	PRC	Malaysia	Singapore	Thailand	Indonesia	Philippines
Raw materials		100	100	100	100	100	100	100	100	100
Taiwan	'96	50	57	49	50	37	19	32	42	50
	'97	46	55	38	45	34	19	27	44	52
Taiwanese subsidiaries in host economy	'96	15	10	16	18	17	10	9	9	10
	'97	18	11	27	21	18	10	10	6	11
Non-Taiwanese firms in host economy	'96	22	25	14	19	33	26	37	29	19
	'97	23	25	14	21	35	26	39	29	18
Imports from elsewhere	'96	13	8	21	13	13	45	22	21	20
	'97	14	9	21	13	14	45	24	21	20
Components, semi-finished products		100	100	100	100	100	100	100	100	100
Taiwan	'96	54	63	52	53	38	40	41	48	65
	'97	50	61	50	48	34	40	39	47	62
Taiwanese subsidiaries in host economy	'96	16	3	20	19	23	0	12	10	2
	'97	18	5	22	22	25	0	13	10	3
Non-Taiwanese firms in host economy	'96	22	23	20	21	33	32	31	28	19
	'97	23	23	20	22	35	32	29	28	22
Imports from elsewhere	'96	9	11	8	8	6	28	15	14	14
	'97	10	11	8	8	6	28	20	15	13

Note: *The effective sample size was 1,264 firms.*

Source: *Ministry of Economic Affairs,* Survey on Foreign Investment by the Manufacturing Industry, *1999.*

Table 9: *Geographic Distribution of Sales by Taiwanese Overseas Subsidiaries (Percent)*

		To Taiwan		To Host Economy		To Elsewhere	
Sales by	**All**	**1996**	**1997**	**1996**	**1997**	**1996**	**1997**
Total	100.00	11.40	10.76	45.98	47.06	42.62	42.18
Small enterprises	100.00	12.26	11.26	42.64	44.31	45.11	44.42
Metal, machinery	100.00	14.15	13.55	58.07	59.94	27.79	26.51
Information, electronics	100.00	16.37	14.70	37.80	41.51	45.83	43.80
Chemicals	100.00	6.57	5.32	45.84	47.11	47.59	47.58
Consumer products	100.00	12.62	11.96	30.63	30.38	56.74	57.66
Large enterprises	100.00	10.03	9.97	51.29	51.44	38.68	38.60
Metal, machinery	100.00	7.85	8.73	57.34	58.55	34.81	32.72
Information, electronics	100.00	12.71	12.57	47.60	47.15	39.68	40.29
Chemicals	100.00	5.25	5.56	56.63	56.92	38.12	37.52
Consumer products	100.00	12.05	10.75	45.81	45.46	42.15	43.79

Source: *Ministry of Economic Affairs,* Survey on Foreign Investment by the Manufacturing Industry, *1999.*

Another category of intermediates suppliers deserving of attention is that of Taiwanese subsidiaries in the PRC and Malaysia.[7] The electronics industry means a trade-off between quantities produced in Taiwan and in the host has set up factories for computers, resistors, passive components, scanners, motherboards, barebones, capacitors, monitors, mice, CD-ROM drives, and so forth in the PRC. Some large companies even have multiple production bases in different Chinese cities. There are also a few related operators such as shipment centers, sales departments, and suppliers of peripherals.[8] In short, Taiwanese business operation abroad has already evolved from the early pioneering, going-alone investment to the herding pattern of investment. An immediate impact of this shift is that local sourcing from Taiwanese companies has been on the rise, and direct sourcing from Taiwan has declined, as is confirmed by 1996–97 data.

Local sales ratios and export ratios have been about the same for small Taiwanese subsidiaries, while local sales ratios have been slightly higher than those of large ones (Table 9). The difference shows that the small subsidiaries, more than large ones, were established with an aim toward export markets. It may be noted that the information and electronics industries have the highest ratio of resales to Taiwan, regardless of firm size, which indicates the presence of global logistics.

How did Taiwan's companies divide their production between parent companies and subsidiaries? In other words, how was intrafirm division of labor taking place in Taiwan's global operations? As is shown in Table 10,

Table 10: *Product Relations between Taiwan Parent Companies and Their Overseas Subsidiaries (Percent)*

Parent	Unrelated	Component[1]	Finished Product	Finished Product-Higher Value[2]	Same Product-Higher End	Same Product-Lower End	Exactly Same
Subsidiary		Finished Product	Component[1]	Finished Product-Lower Value[2]	Same Product-Lower End	Same Product-Higher End	
Total	12.42	11.71	4.59	10.76	21.28	1.90	32.04
Food	15.00	0.00	7.50	5.00	22.50	0.00	47.50
Garment, apparel	0.00	14.29	0.00	17.86	17.86	3.57	46.43
Pulp, paper products	13.64	9.09	0.00	9.09	22.73	0.00	40.91
Precision instruments	7.69	19.23	3.85	13.79	19.23	7.69	26.92
Chemical equipment	21.21	9.09	0.00	12.12	18.18	0.00	36.36
Machinery equipment	22.34	18.09	6.38	8.51	17.02	1.06	22.34
Rubber products	3.45	3.45	0.00	13.79	17.24	3.45	51.72
Wood, bamboo products	38.46	0.00	23.08	11.54	0.00	0.00	23.08
Electronic, electrical	7.81	12.31	5.11	11.71	17.72	2.40	35.14
Plastic products	13.39	10.24	2.36	11.02	18.90	1.57	40.16
Metal products	16.13	9.68	5.38	11.83	19.35	2.15	30.11
Textile	12.68	4.23	1.41	15.49	38.03	0.00	25.35
Miscellaneous	13.89	20.83	4.17	9.72	13.89	2.78	29.17

Note: *1. Components include semi-finished products and parts.*
2. Higher value and lower value mean higher value-added and lower value-added, respectively.

Source: *Ministry of Economic Affairs,* Survey on Foreign Investment by the Manufacturing Industry, 1999.

32.04% of overseas subsidiaries manufactured exactly the same types of products as their parent companies in Taiwan. These firms may represent investment made for reasons of cost or market access. Another 32.04% of Taiwanese companies whose subsidiaries manufactured similar but lower quality products compared to their parent companies invested abroad. Only 1.90% of the subsidiaries produced higher-quality products than the parents. It is more common for the parent company to be the intermediate supplier to the subsidiary than the other way around, except in the food, wood, and bamboo products industries.

It may be concluded that among the overseas operation of Taiwan's multinationals, horizontal division of labor is preferred over vertical division of labor. Nevertheless, cost-down pressure, delivery request, and the ability to meet diversified market demands all work together to push investors to constantly seek a better place for part of if not all the production for diversified markets. When intrafirm production fragmentation intensifies, global logistics serve as both the push and pull for that trend.

6. CONCLUDING REMARKS

Although the world has entered the age of discontinuity, the growth and development of an industry or an economy has never been discontinuous. FDI in the 1960s and 1970s in Taiwan has not only brought in much-needed foreign exchange and technology but has also opened Taiwan's economy by integrating it into the international division of labor. Being friendly to inward FDI proved conducive to Taiwan's rapid industrialization. This strategy so impressed Taiwanese firms that they quickly switched to outward investment when the local business environment changed abruptly after the mid-1980s. Taiwan thereby extended its geographic production line and further deepened its own international division of labor. During the period when Taiwan began to invest significantly abroad, domestic SMEs grew in tandem with the expansion of the economy, including the high-tech sector. Moreover, the strength of Taiwan's manufacturing sector, enhanced by the core-satellite operation between many peripheral SMEs and the core company, continued to develop. This strength in Taiwan's manufacturing sector helped the economy weather the recent regional financial crisis well. The success of SMEs is attributable to the continuously broadening division of labor, which enhanced production efficiency and industrial competition. An important driving force behind this development, indeed, came from a series of market-opening policies that have come into effect since the 1960s (Schive 1999).

As the division of labor is proceeding both domestically and globally, whether the two forces reinforce or offset each other is a crucial and unanswered question. The Taiwan experience indicates that the simultaneous development of these two forces tends to be more positive than negative. Most telling will be the availability of global logistic management among local businesses—high-tech industries in particular. Global logistics emphasize fast delivery to minimize inventory cost, and quick response to market change in order to gain competitiveness, while at the same time taking advantage of global production fragmentation for lower cost. This modern business practice is most clearly by today's IT industry. The practice is based on the idea of ideal geographic location, fast delivery, efficient financial and telecom support, and a well-developed international production network built upon an intensified division of labor, locally and globally. Taiwan has all of these factors in varying degrees. All the factors except the first one are rooted in a strong liberalization movement that was initiated in the early 1990s in related markets.

We may learn a lesson from Taiwan. If you open up a specific market, you know there will be more participants in that market; competition and consumers' surplus will increase. The results of opening up more than one market are harder to predict. It is not unlikely that you will end up in a totally new situation with surprisingly far-reaching benefits.

References

Chang, Shan-Jih (1991), *Estimation of Transaction Costs in Taiwan,* M.A. thesis, in Chinese (Taipei: National Taiwan University, Department of Economics).

Cheng, Leonard K. and Larry D. Qiu and Guofu Tan. (2001), "Foreign Direct Investment and International Fragmentation of Production," in S. Arndt and H. Kierzkowski, eds., *Fragmentation: New Production Patterns in the World Trade* (Oxford: Oxford University Press).

Commercial Times, (1999), October 26, p. 3.

Electronics Research and Service Organization (ERSO) (1998), *Yearbook of Semiconductor Industry,* in Chinese (Taipei: ITRI).

Hagel John III and Marc Singer (1999), "Unbundling the Corporation," *Harvard Business Review*, March–April.

Hu, Ming-Wen (1999), "A Study on the Adaptability of Taiwan's Industrial Size Structure," *IndustryEconomics Conference Proceedings* (Melbourne: Australia: Productivity Commission of the commonwealth of Australia).

Hu, Ming-Wen and Chi Schive (1998), "The Changing Competitiveness of Taiwan's Manufacturing SMEs," *Small Business Economics*, 11:315–326.

Jones, Ronald W. and Henryk Kierzkowski (1990), "The Role of Services in Production and Internatioanl Trade: A Theorectical Framework," in A. Krueger and R. Jones, eds., *The Political Economy of International Trade* (Oxford: Basil Blackwell).

Jones, Ronald W. and Henryk Kierzkowski (2001), "A Framework for Fragmentation," in *Fragmentation and International Trade*. in S. Arndt and H. Kierzkowski, eds., *Fragmentation: New Production Patterns in the World Trade* (Oxford: Oxford University Press), forthcoming.

Ministry of Economic Affairs (1999), *Survey on Manufacturing Activities.* Taipei, in Chinese (Taiwan: OEA).

Ministry of Economic Affairs (1999), *Survey on Foreign Investment by the Manufacturing Industry,* in Chinese (Taipei, Taiwan: MOEA).

Porter, Michael E (1998), "Clusters and the New Economics of Competition," *Harvard Business Review*, November–December.

Schive, Chi (1993), "Cross-Investment in the Asia-Pacific: Taiwan's Inward and Outward Investment," *Business and the Contemporary World*, Vol. V, No. 2, Spring.

Schive, Chi, and Denis F. Simon (1986), "Taiwan's Informatics Industry: the Role of the State in the Development of High-tech Industry," in F. Rushing and C., eds., Ganz Brown *National Policies for Developing High Technology Industries -- International Comparisons* (Boulder: Westview).

Schive, Chi and Sze-Hwa Wu (1989), *Determinants of Industries' R&D – Analysis on the Emerging Automation Industry,* in Chinese, a paper research prepared for Advisory Division of Science and Technology, Executive Yuan (Taipei, Taiwan).

Schive, Chi (1990), *The Foreign Factor: the Multinational Corporation's contribution to the Economic Modernization of the Republic of China*, (Stanford: Hoover Institution).

------------- (1990) "Linkages: Do Foreign Firms Buy Locally?" *Asian Economic Journal*. 1–15.

------------- (1995), *Taiwan's Economic Role in East Asia*. (Washington, D.C.: The Center for Strategic and International Studies).

------------- (1999), "How did Small and Medium Enterprises in Taiwan Survive the Crisis?" In *1999 Industry Economics Conference Proceedings* (Melbourne: Productivity Commission of the Commonwealth of Australia.), 91–110.

Wise, Richard and Peter Baumgartner (1999), "Go Downstream: the New Profit Imperative in Manufacturing," *Harvard Business Review*, September–October.

Wang, Jiann-Chyuan (1998), *SMEs Role in High-Tech Industry*. in Chinese, a research prepared for Small and Medium Enterprise Administration, Ministry of Economic Affairs (Taipei, Taiwan).

Endnotes

[1]The core-satellite relationship is a cooperative relationship between large (core) and small (satellite) enterprises. The cooperating firms must register with the government in order for the SMEs to get loans at preferential interest rates and for the group to obtain technical and other forms of support from the government. However, not all such relationships are registered.

[2]A Foreign business or company is defined as a local company with any amount of foreign capital, not as a company under foreign control.

[3]Only 273 mergers were reported between 1971 and 1986.

[4]According to data provided by the Core-Satellite Development Center, in 1998 there were 184 groupings with 2,850 registered firms among 24 industries, while 17,100 firms were involved with the groupings.

[5]Qume and Wang, for example, both moved part or all of their manufacturing lines to Taiwan with local-based R&D.

[6]Figures for Japan are also quite impressive, but the sample size is too small.
[7]See footnote number 5.
[8]These firms remained strong in the wake of a serious earthquake that struck Taiwan in September 1999. Their PRC subsidiaries took over production that was halted as a result of power supply interruptions.

Chapter 10

EXPORT-ORIENTED FOREIGN DIRECT INVESTMENT IN THE PEOPLE'S REPUBLIC OF CHINA: DIVISION OF VALUE ADDED BETWEEN SOURCE AND HOST ECONOMIES

Yun-Wing Sung

1. INTRODUCTION

Export-oriented foreign direct investment (FDI) has received a great deal of attention in the trade and development literature because it has been important to the growth of many newly industrializing economies (NIEs). By the early 1970s Helleiner (1973) had already pointed out that multinational firms were deeply involved in the expansion of manufacturing for export in Less Developed Countries (LDCs). In many cases such manufacturing was undertaken in export-processing zones or similar enclaves.

In the 1980's, foreign affiliates' share of exports was low for first-tier East Asian NIEs (Hong Kong, South Korea, and Taiwan), with the exception of Singapore (Table 1). However, for the next tier (Malaysia, the Philippines, and Thailand) foreign affiliates accounted for 37% to 52% of exports in the secondary sector FDI also accounted for over half of the People's Republic of China's (PRC) exports, as we will see later.

The costs and benefits of export-oriented FDI have long been discussed in the literature. Helleiner's (1973) seminal article was followed by Sharpston (1975), Watanabe (1976), Keesing (1982), and Helleiner (1988). Helleiner (1973) distinguishes between short-run allocative effects and long-run growth effects of export-oriented FDI.

Allocative Effects

Unlike import-substituting FDI, export-oriented FDI does not result in negative value added (at international prices), as the outputs are sold in the world market, and most of the inputs are imported at international prices (Helleiner 1973, p. 14). Processed exports usually have a low value added because most of the inputs are imported. This may be efficient, as certain economies, especially small ones, may require a substantial amount of imported inputs in order to engage in production.

However, the profits and value added of processed exports are often underreported because of transfer pricing. It may be difficult to prevent transfer pricing in processed exports because the parent firm controls both imported inputs and outputs, which results in twice as many opportunities for transfer

Table 1: *Share of Foreign Affiliates in Total Exports of East Asian Host Countries (%)*

Host Country	Primary Sector	Secondary Sector: Electrical/ Electronic Products	Secondary Sector: Chemical Products	Secondary Sector: Motor Vehicles	Secondary Sector: Total	Tertiary Sector
Hong Kong (1984)	-	-	-	-	16.5	-
South Korea (1978)	-	-	-	-	24.6	-
Taiwan (1986)	18.5	43.6	30.8	-	18.5	3.6
Singapore (1988)	-	-	-	-	86.0	-
Malaysia (1986)	32.6	-	-	-	51.2	15.9
Philippines (1983)	-	-	-	-	51.5	-
Thailand (1980)	-	-	-	-	37.3	-

Note: *Year in brackets refers to year of data.*

Source: *UNTCMD (1992), Table Annex 9.*

pricing as does import-substituting manufacturing (Helleiner 1973, p. 13). While transfer pricing leads to losses in government revenue, the host country earns at least the value of the wage bill from processed exports. As low employment levels in developing economies lowers the opportunity cost of unskilled labor, using unemployed labor to earn scarce foreign exchange is quite attractive (Helleiner 1973, p. 14).

However, host countries usually provide infrastructure to attract FDI, and the large number of low-wage countries competing for "footloose" manufacturing implies that host countries have little bargaining power. It is possible for the costs of infrastructure to exceed its benefits (Helleiner 1973, p. 14).

Growth Effects

Besides discussing the above cost-benefit calculus, which deals with static allocative efficiency, Helleiner (1973) talks about dynamic growth effects, including "linkage," "learning," and "dependence" effects.

As previously mentioned, processed exports usually have low value added because the countries that export them make extensive use of imported inputs, which may be efficient for small economies. However, backward linkage may fail to develop because of the enclave nature of the exporting firms. For example, domestic intermediate inputs into Taiwan's Kaohsiung Export Processing Zone accounted for only 4% of the value of its exports in 1969 (Helleiner 1973, p. 16).

Processed exports, which are intensive in unskilled labor, seem to promise little in terms of learning effects. However, when a country engages in processing for export, it gains experience in modern factory discipline and management skills, since overseas companies will not manage an indigenous labor force for long, because doing so is too expensive. Instead, they will soon turn the task over to local managers. Moreover, the rise in income from processed exports usually raises local demand for education and for sophisticated skills. The first-tier Asian NIEs have been able to upgrade their skills tremendously as a result of their early involvement with processing for export, which illustrates the fact that labor-intensive exporting is not a blind alley for developing economies.

Taking part in processing for export may imply an economy's dependence on foreign investors. Processed exports may constitute an enclave, as they are dependent on neither local resources nor local markets. The host country has very little control over sources of inputs and market outlets. It has little bargaining power, since processed exports are footloose.

Empirical Studies of Costs and Benefits of Processed Exports

Despite the importance of processed exports, undertaking empirical studies of their costs and benefits is difficult because of lack of data. Data on transfer earning are very difficult to obtain.

Exports of foreign affiliates include both processed exports, which mostly use imported inputs, and exports that mostly use local inputs. There are usually no data that distinguish between the two.

Data on the benefits of exported-oriented FDI for the source country are even more difficult to obtain. FDI earnings are available from balance of payment accounts, but such earnings include export-oriented as well as import-competing FDI, and further disaggregation is usually not possible. Moreover, data on FDI earnings are usually quite unreliable. Transfer earnings further exacerbate the problem. Data on the division of earnings between the host and source countries are all but nonexistent.

The case of the PRC is interesting, as the country has been highly successful in exporting and in attracting FDI, especially export-oriented FDI from Hong Kong. Hong Kong's export-oriented manufacturing firms have been relocating to the PRC, especially to Guangdong, in droves since the mid 1980s to take advantage of lower labor costs there. Hong Kong accounts for the bulk of the PRC's export-oriented FDI and processed exports. Around 60% of the PRC's processed exports originate in Guangdong, the prime site of export-oriented investment from Hong Kong.

From 1978 to 1999 the value of the PRC's exports increased at an average annual rate of 15.3% (in nominal U.S. dollars). The country's processed exports, which at the beginning of the period were nonexistent, now account for over half of the PRC's exports. The PRC's rank in world trade rose from 32nd in 1978 to 11th in 1992 and to 10th in 1997. Since 1993 the PRC has been the second-largest recipient of FDI in the world, after the United States. Moreover, good data on the division of earnings between Hong Kong and the PRC are available, permitting an insightful study of the division of value added between source and host economies.

This paper uses PRC and Hong Kong data to examine the division of value added in export processing between host and source economies. Part of the allocative effects can be gauged from the income (value added) generated by processed exports. Part of the growth effects can be gauged from the change in value added over time. However, the study is preliminary, since a full study of the cost and benefits of processed exports has to address the opportunity costs of the resources used in processed exports. Moreover, value added and the change it undergoes over time do not capture the external benefits of processed exports.

2. FORMS OF PROCESSED EXPORTS IN THE PEOPLE'S REPUBLIC OF CHINA

Growth in the PRC's manufacturing exports was especially rapid after 1984, when the State Council approved two schemes (what we call the contractual processing scheme and the ordinary processing scheme) allowing the duty-free import of components and raw materials for use in export industries. The PRC's high tariffs and deep exemptions for imported inputs used in export processing create strong incentives for this kind of trade. Detailed statistics on process exports are available from the PRC's Customs Administration because its imported inputs are entitled to tariff exemption.

Contractual Processing

In contractual processing, Chinese firms process raw materials or components supplied by foreign firms for a processing fee. The processed outputs belong to the foreign firms and are exported by them. The foreign firms usually provide the required machinery, product design, and technical assistance.

The foreign funds involved in contractual processing are classified as other foreign investment (OFI)[1] rather than FDI. Unlike FDI, OFI does not confer legal control of the enterprise to the foreign investor. The distinction between the two types of processing is not sharp, because the foreign partner in contractual processing often has de facto (though not de jure) control. However, the foreign funds involved in contractual processing constitute commercial credit rather than FDI because the PRC partner legally controls the operation and usually pays for foreign machinery and technical assistance with labor services used in making goods for the foreign partner. In this paper the term "foreign investment" includes FDI and OFI but excludes portfolio investment and loans.

According to Chinese terminology, FDI includes investment in three types of foreign-invested enterprises (FIEs), namely, fully foreign-owned enterprises, equity joint ventures, and contractual joint ventures. In the 1990s many contractual processing operations were converted into FIEs because the latter can market part of their output in the domestic market, while the former cannot. This has led to a decline in OFI and to an acceleration in FDI in the PRC (see Sung, Liu, Wong, and Lau 1995, 1978).

Ordinary Processing

The bulk of ordinary processing in the PRC is undertaken by FIEs. Like contractual processing, ordinary processing involves the processing of imported materials and components for export. Unlike in contractual processing, firms in the PRC rather than those outside it own outputs and imported inputs. Instead of earning a processing fee, as they do in contractual processing, the firms involved sell the processed exports for a profit. However, as these firms are often subsidiaries of foreign firms, the prices of inputs and outputs are frequently determined by the foreign parents. The distinction between contractual processing firms and ordinary processing firms is thus largely legal rather than substantive.

Statistics on Processed Exports

There are two sources of data on Chinese trade: the series compiled by the Ministry of Foreign Economic Relations and Trade (MOFERT) and the series compiled by the General Customs Administration. MOFERT data only record imports that lead to payments to foreigners or exports that generate foreign exchange income[2]. In MOFERT data on contractual processing, imports are not recorded, as no payments are made to foreigners, while only the processing fees (mostly labor costs) of exports are recorded. However, in PRC customs statistics on contractual processing the entire value of such imports and exports are recorded, as is standard international practice. To avoid confusion, the PRC stopped releasing MOFERT data on the national level in 1992, although the data are still released on a provincial level. MOFERT data are useful to us here, as they provide information on the value added (labor costs) of contractual processing.

Exports Under Reciprocal Contracts

In reciprocal contracts, a foreign trade corporation (FTC) of the PRC and a foreign partner sign two separate contracts to engage in contractual processing: one contract for provision of materials or components for the FTC to process, and another contract for the purchase of the processed output by the foreign partner. The prices of materials/components and processed output specified in the two contracts determine in effect the FTC's processing fee (Tan 1996, p. 28). As the FTC technically exports products rather than processing services, it is entitled to an export subsidy, which is otherwise denied to contractual processing.

For Guangdong's data on trade recorded in *China Customs Statistics*, trade under reciprocal contracts is excluded from contractual processing and reported separately from 1987 to 1989. From 1990 onward it is included under contractual processing. For national data on trade recorded in *China Customs Statistics*, trade under reciprocal contracts is excluded from contractual processing until 1992. Though detailed data are lacking, trade under reciprocal contracts appears to be diminishing because export subsidy, the main incentive for such trade, was abolished in the trade reforms of the early 1990s.

Growth of Processed Exports

Table 2 shows the growth of the PRC's processed exports, including both contractual processing and ordinary processing. Both categories are large, and both have grown rapidly. In recent years ordinary processing has grown faster than contractual processing because of the conversion of contractual processing operations into FIEs, which has occurred because the latter can sell part of their output in the domestic market. In 1999 the PRC's processed exports were US$111 billion, or 56.9% of the PRC's exports. The share of processed exports in the PRC's total exports is high and rising.

Table 3 shows the growth of processed exports in Guangdong. Data on exports under reciprocal contracts were available from 1987 to 1989. Guangdong's share of the PRC's processed exports is high but declining. This is due to the spread of export-oriented industrialization northward from Guangdong. From 1992 to 1998 Guangdong's share of contractual processing fell from 85.7% to 56.6%, and its share of total processed exports declined from 63.7% to 52.6%. Guangdong accounted for around half of the PRC's ordinary processing in the period.

Table 4 shows that FIEs accounted for a high (85% in 1999) and rising share of ordinary processing. Table 5 shows that ordinary processing accounts for the bulk (72% in 1999) of FIEs' exports. FIEs also undertake some contractual processing. Nonprocessing exports constitute a small (15.9% in 1999) but rising share of FIEs' exports. On the whole, FIEs account for the bulk of ordinary processing, and ordinary processing constitutes the bulk of FIEs' exports.

3. EXPORT-ORIENTED FOREIGN INVESTMENT IN THE PEOPLE'S REPUBLIC OF CHINA

Hong Kong accounts for the bulk of export-oriented foreign investment in the PRC. Hong Kong's rapid economic growth starting in the 1950s was based on the export of labor-intensive manufactures. Successful growth led to a severe labor shortage and to rapidly rising wages in the mid-1980s. The human capital that Hong Kong manufacturers had accumulated in the production and export of labor-intensive products would be obsolete if Hong Kong had not relocated the production processes to economies that still have abundant labor. However, most manufacturing firms in Hong Kong are small, and they are ill equipped to operate across geographic and cultural barriers. The PRC has the advantage of geographic and cultural proximity to Hong Kong, and the opening of the PRC presented Hong Kong manufacturing firms with a golden opportunity to relocate their production processes.

Table 2: *The PRC's Processed Exports*

	PRC's Processed Exports			Hong Kong Imports from PRC Related to Outward Processing	
	Contractual Processing	Ordinary Processing	Total		
1985	2,368	-	-	-	
	(8.7)				
1986	3,417	-	-	-	
	(11.0)				
1987	4,733	-	-	-	
	(12.0)				
1988	6,486	-	-	-	
	(13.7)				
1989	8,230	-	-	14,562	
	(15.7)			(27.7)	-
1990	10,454	-	-	18,629	
	(16.8)			(30.0)	-
1991	12,925	-	-	25,400	
	(18.0)			(35.3)	-
1992	15,340	24,322	39,662	32,566	
	(18.0)	(28.6)	(46.7)	(38.3)	82.1*
1993	15,962	28,286	44,248	38,160	
	(17.4)	(30.8)	(48.2)	(41.6)	86.2*
1994	18,153	38,827	56,980	45,925	
	(15.0)	(32.1)	(47.1)	(38.0)	80.6*
1995	20,660	53,043	73,703	51,650	
	(13.9)	(35.7)	(49.5)	(34.7)	70.1*
1996	24,240	60,094	84,334	58,558	
	(16.0)	(39.8)	(55.8)	(38.8)	69.4*
1997	29,446	70,156	99,602	63,439	
	(16.1)	(38.4)	(54.5)	(34.7)	63.7*
1998	30,730	73,741	104,471	61,684	
	(16.7)	(40.1)	(56.9)	(33.6)	59.0*
1999	35,770	75,102	110,872	62,839	
	(18.4)	(38.5)	(56.9)	(32.2)	56.7*

Figures in brackets represent percentage share of the PRC's total exports.

Figures with an asterisk represent percentage share of the PRC's processed exports.

Source: *Chinese data from* China Customs Statistics, *various issues; Hong Kong data from* Hong Kong External Trade, *various issues.*

Historically, Hong Kong and Taiwan have tended to invest in small-scale export-oriented projects in the PRC (including both contractual processing and foreign-invested enterprises undertaking ordinary processing), whereas developed countries such as the United States and Japan have concentrated their investments on large-scale import-competing projects. Hong Kong and Taiwan recently diversified their investments to import-competing projects as well as

Table 3: *Guangdong's Processed Exports (US$mn)*

	Contractual Processing				
	Reciprocal Contracts	Other	Subtotal	Ordinary Processing	Total
1987	1,302	3,884	5,186	1,565	6,751
	-	(82.1)	-	-	-
1988	1,333	5,777	7,110	2,760	9,870
	-	(89.1)	-	-	-
1989	966	7,284	8,250	4,560	12,810
	-	(88.5)	-	-	-
1990	-	-	9,166	6,842	16,008
	-	-	-	-	-
1991	-	-	11,087	9,357	20,444
	-	-	-	-	-
1992	-	-	13,147	12,115	25,262
	-	-	(85.7)	(49.8)	(63.7)
1993	-	-	13,649	15,487	29,136
	-	-	(85.5)	(54.8)	(65.8)
1994	-	-	15,078	21,136	36,214
	-	-	(83.1)	(54.4)	(63.6)
1995	-	-	15,936	26,340	42,276
	-	-	(77.1)	(49.7)	(57.4)
1996	-	-	16,858	30,298	47,156
	-	-	(69.5)	(50.4)	(55.9)
1997	-	-	19,257	35,574	54,831
	-	-	(65.4)	(50.7)	(54.6)
1998	-	-	20,255	38,106	58,361
			(56.6)	(50.7)	(52.6)

Figures in brackets represent percentage share of the national total.

Source: *Data on Reciprocal contracts are from* The Statistics of the External Economy, Trade, and Tourism of the Guangdong Province *(internal document), 1987, 1988, and 1989. Other data are from the* Statistical Yearbook of Guangdong, *various issues.*

Table 4: *Ordinary Processed Exports from Foreign-Invested and Domestic Enterprises*

	Ordinary Processed Exports from		
	Foreign-Invested Enterprises	Domestic Enterprises	Total
1995	39,177	13,866	53,043
	(73.9)	(26.1)	(100)
1996	48,596	11,498	60,094
	(80.9)	(19.1)	(100)
1997	57,663	12,493	70,156
	(82.2)	(17.8)	(100)
1998	61,958	11,783	73,741
	(84.0)	(16.0)	(100)
1999	64,161	10,941	75,102
	(85.4)	(14.6)	(100)

Figures in brackets represent percentage share of the total.

Source: China Customs Statistics, *various issues.*

Table 5: *Exports of Foreign-Invested Enterprises*

	Exports of Foreign-Invested Enterprises			
	Contractual Processing	Ordinary Processing	Others	Total
1995	2,870	39,177	4,829	46,876
	(6.1)	(83.6)	(10.3)	(100)
1996	4,487	48,596	8,423	61,506
	(7.3)	(79.0)	(13.7)	(100)
1997	6,144	57,663	11,093	74,900
	(8.2)	(77.0)	(14.8)	(100)
1998	7,223	61,958	11,781	80,962
	(8.9)	(76.5)	(14.6)	(100)
1999	10,378	64,161	14,089	88,628
Jan – Aug	(11.7)	(72.4)	(15.9)	(100)

Figures in brackets represent percentage share of the total.

Source: China Customs Statistics, *various issues.*

services and infrastructure, and the United States and Japan have also tapped the PRC's potential as an export base.

Hong Kong's investment in the PRC is substantial. By the end of 1999 the stock of Hong Kong FDI in the PRC was US$16.4 billion, or 51% of the PRC's stock of inward FDI. As mentioned previously, since 1993 the PRC has been the second-largest recipient of FDI in the world after the United States, and Hong Kong has become the world's fourth largest source of FDI after the United States, the United Kingdom, and Germany, but ahead of France and Japan.

As Hong Kong tends to concentrate on export-oriented investment, its share of such foreign investment in the PRC should even be higher than its 51% share of total inward FDI in the PRC. Unfortunately, statistics on investment do not distinguish between export-oriented and domestic market–oriented investments, so we are unable to verify whether this assumption holds true.

From figures on the PRC's processed exports, we can see that Hong Kong clearly accounts for the bulk of export-oriented foreign investment in the PRC. Table 2 shows the PRC's processed exports as well as Hong Kong's imports from the PRC related to outward processing. Owing to the importance of outward processing to Hong Kong trade, the Census and Statistics Department of the Hong Kong government has been conducting quarterly surveys on trade declarations since 1989, asking traders if the trade flows are part of an outward-processing arrangement. The bulk of Hong Kong's outward-processing activities in the PRC are undertaken in Guangdong, and the processed output is imported into Hong Kong for reexport to other markets.

Table 2 shows that, in 1992, Hong Kong's outward-processing-related imports from the PRC amounted to US$32.6 billion, or 82.1% of the PRC's processed exports.[3] The percentage has declined since 1993; it fell to 57% in

1999. The decline can be attributed to two factors. First, export-oriented industrialization has spread north from Guangdong along the PRC's coast, and processed exports in the north cannot be exported via Hong Kong. Second, the PRC has constructed more ports in the south, and even processed exports in the south can bypass Hong Kong. However, some of the PRC's processed exports bypassing Hong Kong are still produced with the help of Hong Kong's investment. The 1999 57% share is thus the lower bound of Hong Kong's share of PRC- processed exports. In a nutshell, Hong Kong clearly accounts for the bulk of the PRC's processed exports.

4. VALUE ADDED IN THE PEOPLE'S REPUBLIC OF CHINA

China Customs Statistics contain data on imports of raw materials and components used in processed exports. It is thus possible to compute the rate of processing margin (the percentage by which the value of processed exports exceeds the value of imported inputs, using the former as the base). This assumes that imported inputs in a given year are all used to produce goods exported in that year. In actual fact, processing takes time, and there is some lag time between imports and exports. When export processing is growing rapidly, the above method of computation will understate the processing margin, as the value of imported inputs in a given year will overstate the amount of inputs actually used to produce exports in that year. However, the problem is not very serious, as the production cycle for labor-intensive processing is quite short (a couple of months or so).

Conceptually, the processing margin is roughly the same as total value added,[4] which gives the total income generated by processed exports.

Rate of Processing Margin

Table 6 shows the rate of processing margin for the PRC. From 1985 to 1991 data were only available for contractual processing. The rates were around 15%, except in 1987 and 1988, when they were negative. It is known that the actual rates in 1987 and 1988 were positive, as positive processing fees were earned. This is not a case of negative value added (at international prices). Though the time lag between imports and exports would bias the rate of processing downwards, it is doubtful that time lag is a major explanation, as the rates in the other years were normal.

The illegal sale of tariff-free imported materials in the domestic market (i.e., smuggling) appears to be the major reason for the negative rates. Some processing operations turn out to be mere shelters for smuggling (Sung 1991, p. 7). Such smuggling exaggerates the amount of imported inputs and can easily give rise to negative rates.

Table 6: *Rate of Processing Margin for the PRC's Processed Exports (%)*

	Processed Exports from		
	Contractual Processing	**Ordinary Processing**	**Overall**
1985	16.6	-	-
1986	12.8	-	-
1987	–2.9	-	-
1988	–2.7	-	-
1989	14.9	-	-
1990	16.7	-	-
1991	15.4	-	-
1992	17.4	22.4	20.4
1993	18.8	17.3	17.8
1994	16.7	16.4	16.5
1995	21.5	20.5	20.8
1996	26.6	25.8	26.2
1997	31.6	32.5	32.3
1998	35.3	34.0	34.4
1999	34.1	33.4	33.6

Source: China Customs Statistics, *various issues.*

Data on contractual processing and ordinary processing are available from 1992, and the two forms of processing do not differ much in their rates of processing margin. We expect the rate for ordinary processing to be higher. Firms undertaking ordinary processing bear more risks because they own the processed output. The fact that the rates are similar may indicate that such profits are channeled to the parent firm through transfer pricing. Another explanation is that such firms do not earn significantly higher profits. Their risks are minimal because the parent firms are responsible for supplying the materials and marketing the output.

Though the relevant tax rates in Hong Kong and the PRC are similar, there is significant incentive for transfer pricing, as Hong Kong has no foreign exchange controls and has better protection of property rights. Funds can be used more flexibly in Hong Kong than in the PRC.

The rate of processing margin for all processed exports was 20% or less from 1992 to 1995; it rose to 26% in 1996 and to over 32% in 1997. The rate stabilized at around 33% in 1998 and 1999. There has been a clear rise in the rate of processing margin since 1995.

Table 7 gives the rate of processing margin for Guangdong's processed exports. Since Guangdong accounts for the bulk of the PRC's processed exports, the pattern for Guangdong should be similar to that of the PRC. This is indeed the case. As in the case of the nation as a whole, there are some negative rates in 1987 and 1988. There has been a clear rise in the rate of processing margin since 1995.

Table 7: *Rate of Processing Margin for Guangdong's Processed Exports (%)*

	Processed Exports from		
	Contractual Processing	Ordinary Processing	Overall
1987	–7.3	23.4	–0.2
1988	–9.2	18.9	–1.5
1989	16.9	21.7	18.6
1990	16.8	24.9	20.3
1991	16.0	20.6	18.1
1992	17.8	15.4	16.6
1993	20.7	18.3	19.4
1994	21.6	16.2	18.5
1995	25.4	22.2	23.4
1996	28.9	28.1	28.4
1997	29.2	28.8	29.0
1998	34.3	31.1	32.2

Source: Statistical Yearbook of Guangdong, *1997 and 1998.*

Table 8: *Rate of Value Added for Contractual Processing in Guangdong (%)*

	Exports of Contractual Processing		
	Reciprocal Contracts	Others	Total
1987	-	7.2	-
1988	31.6*	6.0	9.1
1989	33.9	7.9	11.0
1990	-	-	10.0
1991	-	-	10.3
1992	-	-	10.2
1993	-	13.8*	-

*Both the value-added and exports data are from MOFERT statistics. For the other figures, the value-added data are from MOFERT statistics, while the exports data are from PRC customs statistics.

Source: *See Table 3.*

Rate of Value Added

Table 8 gives the rate of value added for contractual processing in Guangdong, which is equal to the processing fee (obtained from MOFERT statistics) divided by the value of exports (mostly obtained from *China Customs Statistics*). As imports do not enter into the calculation of the rates, they are not affected by smuggling. Indeed, the rates of value added in 1987 and 1988 are positive, showing that the negative rates of processing margin obtained in Tables 7 and 8 are a result of smuggling.

Conceptually, processing fees should roughly be the same as total value added (except for the small leakage caused by the indirect use of imported inputs pointed out in footnote 4). The rates of value added should thus be about the same as the rates of processing margin. However, the rates of value added were lower. From 1989 to 1992 the rates of value added for all contractual processing in Guangdong were around 10% (Table 8), while the corresponding rates of processing margin were around 16 to 17% (Table 7). One explanation for this is that processing fees are less than value added because of the deduction of payments for foreign machinery and technical assistance used in making goods for the foreign partner. Another explanation is that processing fees were paid for goods delivered ex-factory, and the foreign partner has to transport the goods to the border for exports. The value of goods ex-factory is of course lower than the value of goods loaded at the border for exports. Likewise, the foreign partner may have to transport the raw materials and components for processing from the border to the factory. In any case, the fact that the rates of processing margin are higher than the rates of value added shows that the time lag between imports and exports does not lead to significant biases in the rates of processing margin.

Table 8 shows that the rates of direct value added for exports under reciprocal contracts were very high: 31.6% in 1988 and 33.9% in 1989. This indicates that exports under reciprocal contracts use more local inputs, including primary and/or intermediate inputs. This is plausible, as reciprocal contracts involve FTCs, which have their own production, supply, and export network. They are less dependent on the foreign partner for providing machinery, expertise, services, and intermediate inputs. Another possibility is that they have more bargaining power vs. the foreign partners, since exports under reciprocal contracts are substantial.

5. VALUE ADDED IN HONG KONG

Owing to the importance of entrepôt trade in Hong Kong, the Hong Kong government has conducted annual surveys to ascertain the rate of report margin (the percentage by which the value of reexports exceed the value of imports for reexports, using the latter as the base). The reexport margin is roughly the same as total value added generated by reexporting.[5] While the processing margin gives the value added of processing exports for the PRC (the host of FDI), the reexport margin shows the value added for Hong Kong (the source of FDI).

Table 9 shows the rates of reexport margin in Hong Kong's reexports. The rate of reexport margin is much higher for reexports of Chinese goods than for reexports of other goods. This is because the bulk of Hong Kong's reexports of PRC goods involve outward processing, and Hong Kong investors engage in many value-adding activities in outward processing.

Table 9: *Hong Kong's Rate of Reexports Margin by Origin (%)*

	PRC	Others	Overall
1989	13.0	11.5	12.2
1990	21.1	12.7	17.4
1991	25.8	10.3	18.9
1992	29.7	10.3	20.8
1993	35.3	8.5	22.4
1994	33.2	6.0	20.0
1995	32.8	5.9	19.8
1996	34.4	6.6	21.4
1997	34.6	7.4	21.7
1998	35.1	8.6	23.0

Source: Hong Kong Monthly Digest of Statistics, *Census and Statistics Department, Hong Kong, May 1997.*

The rate of reexport margin for Chinese goods rose from 13% in 1989 to a record 35.3% in 1993 and remained at around 33% to 35% thereafter. The increase in the rate for Chinese goods is largely due to the rise in the share of products related to outward processing in Hong Kong's reexports of PRC goods. The share rose from 74.1% in 1991 to 88.4% in 1997 (Sung 1998, Table 4.3).

In fact, Hong Kong's reexport margin for outward processing is likely to be higher than its reexport margin for PRC goods that do not involve outward processing, because Hong Kong provides less value-adding activities for the latter. However, in the absence of more accurate data, the reexport margin for PRC goods is taken as a lower bound of the reexport margin for outward processing.

6. DIVISION OF VALUE ADDED BETWEEN THE PEOPLE'S REPUBLIC OF CHINA AND HONG KONG

Table 10 shows the "take" of Hong Kong investors relative to PRC producers. The take of Hong Kong investors, as indicated by the reexport margin of PRC goods, is higher than the PRC's processing margin,[6] although the gap between the two has narrowed since 1996.

The higher take of Hong Kong relative to the PRC implies that much of what are regarded as the PRC's exports are in fact Hong Kong exports, if value added rather than assembling or fabrication is used to determine country of origin. This is particularly pertinent in the case of Sino–United States trade, as 70% of the PRC's exports to the United States are reexported via Hong Kong, and most of these are processed exports. It is well known that the United States' bilateral trade deficit with the PRC is very large, a fact that generates considerable friction between the United States and the PRC. The PRC takes the heat of U.S. trade protectionism, while Hong Kong captures the lion's share of the benefits.

It appears that the unfavorable division of benefits for the PRC cannot be attributed to a weak bargaining position. The PRC's investment environment

Table 10: *The PRC's Processed Exports: Ratio of Value Added in Hong Kong Relative to Value Added in the PRC**

	1992	1993	1994	1995	1996	1997	1998
Ratio	1.46	1.98	2.01	1.58	1.31	1.07	1.02

**Computed from Tables 5 and 6.*

improved greatly in the 1990s, and many multinationals are rushing into the country. Moreover, for Hong Kong investors, the PRC has the advantage of geographic and cultural proximity, which other low-wage countries cannot offer. Hong Kong's export-oriented investment in the PRC came mostly from small firms with little market power.

It appears that Hong Kong's take is relatively high because Hong Kong performs many value-adding services for processing operations in the PRC, including product design, production management, marketing, order-taking from importers, sourcing, quality control, trade financing, coordination of shipping, and so forth. Production involves a long chain of steps, and Hong Kong performs many initial and final steps in that chain, while the PRC only does the assembly or fabrication. Another likely contributor to the situation is the fact that Hong Kong's reexport margin is inflated by transfer pricing.

The fact that the PRC is dependent on Hong Kong for so many of the above services shows the weakness of the Chinese system. The PRC's processing operations are dependent on Hong Kong's trade financing because of the system of credit rationing in the PRC, which favors state-owned enterprises. The PRC depends on external investors to take care of international marketing and to take orders from importers, partly because it is cumbersome for Chinese nationals to get passports to travel overseas. Processing operations tend to import most of their raw materials, because the quality of local supplies and the reliability of local suppliers are undependable. The rigidity of the PRC's economic system hampers the development of forward and backward linkages to local enterprises.

7. RISE IN THE RATE OF PROCESSING MARGIN

There has been a substantial rise in the rate of processing margin in the PRC since 1995. This may be a statistical artifact caused by a decrease in smuggling, or it may be the result of an increase in backward and forward linkages, indicating that economic reforms have increased the flexibility of the Chinese economy. Data on smuggling is difficult to obtain. From anecdotal evidence it would seem that smuggling was rampant until the crackdown in China in the fall of 1997. In the absence of better data it is difficult to attribute the rise in the rate of processing margin from 1995 to 1997 to a decline in smuggling.

We do have evidence of an increase in backward and forward linkages. Initially the PRC's FIEs consisted largely of relocated labor-intensive downstream operations involving processing and assembling. This created a demand for intermediate inputs, which was partly met by the relocation of midstream and upstream operations from elsewhere, and also by local enterprises (Naughton 1997, p. 296). While the increased use of local inputs decreased the demand for imported inputs made elsewhere and raised the processing margin, it did not directly diminish the value-adding contributions of Hong Kong, because Hong Kong supplies the PRC with services rather than intermediate goods. This is consistent with our finding of a rise in the rate of processing for the PRC and a roughly constant rate of reexport margin for Hong Kong.

There has also been an increase in forward linkage, as more processed goods are exported via Chinese ports rather than via Hong Kong (Sung 1997, pp. 74–76). The share of the PRC's processed exports reexported via Hong Kong fell from a record 50% in 1993 to 35% in 1999. When processed goods are exported via Chinese ports rather than via Hong Kong, the rate of processing margin for the PRC rises. However, the rate of reexport margin for Hong Kong would not fall, as this rate is calculated from goods that are still reexported through Hong Kong. This is again consistent with our finding of a rising rate of processing margin and a constant rate of reexport margin.

It should be noted that Hong Kong has invested heavily in PRC ports. Even for goods that are exported through PRC ports, Hong Kong continues to play an important role, since it provides services such as transshipment, financing, and marketing (Sung 1998, pp. 89–91). Such activities are reflected in service exports instead of in the reexport margin for goods.

Another factor contributing to the increase in the PRC's value added may be the improved quality of the PRC's processed exports. Table 11 shows the unit

Table 11: *Unit Value Index of Hong Kong's Imports from the PRC*

Annual Data	1990	1991	1992	1993	1994	1995	1996	1997	1998	1999
Index	100	101.3	102.4	99.8	102.8	107.6	110.0	110.6	107.3	104.6

Monthly Data	1997	1998	1999	2000
January	110.1	109.5	105.6	103.3
February	111.2	109.0	105.9	
March	110.1	107.8	104.9	
April	109.5	107.7	104.5	
May	110.2	106.9	104.1	
June	110.2	107.2	104.3	
July	111.7	107.6	104.4	
August	111.0	107.7	104.9	
September	111.4	107.1	105.0	
October	110.6	106.0	104.3	
November	110.5	105.8	104.2	
December	110.1	105.6	104.5	

Source: Hong Kong Trade Number, *Census and Statistics Department, Hong Kong, various issues.*

value index of Hong Kong's imports from the PRC, the bulk of which are the PRC's processed exports. The rise in the unit value index from 1994 to 1997 is consistent with increased quality.

Table 11 also shows the monthly unit value index since 1997. The index reached a peak in July 1997, when the Thai baht was devalued, triggering the Asian financial crisis. Since July 1997 the unit value index has declined, first slowly in late 1997 and then sharply in 1998, reflecting the effect of the devaluation of the currencies of the PRC's competitors. However, the decline in the index slowed in 1999, the year of East Asia's initial recovery.

Other things being equal, the fall in the price of processed exports should lead to a decline in the processing margin. However, it seems likely that the crackdown on smuggling in the fall of 1997 has raised the processing margin. The continual increase in backward and forward linkages also works in the same direction. Since the price of processed exports stabilized in 1999, the processing margin may resume its upward trend soon.

8. CONCLUSION

Export-oriented foreign investment in the PRC, the bulk of which is from Hong Kong, has been very important to the PRC's spectacular export drive. The rate of value added for the PRC is relatively low compared with that of Hong Kong investors, indicating a lack of linkages.

Despite a low rate of value added, the rapid rise in the volume of processed exports in the PRC has generated substantial income and tax revenue, which has contributed to prosperity in the country, especially in Guangdong. Unlike the Special Economic Zones, which receive financial assistance from the state to build infrastructure to attract foreign investment, most local governments in Guangdong do not have financial assistance from the state. They sold bonds to the public or borrowed from Hong Kong to finance their initial infrastructure, and they have been able to pay back their loans from tax revenue as a result of the prosperity generated by export processing (Sung, Liu, Wong, and Lau 1995, pp. 28–29). It should be noted that Special Economic Zones only accounted for 23% of Guangdong's exports. The bulk came from former rural municipalities that had little state financial assistance.

The experience of the first-tier Asian NIEs shows that export-oriented industrialization is not a blind alley. These economies have been able to upgrade their skills and diversify their exports. Their period of dependence on foreign investors was brief, and they have become important sources of FDI in their own rights. Hong Kong has been the foremost source of FDI among developing economies since 1992 and the world's fourth-largest source since 1993.

The recent dramatic increase in the PRC's rate of processing margin points to an increase in both backward and forward linkages and perhaps to improved quality as well. Export-oriented industrialization in the PRC appears to be following the same path as it did in first-tier Asian NIEs.

References

Census and Statistics Department (1997), "Hong Kong's Imports Measured on Free on Board (f.o.b.) Basis for 1996," *Hong Kong Monthly Digest of Statistics*, December, Appendix FB.

Helleiner, G. K. (1993), "Manufacturing for Export, Multinational Firms and Economic Development," *World Development*, 1/ 7, 13–21.

---------- (1988), "Transnational Corporations, Direct Foreign Investment and Economic Development," in Chenery, H. and T. N. Srinivasan (eds.), *Handbook of Development Economics* (Amsterdam: Elsevier), 1441–1480.

Lardy, Nicholas R. (1994), *China in the World Economy* (Washington D.C.: Institute for International Economics).

---------- (1992), *Foreign Trade and Economic Reform in China, 1978–1990* (Cambridge, U.K.: Cambridge University Press).

Naughton, Barry (1997). "The Future of the China Circle," in Naughton, Barry (ed.), *The China Circle: Economics and Technology in the PRC, Taiwan, and Hong Kong* (Washington D.C.: Brookings Institution Press), 289–303.

Sharpston, M. (1975), "International Subcontracting," *Oxford Economic Papers*, 27, 94–135.

Sung, Yun-Wing (1997), "Hong Kong and the Economic Integration of the China Circle," in Naughton, Barry (ed.), *The China Circle: Economics and Technology in the PRC, Taiwan, and Hong Kong* (Washington D.C.: Brookings Institution Press), 41–80.

---------- (1998), *Hong Kong and South China: The Economic Synergy* (Hong Kong: City University of Hong Kong Press).

Sung, Yun-Wing, Pak-Wai Liu, Yue-Chim Richard Wong, and Pui-King Lau (1995), *The Fifth Dragon: The Emergence of the Pearl River Delta* (Singapore: Addison Wesley).

Tan, Yinbo (1996), "Utilization of Foreign Capital and the Economic Development of Guangdong," M.Phil. Thesis, Chinese University of Hong Kong.

UNTCMD (United Nations, Transnational Corporations and Management Division) (1992), *World Investment Report 1992: Transnational Corporations as Engines of Growth* (New York: United Nations).

Watanabe, Susumu (1976), "International Subcontracting, Employment and Skill Promotion," *International Labour Review*, 105, 425–449.

Endnotes

[1]The other forms of OFI are leasing and compensation trade.

[2]The two series differ in coverage, valuation, and recording times (Lardy 1992, pp. 12–13).

[3]Strictly speaking, Hong Kong's imports are valued cost, insurance, and freight (cif), while China's exports are valued free on board (fob). The former exceeds the latter by the cost of insurance and freight, and we have ignored such costs. However, in the case of Hong Kong imports from China, such costs are only 1% of the value of imports (Census and Statistics Department, Hong Kong, 1997). We ignore such costs in the interest of brevity.
[4]The processing margin overstates total value added slightly, because processed exports usually use some domestic inputs (e.g., transportation, electricity), which in turn may use imported inputs (e.g., imported petroleum used in transportation). The indirect use of imported inputs represents a leakage that should be deducted from the processing margin to arrive at total value added. However, such indirect leakages are small.
[5]The reexport margin again exceeds total value added slightly, owing to the indirect use of imported inputs. See footnote 4 for details.
[6]The cif-fob differential, which is only 1% for Hong Kong's imports from China, is again ignored. See footnote 3 for details.

Chapter 11

THE GLOBALIZATION OF TRADE AND PRODUCTION: A CASE STUDY OF HONG KONG'S TEXTILE AND CLOTHING INDUSTRIES*

Leonard K. Cheng and William K. Fung

1. INTRODUCTION

Production in the world economy has become increasingly globalized. Producers scan the globe to select the distribution of production processes that result in the lowest total production and transportation costs. Significant advances in production, communication, and transportation technologies on the one hand and multilateral liberalization of international trade and investment policies on the other have led to shifts in the world's comparative advantages and consequently in the geographical distribution of production activities.

However, these geographical shifts in production are not simple shifts from one country to another. Very often fragmentation of the production process occurs first, with labor-intensive processes being shifted to countries where labor is more plentiful and cheaper. Capital-intensive processes tend to be shifted later, if at all, and knowledge-intensive processes like design, engineering, and marketing often stay in the original country. In the end production may stabilize into a cooperative system in which several territories take advantage of their respective comparative advantages to collectively produce the most cost-efficient article possible. The history of Hong Kong's textile and clothing industries provides a vivid illustration of this production-fragmentation process.

As a reaction to the fragmentation process, producers in importing countries that are hurt by shifts in comparative advantages often seek protection from their national governments. Even though they are not always successful in reversing the general trend of trade and investment liberalization, local producers seeking relief from global competition remain a powerful counterforce that can bring about various sorts of trade restrictions. Moreover, geopolitical objectives tend to favor producers in some regions but discriminate again those in others.

Regardless of whether protectionist policies are the results of national or regional political economy, they represent deviations from the principle of comparative advantage and cause distortions in efficient global production. One of the tools of trade restrictions is changing the "rules of origin." (For an excellent exposition of the distortions in production and trade caused by rules of origin as they are used, see Lloyd [2000].) This non-cost-based distortion is yet another factor that concerns producers of textile and clothing products and impacts their decisions about which countries to include in the production process. Thus in this paper we also analyze the impact of the rules of origin pertaining to textile and clothing products on Hong Kong's own production, its outward processing scheme, and trade. In addition, we point out the distortions these rules cause.

2. A BRIEF HISTORY OF HONG KONG'S TEXTILE AND CLOTHING INDUSTRIES

Hong Kong's founding in 1842 was initiated by Britain's need for a permanent entrepôt to the People's Republic of China (PRC), and the former British colony was precisely that from 1842 to 1949. However, with the founding of the PRC in 1949, the subsequent embargo of Chinese goods following the Korean War, and the lowering of the "Bamboo Curtain," that role came to an abrupt end. For 30 years, from 1949 to 1979, Hong Kong essentially lost its place as the PRC's entrepôt.

Fortunately, the arrival of over a million refugees from the PRC heralded the onset of Hong Kong's success as a center for labor-intensive industries. Among the first industries to be established in the territory were the textile and clothing industries. These industries were well established in the PRC, especially in Shanghai, and Shanghai industrialists brought capital, expertise, and plants and equipment to Hong Kong after 1949. Thus began Hong Kong's rise as one of the world's premier exporters of clothing from the 1950s to the 1970s.

The textile and clothing industries were largely vertical all the way from yarn spinning to finished garments. The import of raw materials was confined to raw cotton and wool tops. Later, man-made fabrics, silk, and leather would also be imported. Even as Hong Kong flourished thanks to its success with export industries, the seeds of its loss of competitiveness were germinating. The territory's dependence on labor-intensive manufacturing industries, of which the textile and clothing industries were the largest and most important, meant that its competitiveness was being steadily eroded as the standard of living and wages rose. By the 1970s industrialists in these industries found themselves facing the following choices: (1) give up labor-intensive industries to less developed countries (LDCs), or (2) invest in LDCs, and shift labor-intensive production processes overseas while keeping capital- and knowledge-intensive production processes at home.

They chose option 2. In the late 1960s and early 1970s Hong Kong's textile and clothing industrialists began to invest in neighboring economies including Taiwan, Singapore, Malaysia, Thailand, and Indonesia, thus beginning the process of production fragmentation. However, it was not until Deng Xiaopeng's "open-door policy" came into effect in 1979 that Hong Kong's "hinterland" (the PRC) was restored to it, allowing for a true division of labor on a much greater scale.[1]

3. MOVEMENT OF HONG KONG'S MANUFACTURING INDUSTRIES TO THE PEOPLE'S REPUBLIC OF CHINA

At the end of the 1970s the PRC's establishment of an open-door policy sent shock waves through the region and the world. For the first time in decades the PRC started to make its abundant supply of workers available to foreign investors. To Hong Kong's manufacturers of labor-intensive products the PRC's opening could not have come at a better time. Before the PRC's economic reform they had been under increasing competitive pressure from other Asian economies that had followed Hong Kong's "export-led growth" strategy. Rising wages and rents in Hong Kong threatened to price their labor-intensive and land-intensive industries out of the world market. In response to both the push factors from within the territory and the pull factors from South China (where the PRC's open-door policy began), Hong Kong investors started moving across the border to the PRC.

Initially investment took the form of subcontracting arrangements known as "export processing" (or "outward processing" from the point of view of producers based in Hong Kong). It then began taking the form of joint ventures and wholly owned subsidiaries. Owing to cultural and geographical proximity, Hong Kong's investment in the PRC has to a large extent been concentrated in Guangdong Province, in particular in the neighboring Pearl River Delta (PRD). According to a survey conducted by the Hong Kong Federation of Industries (1991), by 1990, 40% of manufacturing firms in Hong Kong had investment in the PRD, and about 75% of their products were either finished or processed in the PRD.

Until the beginning of the 1990s most Hong Kong investors were small firms. However, beginning in the 1990s, especially after Deng Xiaoping's tour of South China in 1992, major listed companies from Hong Kong flocked to the PRC in droves, seeking investment opportunities. The number of Hong Kong–invested enterprises in Guangdong Province rose from 6,438 in 1988 to 49,341 in 1995.[2] The lion's share of Hong Kong's foreign direct investment (FDI) went into the manufacturing sector, especially into export-oriented labor-intensive industries. According to Sung (1998, Table 5.2), between 1979 and 1995 Hong Kong firms invested a total of US$23.3 billion in Guangdong's manufacturing sector, which represented 72.5% of Hong Kong's total FDI in the province. More remarkably, in 1996 and 1997 Hong Kong investors made a further investment of US$19.2 billion into the province.[3]

Given the twin characteristics of much of Hong Kong's investment in Guangdong (namely, export-oriented and unskilled labor–intensive stages of production), it is not surprising that Hong Kong's outward processing trade accounts for the bulk of its trade with the PRC. Sung (1998, Table 4.3) shows that in 1997, 76.1% of Hong Kong's exports to the PRC (at US$6 billion), 44.7% of reexports to the PRC (at US$25.5 billion), and 81% of imports

Table 1: *Percentage of Hong Kong Exports by Origin*

Year	Made in Hong Kong		Made in the PRC		Made in Third Countries
1988	35.8	(50.0)	35.8	(50.0)	28.4
1991	22.1	(27.7)	57.6	(72.3)	20.3
1995	11.2	(16.0)	58.7	(84.0)	30.1
1997	9.5	(13.1)	62.8	(86.9)	27.7

Source: *Hong Kong Trade Development Council (1988, 1991, 1996, 1998).*

from the PRC (at US$63.4 billion) were related to Hong Kong's export-processing activities in the PRC.

The Hong Kong Trade Development Council (HKTDC) conducted surveys of Hong Kong's manufacturers and traders in 1988, 1991, 1995, and 1997 to determine the relative importance of their exports of products made in Hong Kong, the PRC, and third countries. The results are summarized in Table 1. The figures in parentheses represent the share of the products made in Hong Kong and the PRC if those made in third countries are excluded. These figures clearly show that over the period 1988–97 Hong Kong's manufacturers and exporters increased their dependence on products made in the PRC. At the same time, their dependence on products made in Hong Kong decreased quickly. In contrast, their dependence on products made in third countries showed no clear trend.

As Hong Kong moved many of its manufacturing processes, especially the labor-intensive ones, across the border, goods manufactured in the PRC were exported to Hong Kong to be reexported (or for further processing before they were exported) to the world market. As a result of this development Hong Kong's economy experienced dramatic structural changes. As shown in Table 2, in 1987 manufacturing accounted for 22% of Hong Kong's GDP, but by 1997 the share had declined to 6.5%. Total employment in manufacturing peaked in 1987 (in absolute number) at 919,000, or 34.1% of total employment in that year, but by 1998 it had fallen to 264,600, or 8.3% of total employment. Mirroring the decline in Hong Kong's manufacturing activities was the rapid expansion of its trade-related industries. The share in GDP for "transport, storage, and communications" increased from 8.2% in 1986 to 10.1% in 1995, although it fell back to 9.3% in 1997 because of the Asian financial crisis. The share of "import/export trade" in total employment rose from 7.7% in 1986 to 17.6% in 1995 but dropped to 15.5% in 1998; that of "transport, storage, and communication" rose from 8.2% in 1986 to 12.0% in 1998.

Table 2: *Structural Changes of Hong Kong's Economy*

	Share (%) in GDP		Share (%) in Total Employment		
Year	Manufacturing	Transport, Storage, and Communication	Manufacturing	Import/Export Trades	Transport, Storage, and Communication
1986	22.6	8.2	34.7	7.7	8.2
1987	22.0	8.6	34.1	8.3	8.3
1988	20.5	9.1	32.3	9.4	8.7
1989	19.3	8.9	30.4	10.3	9.1
1990	7.6	9.5	27.8	11.4	9.3
1991	15.4	9.6	24.5	12.7	9.4
1992	13.6	9.7	21.5	13.5	10.3
1993	11.2	9.5	18.4	14.9	10.9
1994	9.2	9.7	15.4	17.1	11.4
1995	8.3	10.1	13.4	17.6	11.0
1996	7.3	9.8	11.2	17.5	11.0
1997	6.5	9.3	9.8	16.7	11.3
1998	N/A	N/A	8.3	15.5	12.0

Source: Hong Kong Monthly Digest of Statistics, *various issues.*

4. HONG KONG'S PRODUCTION OF TEXTILE AND CLOTHING PRODUCTS

Under the Multi-Fiber Arrangement (MFA), Hong Kong exporters must ensure that their products abide by the "rules of origin" as defined by the importing countries in order to use the export quotas allocated to them. This means that certain production processes must take place within the boundaries of its economy. Given these constraints, the textile and clothing industries' shift in production from Hong Kong to the PRC was less dramatic than that of other labor-intensive industries. Nevertheless, the shift was sufficiently significant that Hong Kong's domestic production first grew at a slower rate but eventually declined in absolute terms.

The evolution of the clothing industry's employment from 1978 to 1996 and of value added between 1976 and 1996 is depicted in Figures 1 and 2, respectively. Total employment in the clothing industry peaked in 1980 at over 300,000 workers and experienced a gradual decline in the early 1980s but a more rapid decline in the decade between 1987 and 1996. In 1996 the number of employees in the industry stood below 65,000, less than one-quarter of the 1987 employment level. Despite the remarkable decline in its employment level, the industry's value added continued to increase, with a few minor dips, until 1992. After that year the value added started

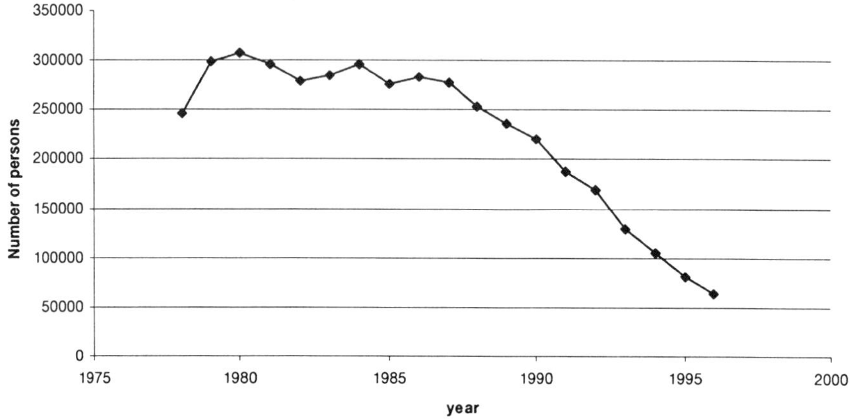

Figure 1: *Number of Persons Employed in the Clothing Industry.*

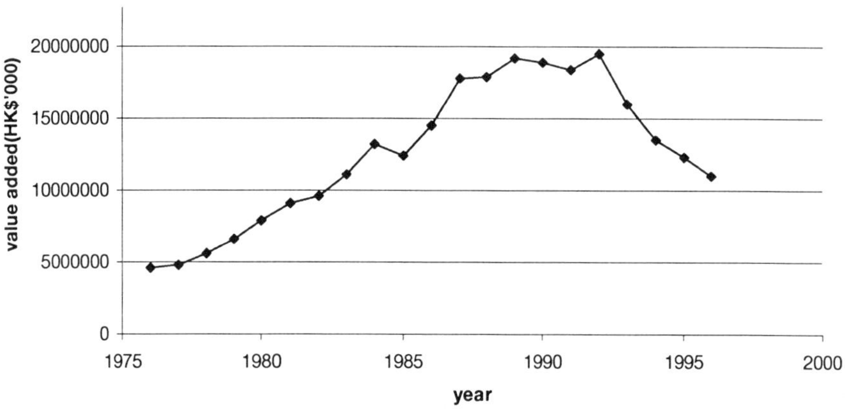

Figure 2: *Value Added Derived from the Clothing Industry.*

to tumble, and by 1996 it was less than 60% of what it had been at its peak in 1989.

Even though the textile industry is regarded as more capital intensive than the clothing industry, the evolution of its employment and value added (depicted in Figures 3 and 4, respectively) was similar. The number of employees in this industry declined fairly drastically between 1979 and 1982. After recovering somewhat between 1983 and 1989, the employment level

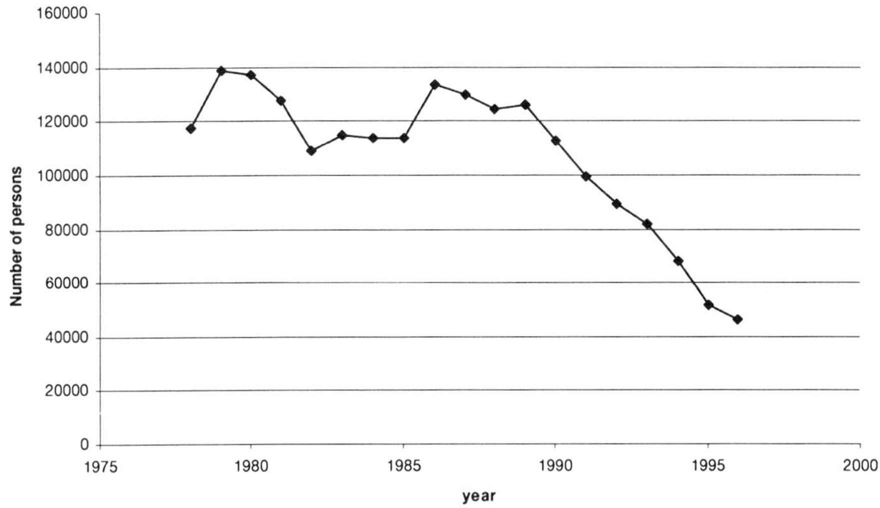

Figure 3: *Number of Persons Employed in the Textiles Industry.*

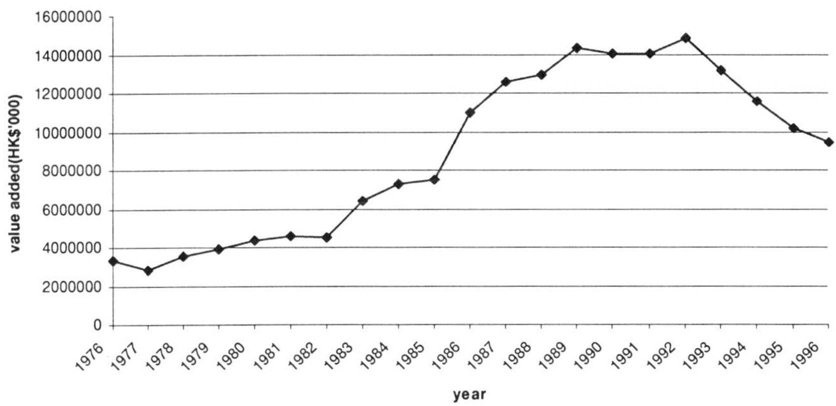

Figure 4: *Value Added Derived from the Textiles Industry.*

again began declining rapidly, falling from over 120,000 in 1989 to about 46,000 in 1996. The decline was only slightly less dramatic than that of the clothing industry. Much as it did in the clothing industry, the value added of the textile industry continued to rise, with one or two minor dips, until 1992. After that it tumbled. Perhaps as a reflection of the textile industry's growth in capital intensity, its value added fared slightly better than the value added of clothing. In 1996 the value added of Hong Kong's textile industry was about 66% of what it had been in 1989 or 64% of what it had been at its peak in 1992.

5. HONG KONG'S TRADE IN TEXTILES AND CLOTHING PRODUCTS WITH THE PEOPLE'S REPUBLIC OF CHINA

Hong Kong's trade with the PRC in clothing and textiles from 1975 to 1997 is given in Table 3. As can be seen, all three categories of Hong Kong–PRC trade (namely, imports from the PRC, domestic export to the PRC, and reexports to the PRC) in these two industries grew fast. However, to put the growth in bilateral trade in perspective we express the trade statistics in Table 3 as a percentage of Hong Kong's total imports, total domestic exports,

Table 3: *Hong Kong's Trade in Textile and Clothing Products with the PRC (HK$ million)**

	Textiles			Clothing		
Year	**Imports**	**Domestic Exports**	**Re-exports**	**Imports**	**Domestic Exports**	**Re-exports**
1975	934	2	14	216		
1976	1218	11	8	345	0	0
1977	1220	13	8	480	0	0
1978	2076	39	20	621	0	0
1979	3001	199	221	1092	9	5
1980	4007	382	1366	2286	58	29
1981	5680	628	3165	3704	163	43
1982	5731	943	2659	4711	218	50
1983	8799	1596	3151	6435	331	72
1984	12862	2023	6492	9289	512	120
1985	12620	2116	9559	10548	489	212
1986	17635	3642	11189	16572	783	355
1987	25497	5430	14764	22143	1298	539
1988	26734	5291	18335	27247	1556	579
1989	30452	6134	23432	38457	1846	861
1990	30019	6715	28153	47134	2020	1095
1991	35646	7518	35713	60191	2572	1530
1992	36925	7523	42629	72201	3329	2006
1993	35244	6920	44860	82554	3752	2187
1994	42019	6654	55689	85966	4702	2210
1995	45444	6485	64561	87347	5899	2011
1996	43400	6799	69452	94186	7424	2049
1997	45745	6726	72540	105383	11364	2973

**Zeros indicate that the value of trade was below HK$1 million.*

Source: *Census and Statistics Department*, Hong Kong Trade Statistics, *various issues.*

Table 4: *Hong Kong's Trade in Textile and Clothing Products with the PRC as Percentage of Its Total External Trade in the Same Products*

	Textiles			Clothing		
Year	Imports	Domestic Exports	Re-exports	Imports	Domestic Exports	Re-exports
1976	18.56	0.38	0.92	48.57	0.00	0.05
1977	18.77	0.48	0.72	50.29	0.00	0.00
1978	24.22	1.40	0.74	47.83	0.00	0.05
1979	25.65	5.03	29.41	55.15	0.04	0.51
1980	27.32	8.73	32.65	65.93	0.25	1.86
1981	29.80	12.16	46.35	70.88	0.58	1.97
1982	32.02	19.12	42.24	72.88	0.76	1.65
1983	37.11	23.10	41.07	76.20	0.96	1.60
1984	39.75	23.97	51.90	80.71	1.10	1.94
1985	38.05	27.63	60.58	80.70	1.09	2.77
1986	41.96	33.88	56.52	84.12	1.50	2.66
1987	45.01	34.45	52.79	85.06	1.99	2.95
1988	43.04	34.47	54.06	85.09	2.31	2.34
1989	42.53	36.90	55.78	86.50	2.57	2.31
1990	38.16	40.05	60.44	87.54	2.80	2.29
1991	38.16	42.93	61.61	90.08	3.39	2.40
1992	36.73	43.78	63.34	90.16	4.31	2.57
1993	35.92	42.83	64.04	90.39	5.22	2.41
1994	35.83	44.29	68.23	89.29	6.43	2.39
1995	35.09	46.29	70.02	89.23	7.99	2.21
1996	34.21	49.72	73.05	89.35	10.69	2.04
1997	36.77	53.20	72.68	90.63	15.73	2.79

Source: *The percentages are calculated by using figures shown in Figure 3 and figures (not shown) of Hong Kong's total trade in textile and clothing products as provided in various issues of* Hong Kong Trade Statistics.

and total reexports, respectively, in these two industries. The results are given in Table 4.

The first column of Table 4 shows that imports of textiles from the PRC accounted for about 24% in 1978 on the eve of the PRC's opening to foreign investment. The ratio rose steadily to about 40% in the late 1980s, but fell back to 36.8% in 1997. The increase in importance of Hong Kong's own exports and reexports of textiles to the PRC was more dramatic. The PRC accounted for about 1% of Hong Kong's domestic export and reexports of textiles in 1978, but the share rose to over 53% and 72%, respectively, by 1997. In absolute value, however, Hong Kong's domestic exports to the PRC declined after 1992, and beginning in 1991 the value of Hong Kong's reexports of textiles to the PRC exceeded imports of textiles from the PRC.

The changes in the relative importance of trade with the PRC in clothing products are described by the last three columns of Table 4. As in the case of textiles, all three categories of trade experienced significant growth. However, relative to trade in textile products, the share of imports of clothing products from the PRC was more important than all categories of textile trade, but the share of domestic exports of clothing products was not very important even with its substantial growth. Reexports of clothing products to the PRC had little significance. The importance of imports of clothing products from the PRC even before its open-door policy came into effect (47.8% in 1978) was a reflection of the PRC's comparative advantage in these products. A share in the range of 90% since 1991 is a reflection of the PRC's comparative advantage in clothing products as well as the role of Hong Kong as the PRC's most important entrepôt in conducting foreign trade.

Since the PRC has a comparative advantage in the production of clothing products, one would not expect it to import many such products from other economies. Thus the volume of reexports of such products to the PRC via Hong Kong would naturally be small. Nevertheless the rise in Hong Kong's domestic exports of clothing products to the PRC accelerated in the early 1990s. This in part reflects exports that went to PRD clothing firms involved in export processing activities for Hong Kong firms. As mentioned in section 3, in 1997, 76% of Hong Kong's exports to the PRC were input components to be processed in the PRC, and the proportion for the textile and clothing industries would have been of a similar order. In addition, some of the increase in Hong Kong's export of clothing products to the PRC went to the final consumers owing to style and brand-name differentiation as the PRC opened its domestic markets wider to Hong Kong clothing producers.

Table 5, which is based on the HKTDC's surveys, provides information about the relative importance in Hong Kong's textiles and clothing exports of products made in the PRC. The figures in parentheses are the percentages of products made in Hong Kong and the PRC, respectively, when products made in third countries are ignored. From these figures we can see that the relative importance of products made in Hong Kong declined from 1988 to

Table 5: *Percentage of Hong Kong's Textile and Clothing Exports, by Origin*

	Made in Hong Kong		Made in the PRC		Made in Third Countries
1988					
Textiles	39.0	(50.5)	38.2	(49.5)	22.8
Garments	50.5	(59.3)	34.7	(40.7)	14.8
1991					
Textiles	15.7	(21.1)	58.7	(78.9)	25.6
Garments	30.5	(34.4)	58.2	(65.6)	11.3
1995					
Textiles	14.6	(21.4)	53.7	(78.6)	31.7
Garments	19.5	(21.8)	70.1	(78.2)	10.4
1997					
Textiles	26.5	(41.4)	37.5	(58.6)	35.9
Garments	17.1	(24.1)	53.8	(75.9)	29.1

Source: *Hong Kong Trade Development Council (1988, 1991, 1996, 1998).*

1991 to 1995, while the relative importance of products made in the PRC rose. Unexpectedly, the relative importance of products made in the PRC in 1997 was below that in 1995, whereas that of products made in Hong Kong rose over the same period. This can be explained by the fact that during the period 1995–97, under pressure from local manufacturers and traders, the Hong Kong Department of Trade relaxed a portion of the rules governing its Outward Processing Approval Scheme (OPA) to allow conformity with the differing rules of importing countries. There was a surge of outward-processing activities because the resulting garments were allowed to be exported as "Made In Hong Kong."

The relative importance of textile products made in third countries continued to rise from 1988 to 1997. The share of garments made in third countries declined from 1988 to 1995 but, surprisingly, jumped by almost 200% in 1997. This increase in third-country production was a result of the constraints imposed by the MFA on clothing exports from Hong Kong and the PRC. As demand, especially from the United States, soared in 1996 and 1997, Hong Kong suppliers were increasingly forced to fill that demand with production in third countries.

The Federation of Hong Kong Industries (1991, pp. 48, 163) found in its 1990 survey that 54.2% of the surveyed firms' clothing products and 69.6% of their textile products were finished or processed in the PRD. These figures are consistent with the figures in Table 5 for 1988 and 1991. According to Sung (1998, Table 4.6), which is based on statistics provided by the Census and Statistics Department, the percentages of textiles and clothing exports of Hong Kong firms produced in Guangdong in 1996 were 68.4% and 52.7%, respectively, of the total exports of goods made in Guangdong and Hong Kong. While the figure for textiles was below that in Table 5 for 1995 but above that for 1997, the figure for clothing was substantially below those in Table 5 for both 1995 and 1997.

6. HONG KONG'S "BORDERLESS" MANUFACTURING OF CLOTHING: FROM "MADE *IN* HONG KONG" TO "MADE *BY* HONG KONG"

At this point, one can see that the Hong Kong manufacturers of clothing and textile products have integrated operations between Hong Kong and the PRC to such an extent that the designation of either "Hong Kong" or "the PRC" as the country of origin has largely lost its meaning.

So long as the border between Hong Kong and the PRC separates their respective labor pools, which are priced at highly disparate rates, the production process and semi-processed goods will flow back and forth across the border. (For an illustration of cross-border flows see section 8 below.) In fact, manufacturing activities have become "borderless," in the parlance of the trade.

The HKTDC used to promote the "Made in Hong Kong" label as a symbol of quality. In recent years, despite the outward shift of production bases, Hong Kong firms have remained firmly in control of the entire manufacturing process. The resultant products manufactured in its offshore production bases are now marketed by the HKTDC as "Made by Hong Kong."

7. IMPACT OF RULES OF ORIGIN ON TRADE AND PRODUCTION IN THE TEXTILE AND CLOTHING INDUSTRIES

The textile and clothing industries are regarded as politically sensitive industries in many developed economies. To avoid undesirable market disruptions in the consuming developed economies, world trade in textile and clothing products is governed by the MFA. The agreement can be regarded as a compromise of the interests of the world's exporters, the producers of importing countries, and the latter's consumers.

To ensure implementation of the MFA's market-sharing arrangement, rules governing the "country of origin" are necessary. The criterion used to determine the country of origin of industrial products like clothing and textiles is "substantial transformation." Three methods are adopted in the application of this criterion: (1) tariff heading, (2) manufacturing or processing operations, and (3) relative value added.[4] Despite these three methods, however, the determination of origin is far from clear-cut because in today's global economy a final product may entail a large number of production processes that take place in many countries.

Given the inevitable ambiguity of the criterion of substantial transformation, different importing countries often apply different rules. These rules are often complicated and arbitrary. For example, according to U.S. rules of origin that were in place from 1984 to June 1996 the origin of garments depended on where the fabric used to make the products was cut; the origin of fabric depended on where it was formed (unless it was dyed and printed and underwent two additional processing operations in another country), and the origin of yarn depended on where the staple fibers were spun or the filaments extruded.

Although complicated, the system would not be so inadequate if it were not subject to arbitrary changes by importing countries. For example, since July 1996 the origin of garments imported into the United States has depended on where they are "wholly assembled" rather than where their fabric is cut. The origin of fabric depends only on where it is formed, regardless of whether the fabric is dyed, printed, or otherwise processed elsewhere.

The confusion is further compounded by the fact that different importing countries having different rules. In the case of sweaters, the United States confers origin based on the knitting of panels, whereas the EU confers origin based on the linking of panels. The change in U.S. rules in August 1984 led to a significant increase in the number of panels being knitted of in Hong Kong and to a reduction in number being imported from the PRC for linking in Hong Kong (see Sung 1991, p.111). In contrast, the change in the rules of origin in July 1996 with regard to garments seemed to have little impact on Hong Kong, partly because Hong Kong's own rules required garment products to be both cut and sewn in Hong Kong to qualify for Hong Kong origin. Nevertheless, it was reported that the number of fabric-cutting jobs in Hong Kong fell and that the number of sewing jobs rose.[5]

8. DISTORTIONS CAUSED BY THE RULES OF ORIGIN

A fundamental problem with any practically implementable system of rules of origin is that it attempts to force an ubiquitous multi-country reality in production into an artificial single-country designation. To a large extent this problem is rooted in the adoption of gross value rather than value added in measuring trade flows and defining trade policies despite the global nature of the production process. When goods are produced efficiently on a global scale, the different stages are distributed to different countries in accordance with their comparative advantages. For example, unskilled labor-intensive stages are performed in countries abundant in unskilled labor; skilled labor-intensive stages are performed in countries abundant in skilled labor; and capital-intensive stages are performed in countries abundant in capital. Single-country rules of origin cause a distortion in efficient production because manufacturers and traders act to maximize their financial interests in response to whichever set of rules of origin they are following.[6] They may change the products they make, the manufacturing processes they undertake, and the magnitude of their products' value added. For example, Hong Kong sweaters exported to the EU are knitted in the PRC but linked in Hong Kong, while sweaters exported to the United States are both knitted and linked in Hong Kong.

Sung (1991, p. 110) points out that the 1984 change of sweaters' origin from linking to knitting removed a distortion (because linking is not any less labor-intensive than knitting) but that the United States' decision not to confer origin on the basis of dyeing and printing (which are capital intensive) created a new distortion by forcing such activities to take place in the country in which the fabric is made.

Another example of how insistence on a "substantial-transformation" country-of-origin conferring process can lead to uneconomic production is the case of jeans manufacturing. U.S. rules of origin require both the cutting of the denim fabric and the sewing of the two main side seams to confer origin. Because of PRC's binding quotas, Hong Kong non-binding quotas, and cheaper labor and production costs in the PRC, at one point jeans were manufactured as follows:

a. Fabric shipped from the PRC to Hong Kong
b. Fabric cut in Hong Kong
c. Panels sent to the PRC, where pockets are made and embroidery and other labor-intensive processes take place; new items attached to the panels
d. "Decorated" panels sent back to Hong Kong
e. Main seams sewn together in Hong Kong
f. Jeans sent back to the PRC for washing, trimming, ironing, and packaging
g. Finished products sent to Hong Kong for reexport to the United States as products with a Hong Kong origin

In this example, the semi-finished products cross the PRC–Hong Kong border a total of five times in order to conform to U.S. rules of origin and to adhere to the Hong Kong–administered OPA Scheme.

9. OTHER BARRIERS TO EFFICIENT GLOBAL PRODUCTION: ENVIRONMENT, LABOR STANDARDS, AND SAFETY STANDARDS

Many factors other than the MFA quota system now affect countries' choice of how to organize what would otherwise be a rational clothing- and textile-manufacturing process. The advent of consumer concerns about non-economic areas like the environment, human rights, labor standards, and so forth, which sometimes coalesce with protectionist interests, is forming what is called in the trade the "ethical sourcing" of products. The presence of child labor, prison labor, lack of environment controls, excess overtime work, and the like may disqualify certain countries from participating in the production process.

This movement toward "ethical sourcing" is very real. It will change and temper in important ways what would otherwise would have been a rational distribution of the production process based on comparative advantages.

10. CONCLUDING REMARKS

During the past 20 years the world has been moving toward a model of globalized and fragmented production, especially of labor-intensive products. Hong Kong's textile and clothing industrialists' dispersion of production, first to Southeast Asia and then to the PRC, with the subsequent development of control systems to manage geographically dispersed manufacturing, vividly illustrates the production-fragmentation process.

The example set by Hong Kong's textile and clothing industries was subsequently widely copied by manufacturers in Taiwan, Korea, and other economies. Today export-oriented, dispersed manufacturing is in fact the norm rather than the exception in the production of clothing and textile products. This system of manufacturing has produced products whose retail price level in the importing economies of the world has not changed substantially in 20 years, to the great benefit of consumers there. This type of export-oriented, dispersed manufacturing that combines and coordinates the comparative advantages of several countries to produce a globally competitive product is an important contribution to the economics of globalization.

Owing to the MFA quota system, however, there are obstacles to the efficient functioning of this system. Although as part of the Uruguay Round trade agreements MFA quotas are supposed to be phased out by 2005, questions remain about whether this deadline will be met. In addition, other non-quota concerns and restrictions are beginning to surface to substitute for the earlier protectionist instruments, and these concerns will likely have detrimental effects on economic efficiency. Nevertheless, despite these countercurrents, the globalized system of manufacturing clothing and textile products seems destined to remain.

References

Cheng, Leonard K., Larry D. Qiu and Guofu Tan (2001), "Foreign Direct Investment and International Fragmentation of Production," in Arndt, Sven and Henryk Kierzkowski (eds.), *Fragmentation: New Production Patterns in the World Trade* (Oxford University Press).

Ho, Lok Sang (1998), "Country-of-Origin Rules: Its Origin, Nature, and Directions for Reform," *Pacific Economic Review*, 3/ 2, 161–166.

Federation of Hong Kong Industries (1991), Hong Kong's Industrial Investment in the Pearl River Delta, Hong Kong.

Hong Kong Trade Development Council (1998), *Survey on Hong Kong Re-Exports – Summary Report*, November, Hong Kong.

Hong Kong Trade Development Council (1991), Survey on Hong Kong Domestic Exports, Re-Exports, and Triangular Trade, November, Hong Kong.

Hong Kong Trade Development Council (1996), *Hong Kong's Trade and Trade Supporting Services*, April, Hong Kong.

Hong Kong Trade Development Council (1998), *The Rise in Offshore Trade and Offshore Investment*, September (2nd edition), Hong Kong.

Lloyd, P.J. (1993), "A Tariff Substitute for Rules of Origin in Free Trade Areas," *The World Economy*, 16, November, 719–735.

Lloyd, P.J. (1999), "Rules of Origin and Fragmentation of Trade," this volume.

Spinanger, Dean (1994), "Profiting from Protection in an Open Economy – Hong Kong's Supply Response to EU's MFA Restrictions," Kiel Working Paper No.653, Intitut fur Weltwirtschaft, Kiel, Germany.

Sung, Yun-Wing (1991), *The China-Hong Kong Connection: The Key to China's Open-Door Policy* (Cambridge, U.K.: Cambridge University Press).

Sung, Yun-Wing (1998), *Hong Kong and South China: The Economic Synergys* (Hong Kong: City University of Hong Kong Press).

World Customs Organization (1985), *The Kyoto Convention*, Amending Supplement Annex D.1. (concerning rules of origin), September.

Endnotes

[*]The authors are grateful to conference participants for their comments and to Zhou Hailan and Tse Tim for research assistance.

[1]Spinanger (1994) documents two key ways in which Hong Kong textile and clothing producers circumvented the Multi-Fiber Agreement (MFA) quota restrictions, namely, quality upgrading and conducting production activities in neighboring economies. He has estimated the value of textile and clothing exports from these economies that were controlled by or connected to Hong Kong companies.

[2]See Table 3 in the appendix to Zhang, Bojian, Wang Liwen, and Chen Shengho, "The Development and Prospect of Hong Kong-Invested Enterprises in Guangdong," in *The Ninth Five Year Plan, the Long-Term Target for the Year 2010, and the Hong Kong Economy* Vol. 1, Hong Kong: One Country Two Systems Economic Research Institute, 1997, 95–124.

[3]*Statistical Yearbook of Guangdong*, 1997 and 1998.
[4]World Customs Organization (1985). For a discussion of their advantages and drawbacks see Lloyd (2000).
[5]According to "New United States Rules of Origin for Textiles and Apparel," *Trade Watch*, Hong Kong Trade Development Council, June 1996, the new rules of origin for textiles and apparel may "cost Hong Kong 4,000 jobs—mainly among workers currently cutting fabric into component parts for apparel and other products made of textiles. At the same time, there has been some speculation that the new rules of origin could foster an increase in sewing operations in Hong Kong."
[6]or a criticism of single-country rules of origin and an argument in favor of multicountry rules of origin, see Ho (1998) and Lloyd (1993, 1999).

Chapter 12

ARMS LENGTH TRANSACTIONS VS. AFFILIATES: A STUDY OF TWO ELECTRONIC COMPONENT FIRMS IN SINGAPORE

Hian Teck Hoon and Kong Weng Ho

1. INTRODUCTION

Key to understanding Singapore's outward-looking development experience is the question of why international production fragmentation there has taken the form of subsidiaries set up to produce a part of the value chain rather than of purchases made from contract manufacturers in arms-length transactions. Singapore is of interest in this regard because, whereas it once served largely as a production manufacturing base for many multinational corporation (MNC) subsidiaries that set up factories there to produce their own parts and components, in the 1990s the number of local contract manufacturers, which were set up to serve MNCs, began to grow. In our interviews with two firms—one a major local contract manufacturer and the other an MNC subsidiary—our aim was to gain an understanding of how firms decide whether to rely on arms-length transactions vs. affiliates for the supply of parts and components.

2. PROFILE OF THE TWO FIRMS

Table 1 provides a summary of the profile of the two firms we interviewed. Both firms produce electronic components, and both were set up in Singapore in the early 1990s. The local contract manufacturer (Firm A) produces electronic components and parts for computers to serve MNCs and has plants in Singapore, Malaysia, and Shanghai. The foreign MNC subsidiary (Firm B) produces a vital electronic component used in television to serve its parent firm. It is interesting that both firms saw their share of intermediate inputs supplied by local manufacturers increase over the years, reflecting the increased reliability of local suppliers. Another interesting feature of the local contract manufacturer is that although it has a plant that produces parts in Singapore, its main production activity takes place in plants based in neighboring countries where labor is cheaper. An important operation handled in Singapore is order fulfillment, that is, the receipt of detailed order specifications to meet the needs of different customers, and the communication of that information to neighboring plants. One more point worth highlighting is that both firms chose Singapore as their operations base because of the city's reliable infrastructure (i.e., logistics, transportation, and communications) as well as its relatively solid supply of skilled personnel.

As a result of the increased international economic integration of Singapore's labor-abundant neighbors, combined with the city-state's higher level of skill endowment and reliable infrastructure, firms are setting up

Table 1: *Profile of the Two Firms*

	Firm A	Firm B
Ownership	Local contract manufacturer	Foreign MNC subsidiary
Product line	Electronic components	Electronic components
Establishment year	1991	1990
nature of business	Subcontractor for MNCs	Produce own intermediate product
Major markets	MNCs in Singapore	Asia
Intermediate inputs	Inputs from foreign suppliers embody advanced technology. Share of inputs from domestic suppliers has increased over time, with very good quality and easy availability, compared to foreign competitors. Intermediate inputs constitute 90% of cost.	Inputs from foreign suppliers are mainly manufactured goods. Share of inputs from domestic suppliers has increased over time. Reliability of domestic supplies has improved as a result of firm's efforts. Intermediate inputs constitute 60% of cost.
Reason Singapore was chosen	Cost and proximity to customers (MNCs). Opportunity is shrinking in Singapore.	Good infrastructure and skilled labor. Reliable financing and tax incentives. Logistics, transportation, energy, and other infrastructure provision by government.
Overseas subsidiary	Firm A has an overseas subsidiary. "Extra payments" to regulators. Burdensome local material content requirement. Poor to fair communication infrastructure. Cannot sell to MNCs located in Singapore.	Not applicable.
Management employees	Fairly easy to retain good managers and to find local candidates. No training for management employees.	Somewhat difficult to retain good managers and to find local candidates. Training available for management employees.
Nonmanagement employees	Training provided. No specific skill required for operators.	Training provided. Employees must be able to program electronic gadgets.

headquarter and product design activities in Singapore and conducting labor-intensive assembly activities elsewhere in the region. The MNC subsidiary we interviewed told us that when it opened its factory in Singapore in 1990 the design and product development activity was conducted entirely by its parent firm. Since 1998, however, some of that design activity has been relocated to Singapore, particularly after the Singapore plant came out with a certain innovation in its product. The parent firm has since increased the Singapore plant's budget for product design engineers. The region's opening to international trade and commerce has not only enabled local contract manufacturers to base more labor-intensive production activities in the region but also, with more MNCs in the region becoming potential customers, caused the market for arms-length transactions to thicken, to use McLaren's (1998) terminology.

3. MNCS VS. ARMS-LENGTH TRANSACTIONS

From our interview with the MNC subsidiary we gained a sense of how the decision to buy components from subcontractors rather than to produce them using a vertically integrated process was made, on the one hand, to lower governance costs. On the other hand, the desire to maintain proprietary rights over certain technologies limited the range of intermediates that could be bought at arms length. The MNC subsidiary we interviewed explained that the technology for a certain part had been transferred to a local subcontractor but that the production of another very critical component was unlikely to be outsourced. To maintain its proprietary knowhow over the production of this particular intermediate product the firm is carefully considering the option of using subcontracted labor at its plant, that is, hiring workers from a firm that offers labor services to its clients. This way the MNC subsidiary would maintain its proprietary knowhow over the production process but would turn over to a subcontractor the problems of absenteeism, worker turnover, and workers' medical benefits.

Grossman and Helpman (1999) recently developed a model to explain when arms-length transactions rather than vertically integrated production processes occur. They view the emergence of arms-length transactions as arising from a trade-off between the costs of governance and the hold-up problem. In our interview of the local contract manufacturer, we found it interesting that many of the components it produces can actually be sold to other customers with minimal cost in changing specifications. As Jones and Kierzkowski argue in this volume, the horizontal aspects of vertical integration are important. The fact that components produced by local contract manufacturers can be sold to other customers means that the outside options of such manufacturers increase, which increases contract manufacturers' incentive to emerge and causes more arms-length transactions to occur. In other words, a wider horizon of markets reduces the hold-up problem and hence encourages the emergence of more contract manufacturers. As we noted earlier, the fact that neighboring regional economies also began to attract MNCs in the 90s has helped to thicken the market for arms-length transactions as well, as the local contract manufacturer now has a larger downstream market to serve.

The managing director of the local contract manufacturing firm we interviewed spoke of a "portfolio of customers." It is interesting that he emphasized the need to serve the global market and the challenge his firm faces in finding customers from around the world. His plant based in Shanghai, for example, principally seeks to serve Shanghai-based MNCs. He also spoke of how shorter product cycles have driven MNCs, which traditionally produced parts and components through their affiliates, to rely increasingly on contract manufacturers. Since they serve a wider portfolio of customers, he feels that contract manufacturers are better placed to absorb the risk of product obsolescence. For this reason, too, he mentioned that competition among contract manufacturers has increased as the contract-manufacturing sector has expanded, drawing employees who formerly worked in MNC subsidiaries. He saw the global shortage of computer product designers as contributing to the future growth of the design subcontracting field.

References

Grossman, Gene M. and Elhanan Helpman (1999), "Incomplete Contracts and Industrial Organization," manuscript, Princeton University.

McLaren, John (1998), "'Globalization' and Vertical Structure," manuscript, Columbia University.

Chapter 13

THE IMPLICATIONS OF INCREASING FRAGMENTATION AND GLOBALIZATION FOR THE WORLD TRADE ORGANIZATION

Robert E. Baldwin

1. INTRODUCTION

In a series of papers over the last 10 years, Ronald Jones and Henryk Kierzkowski (1990, 2001a, 2001b, 1999) have significantly improved our understanding of the process by which the production of final consumption goods becomes divided into an increasing number of vertical production blocks connected by service links. The fragmentation of production at the international level has become especially noteworthy in recent years. As these authors point out, this process is driven both by the increasing returns stressed by Adam Smith that result from the increasing division of labor permitted by economic growth and from freer trade among countries and by technological progress in both production processes and the delivery of such services as transportation, communication, coordination, and accounting. In their paper in this volume Jones and Kierzkowski describe how reductions in the costs of services and advances in technology not only spread horizontally across sectors but often lead to interactions between technology and fragmentation whereby improvements in technology encourage both further fragmentation and additional fragmentation acts as an incentive for further technological progress.

In view of the tremendous increase in the fragmentation of production at the international level and of the rapid spread of globalization generally in recent years, this paper briefly considers whether the rules of the World Trade Organization (WTO) have been appropriately adapted to reflect these circumstances. Section 2 focuses on whether the greater fragmentation of production calls for modifications to existing WTO rules, and Section 3 appraises some of the recent proposals (the purpose of which is to better adapt world trading conditions to the broad forces of globalization) for introducing entirely new rules covering such issues as competition policy, investment policy, and labor and environmental standards. A final section summarizes the various conclusions reached.

2. THE IMPLICATIONS OF INCREASED FRAGMENTATION FOR EXISTING WORD TRADE ORGANIZATION RULES

As Jones and Kierzkowski emphasize in their writings, increased vertical fragmentation of production at the international level can be an important component of economic developments that promote more rapid economic growth and higher per capita income levels. Consequently, it is incumbent on trade negotiators to ensure that WTO rules facilitate this fragmentation. Important progress in implementing this goal was made in the Uruguay Round of multilateral negotiations (1986–92). The most significant development was the negotiation of the General Agreement on Trade in Services (GATS), which extended WTO rules to cover trade in services (broadly defined to cover the supply of a service by one WTO member through its commercial presence in the territory of another member, the supply of a service in a member's territory to service consumers of another member, and traditional cross-border trade in services) in addition to trade in goods. Given the key role of service links in the fragmentation process, it is crucial that various government rules or private practices that serve to impede foreigners from establishing service links among their international productive activities be minimized. Since GATS came into effect, sector-specific agreements covering financial services and telecommunications have been negotiated and, as part of the built-in agenda included in the Uruguay Round, further negotiations on services are currently under way among WTO members, despite the failure of the Seattle meeting in late 1999, which aimed at starting a new multilateral round of trade negotiations.

The Uruguay Round agreement on trade-related investment measures represents another positive response to technological advances in the services sector and to the increasing international mobility of capital. While most countries welcome foreign direct investment (FDI), many impose discriminatory restrictions on foreign-owned firms. The new agreement prohibits the requirements that foreign firms purchase a certain proportion of domestically produced products or export a certain proportion of their output, and it represents an important first step in treating foreign investors the same as domestic firms. Other aspects of investment policy are discussed in Section 3.

A. Tariff Escalation and Rules of Origin

Although vertical fragmentation is an important force for increased economic growth, the division of the production process for goods and services into more and more component parts also provides new opportunities for introducing additional trade-distorting measures. Two actions related to the fragmentation process that have long been of concern to trade policy officials are increases in the degree of tariff escalation and the introduction of more restrictive rules of origin.

i) Tariff Escalation

Tariff escalation refers to most countries' common practice of imposing increasingly high duty rates on imported inputs as the production process proceeds from the raw-materials stage to the final-processing stage. It is common to permit duty-free (or very-low-duty) imports of such raw materials as cotton, for example, but to impose progressively higher tariffs on imports of cotton yarn, cotton fabric and, finally, clothing made from this fabric. Taking into account the concept of the effective rate of protection or protection measured on a value-added basis brings out the reason that tariff escalation makes it even more difficult for industrializing developing countries to penetrate the final goods markets of developed countries than it appears if one considers only the nominal rates of protection on final goods.

To take a simple example, suppose that the per-unit cost of the intermediate product used in producing a final good is $50, that the final product sells for $100, and that both of these goods are traded internationally at these prices. If these prices are fixed in the free international market, a 10% duty on the final product imposed by a country that both produces and imports the final good will increase its domestic price to $110 but will not affect the price of the intermediate input. Therefore the value added in the final manufacturing process will increase from $50 to $60, or by 20%. The effective protection is, consequently, 20%, while the nominal rate of protection is only 10%.[1] Potential exporters of the final product to the tariff-imposing country will have to overcome a 20% cost advantage given to the primary factors used in producing the final product in the tariff-imposing country in order to enter this market successfully.

Now suppose that technological progress uniformly reduces costs at all stages of production but that in doing so it vertically fragments the first stage of production into two parts, so that instead of two stages in producing the final product there are three. The first stage produces a good that is used as an intermediate input in the production of a second good, which, in turn, becomes an intermediate input into the final good. The past history of the political process of setting import duties indicates that there is likely to be political pressure to protect the domestic production of at least the second intermediate good, particularly if it has a comparatively high value-added component. If this does occur, it will trigger additional protectionist pressure from final-stage producers who will find their effective rate of protection reduced owing to a relative cost increase in their intermediate inputs. The outcome could well be a tariff increase on the final good that maintains or increases the initial nominal tariff gap between the final product and its final intermediate input. For example, in the illustration above, if the vertical fragmentation results in a 5% duty (instead of the initial zero duty) on the second intermediate good, the final-good producers may be able to raise their import duty from 10% to at least 15%. In both cases the effective rate of protection on the final good increases, making it more difficult for industrializing developing countries to penetrate the domestic market for this good as well as, of course, to penetrate the domestic market for the newly protected intermediate good.

If the vertical fragmentation process leads to the development of intermediates that can take advantage of relatively cheap foreign productive factors (e.g., labor), there may also be domestic pressures to lower tariff rates on these goods. The pressure would come from domestic multinational firms who wish to outsource the production of the goods to their foreign affiliates. The main point is that by creating new products that increase the number of production stages, vertical fragmentation offers the opportunity to impose new tariff rates that can increase the degree of effective protection on many products. In tariff negotiations with developed countries, the developing countries, such as those in East Asia and Southeast Asia, should focus on the effective rates of protection in markets into which they wish to gain greater market access and should also be on guard that effective rates do not rise as a result of the vertical fragmentation that characterizes the globalization process.

Fortunately, in both the Kennedy Round (1964–67) and the Tokyo Round (1973–79) of multilateral trade negotiations some progress was made toward reducing the gap between high and low duties on manufactured goods in the developed countries, but this effort did not focus specifically on the tariff escalation issue. The Uruguay Round (1986–92) of negotiations on tariff reductions did not make this a major objective, but I would urge that the next multilateral round make reducing the degree of escalation in tariffs on inputs into the early and later stages of the production of goods and services a top priority.

ii) Rules of Origin

Rules of origin are the criteria countries establish to determine for customs purposes the country in which a particular product originates. Rules of origin have becoming increasingly important because of the proliferation of free-trade agreements in which duties among signatories to the agreement are reduced to zero, whereas individual members still maintain duty levels on imports that are separate from nonmembers' duty levels. Since the fragmentation process has meant that more and more traded goods contain components produced in more than one country, determining which products exported from one member to another member should receive duty-free treatment when such products use of intermediates imported from nonmember countries becomes a major concern for exporters and domestic competitors alike. Typically such products receive duty-free treatment if they undergo "substantial transformation" in the exporting member of the agreement (usually indicated by a change in the tariff classification of the product) or contain at least a certain minimum percentage of value added by the member country.

While rules of origin are necessary to maintain the preferential treatment inherent in free-trade agreements and customs union, they are sometimes used for protectionist purposes. For example, in the North American Free Trade Agreement (NAFTA), imports of apparel generally qualify for duty-free treatment only if the apparel is cut and sewn from fabric spun from North American fibers. Consequently, Mexico, which has the potential to significantly expand apparel exports to the United States and Canada because of its relatively inexpensive labor, finds its competitiveness in this product area undercut because it is required to source the key intermediate inputs into apparel from less-efficient producers located in the region. The House and Senate have been deadlocked for several months over this issue with regard to a bill designed to expand trade preferences for sub-Saharan Africa. The Senate bill permits duty-free imports of apparel only if U.S.-produced thread, yarn, and fabric are used, whereas the rules of origin in the House legislation are more liberal. Critics of the Senate version argue that the potential export benefits to the developing countries of Africa will be largely lost because of the higher costs associated with the requirement that the fabric used to make clothing be manufactured in the United States and then shipped to Africa. Still another example of the use of rules of origin for protectionist purposes is the increase in the domestic content requirement for automobiles in NAFTA to 62.5%, compared with the 50% level that existed in the Canada-U.S. Free Trade Agreement.

Part of the problem is the failure of any WTO committee to carefully scrutinize the rules of origin and other arrangements agreed upon by countries when establishing customs unions and free trade agreements. As Srinivasan (1998) has pointed out, although WTO working parties have been established to examine these agreements, only 6 of the 98 groups established between 1947 and 1994 explicitly acknowledged conformity of the agreements with the WTO rules established for the formation of customs unions and free trade areas. It seems clear that the WTO must provide more oversight to ensure that rules of origin in these agreements do not themselves create restrictive and distorting effects on international trade. One modest step in this direction would be for the WTO secretariat itself to undertake periodic studies that compare rules of origin for particular products in different free trade agreements. Such an action might result in international pressures to liberalize those origin rules that most WTO members would regard as overly trade distorting.

B. Fragmentation and the Traditional Rules of Administered Protection

In the absence of ongoing trade negotiations, the day-to-day administration of a country's trade policy consists largely of responding to petitions from various industries and groups of workers that believe they are being injured or otherwise disadvantaged by the trade actions of other countries. These claims range from charges that the industries or groups are being injured by increased imports that are being either fairly or unfairly traded, to charges that other countries' policies are nullifying or impairing the benefits of previous trade agreements. Unfortunately, the procedures set forth in the WTO for dealing with these petitions are based on a trading world in which all the intermediate inputs into the production of a final good are produced domestically. The fact that production is internationally fragmented, with many intermediate inputs being imported, is not considered.

i) Safeguards

The problems produced by the failure to recognize international fragmentation can be illustrated by the WTO safeguards agreement. Article XIX of the GATT/WTO Agreement provides for the imposition of temporary trade-restricting measures if a product is being imported in such increased quantities "as to cause or threaten to cause serious injury to the domestic industry that produces like or directly competitive products" in order to "prevent or remedy serious injury and to facilitate adjustment." Economists have long criticized these provisions for their failure to consider the adverse price-increasing effects on consumers. However, the provisions also fail to take into account the adverse employment and profit effects of the temporary protection permitted on industries using the product as an intermediate input—considerations that traditionally have been of considerable concern to WTO/GATT rule makers.

A simple diagram devised by Corden (1971) to explain the concept of the effective rate of protection can be used to demonstrate these detrimental effects. Assume that the chips used in making personal computers (PCs) can be purchased either domestically or from abroad (perhaps from affiliates owned by the makers of the PCs). It is assumed that other intermediate products as well as the labor and capital needed to produce PCs is supplied only domestically. The PCs are both exported and sold domestically. In Figure 1 let the quantities of chips and PCs be measured along the horizontal axis, with the units being selected so that one chip is required to make one PC, as indicated by the fixed input coefficient derived from the production function. Assume for simplicity that the country is small and that it faces an infinitely elastic import supply curve for chips, GG', and an infinitely elastic foreign demand curve for PCs, SS'. In addition, let EE' be the domestic supply of chips and DD' the domestic demand for PCs.

Under free trade in chips the domestic production of chips will be OK, with any further demand for chips being satisfied from imports at the price of OG. The domestic supply curve of PCs, HJ'H', equals the per-unit cost of the chips plus the additional costs of the other intermediate inputs and direct factors used in producing the PCs. It has a kink at J' because whereas both the per-unit costs of chips and the per-unit costs of other necessary inputs rise until the output, OK, only the per-unit costs of the other inputs rise thereafter,

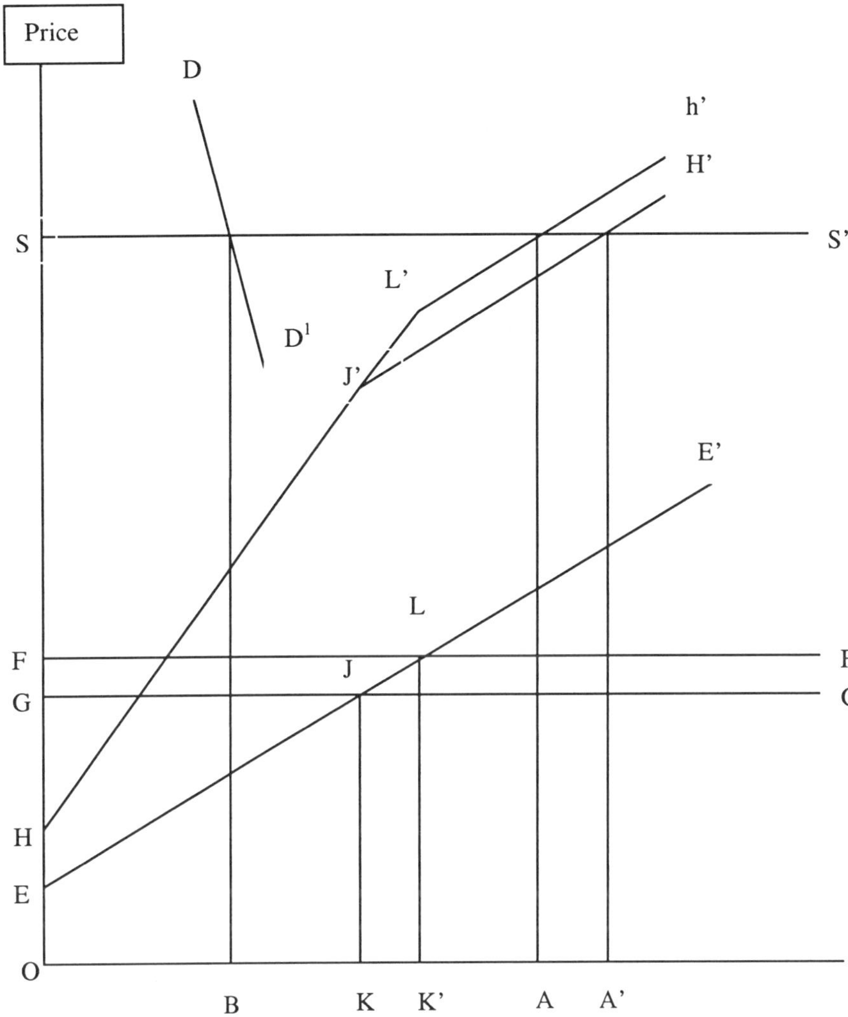

Figure 1: *Quantities of Chips and Computers.*

since the additional chips needed are imported at a constant price. Under free trade conditions domestic PC production will be OA, with OB being purchased domestically and BA being exported. Of the OA of chips required for this production, OK will be produced domestically and KA will be imported.

Now suppose that the domestic chip industry successfully petitions for a protective tariff on imports of chips under the safeguard provisions of the WTO and U.S. trade laws because imports of chips have recently increased. An import duty of GF on these imports shifts the supply curve of chips to

ELF' and the supply curve of PCs to HL'h', thereby reducing imports of chips from KA to K'A' but also reducing exports of PCs from BA to BA'. It is possible that the employment gain in the chip industry could be more than offset by the employment loss in the PC industry as the cost of its inputs of chips increase. However, in determining whether to grant protection, existing WTO rules make no provision for considering the possible injury to other industries caused by the protection, and there are no WTO articles under which such injury can be addressed.

If PCs are being imported rather than exported, the increase in the cost of producing them because of the protection of the chip industry causes injury to the PC industry by increasing imports of PCs. In this case the PC industry could also petition for protection under the safeguard provisions. However, chips are used as an input in thousands of different products where the chips are only a small fraction of unit costs. Consequently, the effect of the increased cost of chips in increasing imports (and reducing domestic production) is likely to be too small for these industries to claim that they are being seriously injured. Yet the sum of the loss in employment in all the industries could exceed the gain in employment in the chip sector.

Protection granted solely to final goods can also have adverse effects on producers of intermediate goods that are not taken into account under WTO rules. For example, suppose that both PCs and chips are being imported and that the PC industry is successful in bringing a safeguard action resulting in an additional tariff on imported PCs. As the demand for PCs decreases in response to the price increase brought about by the increased protection, the demand for imported chips will also decrease, thereby adversely affecting foreign producers. If chips produced abroad themselves contain imported inputs from the country imposing the additional protection of PCs, domestic producers of these inputs will be adversely affected too. Whereas the governments of foreign producers of PCs, whose exports to the country directly decrease as a consequence of the additional duty on PCs, can seek compensation under GATT/WTO rules from the duty-increasing country in the form of tariff cuts on other products, the governments of the foreign producers of chips cannot, since their exports have not been subject to a safeguard action. Their only recourse is to claim nullification or impairment of the benefits of increased exports of chips that they had previously secured by reducing tariffs on some of their own products. However, it is not easy to bring successful WTO cases under these provisions.

ii) Antidumping Duties

The two other main forms of administrative protection, namely, the imposition of antidumping (AD) and countervailing duties, also fail to take into account the fragmented nature of world input production. Countries are permitted to impose AD duties on imports if they are being sold by foreign firms at prices that either are lower than these firms' prices in their own their home markets or are below their average production costs and, in addition, if this pricing behavior causes or threatens to cause material injury to domestic industries in the importing countries. The AD duties cannot exceed the margin of dumping (i.e., the difference in the price foreigners' charge for imports and sales in their home markets or the extent to which the prices are below costs), however. Similarly, countries can impose countervailing duties on imports if the foreign producers of these imports are being subsidized by their governments and if this action causes material injury to domestic producers in the importing countries.[2]

Suppose that the domestic chip industry claims that chips are being dumped by foreign producers and/or are being subsidized by foreign governments and that the industry is being materially injured by these actions. If the domestic administrators of the AD and countervailing duty laws find in favor of the chip industry, the additional duties imposed can have the same unfavorable effects on the PC industry that are shown in Figure 1. The increased costs of chips for the PC industry can reduce employment in that industry by either increasing imports or reducing exports. Moreover, if PCs are being imported, the PC industry would not be able, as in the safeguards case, to use the same laws to offset its losses, unless it can make a case that PCs are also being dumped or subsidized.

In a world of fragmented production, still another problem with existing rules on administered protection is that protection may fail to provide the expected benefits to the domestic industry receiving this protection. For example, if the PC industry secures protection, foreign producers may react by stopping their shipments of PCs and instead may ship all the parts besides chips needed to assemble a PC. These inputs are likely to face much lower import duties. While this activity is presumably more costly than exporting assembled PCs (otherwise foreign producers would have previously shipped the components), the benefits to domestic PC producers of the protection may well be small. In the case of dumping, some countries (e.g., the United States) have tried to deal with this problem by changing their domestic AD laws to permit the imposition of appropriate duties on imported inputs as well as on the final product. But with many imported inputs being used to produce many different products, such a rule can cause significant job and profits losses in other industries that far outweigh any benefits to the sector protected by the AD duty. Fortunately, the WTO AD rules do not as yet include this provision.

The conclusion to be drawn from the preceding analysis is that the administered protection rules of the WTO are seriously flawed in a world-trading system characterized by extensive vertical and horizontal international fragmentation of production, even judging by the merchantilistic objectives of typical trade negotiators. In more and more situations their implementation is likely to result in net job and profit losses in the economy. Consider, for example, the recent effort by the U.S. steel industry to gain protection. Steel producers claimed they had been seriously injured by increased imports, particularly from countries adversely affected by the Asian financial crisis. However, the industry bypassed the procedures for gaining protection under the safeguards law and went directly to Congress with a request for the imposition of import quotas.[2] A study by Hufbauer and Wada (1999) found that, because steel is widely used as an intermediate input into many final goods, the economic costs to other industries and American households for reducing imports by the amount stipulated in the bill passed by the House of Representatives would have been $800,000 for each job retained in the steel industry. While the House of Representatives passed a steel quota bill, the Senate, fortunately, appreciated the absurdity of incurring these costs and rejected the bill. In another recent case a group of independent U.S. oil producers claimed that they were being materially injured by the dumping of oil into the U.S. market. The petition was dismissed on a technicality (the failure of a majority of U.S. oil producers to support the petition), but if the case had been successful the resulting increase in the price of imported oil would undoubtedly have caused the loss of many more jobs in other industries than would have been gained in the domestic oil industry. Yet, because WTO AD provisions take no account of the indirect effects of AD duties, such outcomes are quite possible.

How should WTO rules be changed to prevent such outcomes? With regard to the safeguards rules, a minimum change would be to introduce a "national interest" provision into the WTO safeguards agreement. The United States actually has such a provision in its safeguards law. The International Trade Commission determines whether the industry is seriously injured and recommends specific responses, but the President makes the final decision about whether to grant greater import protection. In reaching this decision, the President is directed "to take all appropriate and feasible action within his power which the President determines will facilitate efforts by the domestic industry to make a positive adjustment to import competition and *provide greater economic and social benefits than costs*." (Italics added by author.) In following this objective, the law specifies that the President shall consider the effect of implementation "on consumers and on competition in domestic markets for articles." It is through these provisions that other producers and final consumers are able to bring pressure against protective action, and the provisions are in part responsible for the very small number of safeguard actions in the United States in recent years. The WTO safeguards agreement also needs a provision that directs administrating authorities to balance the benefits of additional industry import protection against the losses of this increased import protection to other industries as well as consumers and that imposes protective measures only if there will be a net gain in national social welfare.

Any additional protection at all is a second-best response to increased imports that are "fair" as defined by the WTO but that cause serious injury to an industry. The best government response is to provide adjustment assistance that enables workers to find alternative employment at minimum personal economic and social costs yet neither restricts trade in the affected product nor distorts trade in other products. One adjustment-assistance proposal that seems worthy of implementation is the earnings-insurance scheme suggested by Lawrence and Litan (1986). Under this program displaced workers who move to new jobs would receive a certain fraction (e.g., 50%) of the difference between their former and current earnings. The proportion of income replaced would decline over time as the workers acquired additional skills in their new jobs and, consequently, moved up the wage ladder. Another key component of any adjustment-assistance program should be job retraining for displaced workers and also education allowances that permit their children to gain the skills that are in demand in the global economy. Employers should be encouraged through loans with low interest rates to move into new sectors that are less vulnerable to import competition. Community development programs such as those implemented in the United States to cope with the reduction in military spending after the end of the Cold War should also be part of adjustment-assistance efforts when appropriate. Thus the WTO safeguard agreement should also be amended to include a statement to the effect that the preferred response to successful safeguard petitions is domestic adjustment measures and to include an illustrative list of such measures.

Existing WTO AD provisions not only do not deal with the indirect consequences discussed above but are also being misused both to gain import protection and for price-colluding purposes.[3] The case for AD provisions on economic efficiency grounds is based on predatory dumping. Predatory dumping occurs when a firm sells a product in a market at below its marginal production costs with the intention of driving competitors out of business and then raising the price to its monopoly level. This action leads to a distorted use of resources that reduces aggregate real income. There is, however, no general efficiency basis for opposing the sale of a product at a lower price abroad than a home or at a price below average costs. Foreign dumping is "condemned" under WTO rules because it is regarded as an unfair trade practice, and then only if it causes or threatens to cause material injury to a domestic industry. A strong indication that the main users of the AD provisions administer them in a manner that discriminates against foreign countries is the finding from an Organization for Economic Cooperation and Development (OECD) study cited by Finger (1998) that 90% of the instances of import sales found to be unfair under AD actions brought in Australia,

Canada, the European Union, and the United States would have been considered fair if they had taken place in domestic commerce.

The best way for the WTO to deal with dumping is to drop the present agreement and replace it with an agreement on competition policy. This would be used to police predatory dumping. However, as is discussed in more detail in the next section, such a change does not appear politically feasible at this time. Too many politically well-organized producers strongly support the existing AD agreement and, moreover, want to make it even easier to secure AD duties. Just modifying the existing AD agreement to prevent the unfavorable effects on other industries discussed above will be very difficult to achieve politically. However, one reform that might be feasible is to introduce the type of national-interest provision proposed for the safeguards agreement into the AD agreement. This would at least prevent some of the worst socially adverse outcomes in AD cases. In determining the net benefits of any AD action, the administering authorities would be directed to consider whether the dumping rose to the level of being predatory.

Another modest reform would be to require that the level of AD duties or price increases under price undertakings be no higher than necessary to remove the injury to the domestic industry and in no case higher than the margin of dumping. The current WTO provisions state only that "it is desirable" that the AD duties or price increases be less than the dumping margin if this would be sufficient to remove the injury. But, while some WTO members (e.g., the EU) follow this procedure, others (e.g., the United States) do not, but instead automatically impose duties equal to the dumping margin. Since WTO dumping provisions condemn dumping only if it causes material injury, it seems appropriate that any remedial action be sufficient only to remove the injury.

In using public policies to remove the injury from dumping, governments should also be sure that the domestic firms alleging injury are not themselves able to eliminate the dumping. It is only possible for foreign producers to maintain the margin between the higher price they charge at home and the lower price they charge in another country's domestic market if the costs of shipping the product back to the foreign country are greater than this difference. Otherwise the dumped good or a comparable one can be profitably shipped to the dumping country, thereby driving down its price. A tariff imposed by the foreign country on the product, various nontariff barriers, and transportation costs can make the cost of shipping the good back greater than the dumping margin. But if the price gap is greater than the cost of exporting into the dumping country, there seems little reason to publicly assist an industry more than can be justified by the cost gap. In cases where the price gap exceeds the cost gap, it may be that the domestic industry is using the AD laws to enforce a cartel-like price agreement with foreign producers.[4] Consequently, under WTO AD rules, petitioners for protection should be required to provide evidence on costs of shipping the product into the dumping country's home market. The administering authorities should also collect information on these costs. If the cost margin is less than the dumping margin, these authorities should implement AD measures that only remove the injury related to this cost margin. Furthermore, when the administering authorities find the cost margin to be less than the dumping margin, these findings should be transmitted to the antitrust authorities in both the domestic and the foreign countries for a determination of whether national competition rules are being violated.

iii) Countervailing Duties

The countervailing duty provisions of the WTO are much more defensible on economic efficiency grounds than are the organization's AD provisions, especially in light of the fact that the Uruguay Round agreement on subsidies and countervailing measures recognizes that some subsidies can be justified on these grounds and, therefore, should not be countervailed. However, most government subsidies benefit a special-interest group within a country at the expense of the rest of the economy and distort the efficient use of world resources. But permitting the imposition of countervailing duties equal to the margin of subsidization is a clumsy means of trying to offset the injury caused to domestic producers by foreign subsidization. If foreign governments are just subsidizing exports, countervailing duties will restore foreign exports to their pre-subsidy levels, but export subsidies are already banned (except for subsidies of agricultural products) under the WTO. In the case of foreign subsidization of domestic production, some of which is exported, an additional duty equal to the full amount of the domestic subsidy will actually reduce foreign exports to less than their free-trade level and thus will more than make up for the material injury caused by the foreign subsidization. It is for this reason that the WTO countervailing provisions, like the AD provisions, should be changed to specify that countervailing duties be no higher than is necessary to eliminate the material injury caused by the foreign subsidies.

There still remains the problem that a countervailing duty will reduce employment in industries that use the good as an intermediate input because of the increased cost of this input. However, unlike the safeguards case discussed above, the resources previously shifted to these other industries as a consequence of the foreign subsidization do not represent an optimum use of resources. Therefore, implementing adjustment-assistance measures is the appropriate way to deal with the disruption caused by the countervailing duty in these sectors.

3. EXTENDING WORLD TRADE ORGANIZATION RULES INTO NEW AREAS

A. Competition Policy

Most WTO rules are directed at government measures that affect international trade, such as tariffs and quantitative import restrictions, subsidies and countervailing duties, AD duties, government purchasing policies, customs regulations, licensing procedures, rules of origin, technical standards, investment requirements, and so forth. The objectives of these rules range from reducing or offsetting the trade-distorting or "unfair" effects of such government measures to simply ensuring greater transparency in public regulations and requirements. However, it is widely recognized that the actions of private producers can also distort international trade or be "unfair." Dumping by foreign firms is a prime example of the latter type of action, and this is covered by WTO rules. Trade-distorting private activities cover the wide variety of means that businesses can use to restrain competition, limit access to markets, and foster monopolistic control in an effort to increase profits.[5] These activities are not covered by WTO rules.

A number of governments, for example, the EU and Japan, and some private groups (e.g., consumer organizations, academic economists) are urging that the WTO adopt a set of rules covering competition policy. That an international organization devoted to reducing trade distortions has no rules of "good" international trading behavior for the private and public sectors means that it runs the risk, they argue, of ignoring a set of efficiency-reducing practices that may be even more important than the measures introduced by governments. Pointing out that such rules were included in the charter for the ill-fated International Trade Organization (ITO) agreed upon by over 50 countries in 1948, they also contend that the globalization trend of recent years makes these rules even more necessary now than in the past.[6]

It is certainly true that improvements in communications and the creation of world markets for many goods and services have resulted in new opportunities and incentives for producers to collude for the purpose of increasing their collective profits. Moreover, by locating more production activities in different countries, producers may be better able to hide monopolistic actions from the antitrust authorities of any particular nation. Yet globalization has also provided new opportunities for the most efficient producers to increase their profits by competing vigorously in the domestic markets of producers who have become inefficient through long reliance on government protection and subsidization. Thus far, it would appear that the pro-competitive effects of globalization have dominated the pro-collusive effects and have served on balance to promote greater international competition. However, it is clear that collusive arrangements between domestic manufacturers and wholesalers-retailers and input suppliers sometimes prevent the pro-competitive forces from operating effectively, as the U.S.-Japan trade disputes over photographic film and auto parts illustrate.

Although there is considerable support for the introduction of competition rules into the WTO as part of the next negotiating round, opposition to this proposal on the part of governments (e.g., those of the United States and many developing countries) and business groups is widespread. There is much disagreement among governments and private groups about just what acts should be punishable under an antitrust regime, as the Working Group on Trade and Competition Policy (established at the last ministerial meeting in December 1996) has discovered. Countries like the United States, whose competition policy is based on an allocative-efficiency, consumer-welfare principle, fear that they will have to give up the essential and correct nature of U.S. law as part of the compromises necessary to gain agreement in a WTO negotiation.[7] In contrast, countries with weak or no competition laws fear they will be subject to rules that are inappropriate for their economic systems and stages of development.

Because of these divergent views on the desirability of competition rules, most trade and antitrust officials believe that, at best, it is only possible to establish a general framework of competition rules requiring WTO members to adopt minimum competition standards that are widely accepted, such as ones relating to price-fixing and group boycott. Moreover, under this approach WTO dispute-resolution procedures would be available only to decide whether countries were enforcing their domestic competition laws and not to make judgement about any particular case.[8] But that even this modest set of competition rules will be introduced into the WTO in the foreseeable future seems unlikely given the opposition of key trading nations and most business interests. It appears that much more evidence of anti-competitive behavior that harms other producers and consumers is needed before political pressures will be strong enough to bring about the introduction of competition rules into the WTO.

B. Investment Policy

Technological advances that have created new opportunities for vertical and horizontal fragmentation at the international level have stimulated FDI enormously. Although completely eliminating the barriers that prevent investors from taking full advantage of these opportunities promotes the efficient use of the world's productive resources, there are strong political forces in both host and investing countries that operate to prevent this outcome. Most governments now welcome FDI because of its income, employment, and technology-transfer benefits, but some are under domestic political pressures to allow such investment only in product lines in which local capitalists have not already invested significantly. Domestic interests also often pressure their governments to require foreign affiliates to purchase a certain proportion of their intermediate inputs locally, hire a certain proportion of mangers locally, export a certain share of their output, and so forth. In contrast, in countries undertaking extensive FDI, governments face the conflicting pressures of being lobbied by multinational enterprises who want the governments to seek the opening of direct investment markets in other countries for national growth and efficiency reasons, and by domestic labor groups who urge the governments not to pursue such market-opening actions on the grounds that FDI causes domestic job losses.

The chances of achieving reciprocal reductions in barriers that restrict FDI increase if there is a significant volume of both outward and inward direct investment between countries. In this situation the multinational firms of each country can argue that their own governments must remove their discriminatory measures so that other governments will do the same and thereby enable these firms to take advantage of foreign investment opportunities. However, when the direction of FDI tends to be one way, as is the case between developed and developing countries, the absence of a politically active domestic group of investors in the developing countries who wish to gain greater access to FDI opportunities (or to prevent retaliatory action by foreign governments) tends to lessen the extent of investment liberalization these countries are likely to undertake. This difficulty can sometimes be overcome, as in the Uruguay Round negotiations, if the developed countries make other concessions (e.g., the phase-out of the Multifiber Arrangement on textiles and apparel) that are perceived by the developing countries as balancing the package of their concessions on investment policy.

As recent experience with the Multilateral Agreement on Investment (MAI) negotiated under the auspice of the OECD indicates, it is even difficult for developed countries to implement a stand-alone agreement on investment. A wide range of nongovernmental organizations (NGOs) concerned about such matters as the environment, human rights, labor standards, and national sovereignty strongly oppose the key national treatment and most-favored-nation provisions of the MAI (and the WTO). They wish to continue policies that, for example, place special restrictions on what foreign investors can own, maintain economic programs that benefit only domestic companies, or require firms to hire a certain proportion of managers locally. They also favor economic sanctions that punish particular countries for human rights violations by preventing corporations from doing business in these countries. As a consequence of the lobbying campaign by these groups that, among other accomplishments, was successful in getting various municipalities to pass resolutions declaring themselves "MAI free zones," the OECD secretariat has, in effect, abandoned its efforts to secure approval of the draft MAI by member governments.

This experience does not bode well for introducing comprehensive rules on investment into the WTO in the upcoming Millennium Round. On economic efficiency grounds the need for such rules is clear, but sufficient political support among and within countries to achieve this goal seems absent.

C. Labor and Environmental Standards

As world markets become more and more integrated by the forces of globalization and fragmentation, there are increasing political pressures to establish global rules covering not only traditional economic issues such as competition and direct investment policies but also broader socioeconomic subjects such as labor and environmental standards. NGOs concerned about such policies have become increasingly aware of the inter-country differences in government rules and regulations pertaining to these subjects and believe that these differences can significantly undermine their own ability to introduce stricter (or even to maintain existing) government standards within their own countries. For example, labor unions in developed countries have concluded that their efforts to improve the wages and working conditions of their members are being hampered by multinational firms' increased ability to locate the production of intermediate inputs in countries with more lenient government rules on such matters as child labor, workers' rights to organize, and so forth. Similarly, environmental groups in developed countries believe that the existence of weaker environmental standards in developing countries, coupled with the increased fragmentation opportunities in these countries for multinational firms, impair the groups' ability to preserve and strengthen existing standards in their own countries.

While standards covering labor and environmental issues already exist in various multilateral environmental agreements and in the International Labor Organization (ILO), non-adherence to the standards is widespread. In contrast, the WTO has proved surprisingly successful in achieving adherence to its trading rules. An important reason for this has been the ability of members harmed by the actions of rule violators to impose trade sanctions on the violators. Labor and environmental NGOs understandably would like to take advantage of the WTO enforcement process to promote their goals by introducing labor and environmental rules into the WTO. However, they also want new rules, because WTO dispute-settlement panels have concluded in some cases that existing national laws covering these matters violate WTO rules.

An important distinction between labor and environmental standards is that different labor standards among countries, unlike different environmental standards, do not produce spillover effects that directly harm the physical health of humans, animals, and plants in other countries or bring about an inefficient economic allocations of world resources. The existence of poorly enforced child labor laws in particular countries can have adverse income-distribution consequences for some groups in other countries, just as intercountry differences in the distribution of natural resources, skilled workers, and technology can, but these laws do not produce a technological spillover that directly harms workers and consumers in other countries and that distorts the efficient use of resources. In contrast, lax laws that permit factories to pollute the air, thereby causing acid rain to fall in other countries or destroying the ozone layer, do cause such damage and result in a misallocation of resources.

Establishing a set of common labor standards is a matter of adopting a uniform set of social conditions among countries. As such, the motivation for introducing uniform international labor standards should be a concern for the welfare of the workers in countries with weakly enforced laws rather than for the income-distribution consequences in other countries. Otherwise there would be no limit to the types of economic conditions whose differences across countries would be legitimate candidates for imposing international uniformity. Consequently, under the rights of sovereignty (whereby a nation has the right to decide what policies are in its best interests without being coerced by other countries) international rules on such matters as labor standards should apply only to those countries that agree with such rules. This is the arrangement followed in the International Labor Organization (ILO).

Attempting to introduce a comprehensive set of labor rules into the WTO patterned after those in developed countries is unwise for a number of reasons. First, most developing countries strongly oppose such an effort. When it comes to the issue of child labor regulations, for example, the governments of these countries maintain that their levels of development are too low to be able to afford stricter labor standards. They believe their children are likely to end up worse off, not better off, under a set of rules that bans their employment. They also point to the voluntary actions that most countries have taken to improve labor standards as per capita income has risen and argue that this is the proper way to achieve better standards. In addition, some also have expressed their concern that the attempt to tie labor standards to trade policy is motivated more by a desire to reduce imports from developing countries than by human rights considerations. There can be little objection to WTO rules on social issues on which there is unanimity among WTO members, such as in banning imports produced by prison labor, but attempting to force a large group of countries to accept standards they strongly oppose by withholding trade concessions in markets of export interest to these countries threatens to cause a breakup of the organization.

Adding rules on social issues such as labor standards to the WTO is also likely to overburden the organization's ability to enforce both the new rules and its old ones. Determining whether the standards are being observed in producing a particular product is likely to require a firm-by-firm investigation that could overwhelm an organization whose staff and country-representatives are trade rather than labor specialists. Moreover, if comprehensive labor standards become part of WTO rules, it will be difficult to keep a variety of other human rights issues off the rule-making agenda. The WTO could end up trying to use trade policy to force international conformity with regard to such matters as religious and political freedom. Yet we know from past experience that trade policy has not been effective at forcing changes in other countries' social and political behavior. By using one policy tool, namely, trade policy, to achieve multiple and sometime conflicting objectives, the WTO will almost surely become an organization that is ineffective in achieving both its new and its old goals.

As noted above, when economic activity produces adverse technological spillovers that are harmful to other countries, there is a sound economic case for international environmental standards backed by effective enforcement mechanisms. However, in light of the aforementioned points pertaining to overburdening an organization with too many goals, the WTO is not the proper organization in which to introduce these standards, nor is the use of trade policy the best way to achieve conformity with the standards. Negotiation of multilateral environmental agreements (e.g., The Montreal Convention on Fluorocarbons), which rely on environmental specialists, is the appropriate means for developing and enforcing such standards. Furthermore, imposing trade sanctions on countries that do not meet the standards is an inefficient and likely an ineffective means of securing compliance. Since the polluting actions of countries not adhering to the agreements usually harm not only citizens of other countries but also those of their own, the governments of these countries are usually supportive of the standards but, as in the case with labor standards, conclude that their development levels are too low to enable them to afford the costs of adherence. Imposing trade sanctions on such countries does not solve their basic financial problem but merely breeds resentment and non-cooperation in other fields. A more effective means by which to address the problem is for higher-income countries and/or international lending organizations to provide the less-developed countries with financial and technical assistance so that the latter can introduce the capital equipment and technology needed to meet stricter environmental standards. Trade policy should only be used as a last resort and then only when there are adverse technological spillovers. In other situations group pressure coupled with financial and other incentives should be the means by which to effect adherence to the standards of a multilateral environmental agreement.

4. CONCLUSIONS

Technological and political developments that have created new opportunities to fragment production processes into different blocks located in different countries have become a major globalization force. As the major international economic organization devoted to promoting higher living standards through the reduction of barriers and distortions to the exchange of goods and services among nations, the WTO is faced with a difficult task: it must adapt its rules and regulations to meet this and other globalization challenges in order to remain relevant without, however, overextending its range of competency and becoming ineffective and self-destructive. In considering traditional WTO rules from the perspective of the fragmentation phenomenon as well, and in looking at the main proposals for new rules designed to meet the challenges of globalization, this paper reaches mixed conclusions concerning the desirability of changing existing rules and adding new ones.

The basic WTO rules should be modified so that they permit the imposition of import barriers in response to either fair or unfair injurious foreign competition, because the present rules ignore the existence of traded intermediate inputs. In the safeguards agreement, for example, an industry whose output is an important intermediate input for other industries qualifies for import protection if it is seriously injured as a result of increased imports, even if this protection will cause significant output reductions in the input-using industries. As a consequence of the increased costs of the input, the loss in profits and jobs in these latter industries could be greater than the profit and employment gains in the protected sector—a result that runs counter to the basic notion of safeguards. At the minimum a "national interest" provision should be introduced into the WTO safeguards agreement that directs the administering authority to take into account, when making its decision, the impact of protection for a particular industry on the total economy. However, the fundamental implication of fragmentation for the safeguards agreement is that the response to injurious but fair import competition should be adjustment assistance to the injured industry, and not import protection.

The AD and countervailing provisions of the WTO also fail to take into account the existence of traded intermediate inputs. At the minimum these provisions should also be modified to include a national interest clause. However, the AD provisions are also flawed for other reasons and should be completely replaced by rules on competition policy. Since the short-run prospects for introducing competition rules into the WTO are not favorable, however, various suggestions are made for preventing the misuse of the AD laws for straightforward protectionist purposes and for promoting price-increasing collusive actions among domestic and foreign producers.

My conclusion about the desirability of WTO rules on competition policy and of more extensive rules on investment policy is that these are legitimate areas for expanding the scope of the organization's rule making. However, in view of the absence of support for doing so on the part of a many WTO members, competition and investment policies should be set aside as negotiating issues for the time being but continue to be subjects of formal WTO study groups. In contrast, I argue that comprehensive labor standards and environmental standards are not appropriate subjects for inclusion in the WTO. These are important topics for international rules, but they should be dealt with through agreements outside the WTO. Given the magnitude of these subjects and their differences from the traditional market access concerns of the WTO, it seems likely that if the WTO attempts to enforce labor and environmental standards in addition to its existing rules it could become so overburdened that it would become ineffective and inefficient at carrying out either task.

References

Baldwin, Robert E. (1998), "Imposing Multilateral Discipline on Administered Protection," in Krueger Anne O., (1998), ed., *The WTO as an International Organization* (Chicago: University of Chicago Press).

Boltuck, Richard and Robert E. Litan, (1991), eds., *Down in the Dumps: Administration of the Unfair Trade Laws* (Washington, D.C.: The Brookings Institution).

Cordon, W.M., (1971), *The Theory of Protection*, (Oxford: Clarendon Press).

Finger, J. Michael, (1998), "GATT Experience with Safeguards: Making Economic and Political Sense of the Possibilities That the GATT Allows to Restrict Imports" (Washington, D.C.: World Bank).

Fox, Eleanor M., (1999), "Should Competition Join the WTO?" Paper presented at conference on "The Next Trade Negotiating Round: Examining the Agenda for Seattle," Columbia University, July 22–23, 1999.

Graham, Edward M. and J. David Richardson, (1997), eds., *Global Competition Policy* (Washington, D.C.: Institute for International Economics).

Jones, Ronald W. and Henryk Kierzkowski, (1990), "The Role of Services in Production and International Trade: A Theoretical Framework," in R.W. Jones and A.O. Krueger, eds., *The Political Economy of International Trade* (Oxford: Basil Blackwell).

---------- (2001a), "Globalization and the Consequences of International Fragmentation," in R. Dornbusch, G. Calvo, and M. Obstfeld, eds., *Money, Factor Mobility and Trade: The Festschrift in Honor of Robert A Mundell*, (Cambridge, MA: The MIT Press).

----------- (2001b), "A Framework for Fragmentation," in S. Arndt and H. Kierzkowski, eds., *Fragmentation: New Production Patterns in the World Economy*, (Oxford: Oxford University Press).

----------- (1999), "Horizontal Aspects of Vertical Fragmentation," in this volume.

Jenny, Frederic (1999), "Globalization, Competition and Trade Policy: Issues and Challenges," Paper presented at conference on "The Next Trade Negotiating Round: Examining the Agenda for Seattle," Columbia University, July 22–23, 1999.

Hufbauer, Gary C. and Erika Wada (1999), "Steel Quotas: A Rigged Lottery," *International Economic Policy Briefs,* No. 99-5 (Washington, D.C.: Institute for International Economics).

Lawrence, Robert Z. and Robert E. Litan (1986), *Saving Free Trade: A Pragmatic Approach*, (Washington, D.C.: The Brookings Institution).

Messerlin, Patrick A. (1989), "The EC Antidumping Regulations: A First Appraisal," *Weltirtschaftliches Archiv*, 125:563–87.

Srinivasan, T. N. (1998), "Regionalism and the WTO: Is Nondiscrimination Passe?" in Anne O. Krueger (ed.), *The WTO as an International Organization* (Chicago: The University of Chicago Press).

Endnotes

[1]Assuming fixed international prices of intermediate and final goods and fixed intermediate inputs, the formula for the rate of effective protection on good *j*, namely, g_j, can be written as: $g_j = (t_j - E_i\, a_{ij}t_i)/(1 - E_i\, a_{ij})$, where t_j is the nominal rate of protection on the jth good, a_{ij} is the share of input i in the cost of *j* at free trade prices, and t_i is the nominal tariff rate on input *i*. Notice that the effective rate of protection on good *j* equals the nominal rate on this good if the nominal rate of protection on good *j* and the nominal rates of protection on all the *i* inputs are the same.

[2]Certain types of government subsidies (e.g., those for research purposes that do not exceed a particular level) are not subject to countervailing duties.

[3]The reason may have been the fear that the President, who makes the final decision in safeguard cases and is guided by a national interest clause, would have rejected the industry's petition.

[4]For an elaboration of the manner in which U.S. antidumping laws are being misused, see the essays in Boltuck and Litan (1991). Messerlin (1989) points out the deficiencies of the EU's antidumping laws.

[5]For a discussion of the manner in which the antidumping provisions can be used to promote price collusion among domestic and foreign producers, see Baldwin (1998).

[6]See the essays in Graham and Richardson (1997) for a detailed discussion of the many types of trade-distorting actions by private firms as well as the nature of competition policies in the major industrial countries.

[7]Only the commercial policy sections of the ITO were actually implemented (as the GATT), because of the opposition of business interests, especially in the United States.

[8]American antitrust officials also maintain that existing U.S. laws in conjunction with antitrust cooperation agreements negotiated bilaterally with other countries have proved quite effective in handling foreign anticompetitive actions directed against U.S. producers and consumers.

[9]For more details on such proposals, see Graham and Richardson (1999, chapter 17), Jenny (1999), and Fox (1999).

Chapter 14

RULES OF ORIGIN AND FRAGMENTATION OF TRADE

Peter J. Lloyd

1. RULES OF ORIGIN

In the last decade or so economists and multilateral organizations have devoted more attention to rules of origin than ever before. (For a comprehensive review of the topic see Vermulst, Waer, and Bourgeois 1994). Although attempts to address the issues on the rules of origin had been made in the General Agreement on Tariffs and Trade (GATT) the Organization for Economic Cooperation and Development (OECD) and the United Nations Council on Trade and Development,[1] the Uruguay Round Agreement on Rules of Origin was the first binding multilateral agreement on the subject. This agreement relates to non-preferential rules of origin. It imposes some discipline on these rules in terms of their transparency and consistency of application, and it clarifies the objective of the harmonization of the rules for all non-preferential policy instruments. Negotiations under the joint auspices of the World Trade Organization (WTO) and the World Customs Organization (WCO) were not concluded by the July 20, 1998, deadline and are still ongoing. Preferential rules of origin are covered by a non-binding common declaration with regard to such rules; this declaration was annexed to the WTO agreement. Preferential rules are not subject to the agreement's harmonization objective. Consequently countries are free to determine which rules of origin will be applied to both preferential and non-preferential trade.

In practice a wide variety of methods are used to determine the country of origin of imported goods. The Kyoto Convention established that origin should be determined by the last/final country in which "substantial transformation" of the goods took place and indicated that the meaning of "substantial transformation" could be determined with one of three different methods. These are a change of tariff heading (CTH) rule, a percentage of value added rule, or a specified process of manufacture rule. The European Union (EU), which is the largest regional trading arrangement (RTA) in terms of its import trade, uses the CTH rule. The percentage rule is used by a number of RTAs including the European Free Trade Area, the Association of Southeast Asian Nations (ASEAN) Free Trade Area, the Closer Economic Relations (CER) agreement between Australia and New Zealand, and the Andean Community. The specified process of manufacture rule is used only for particular product groups in some regional agreements. For non-preferential rules, a wide variety of rules are also used. Non-preferential rules tend to be less formal than preferential rules of origin; the former are based on substantial transformation and are often applied on a case-by-case basis and couched in terms like "the country in which the last significant process of manufacture was performed." The remainder of this paper concentrates on preferential rules of origin.

Where the CTH rule is used it may be applied in one agreement at different levels of the tariff classification for different groups of goods. Where the percentage rule is used it is usually uniform within an agreement, but the percentage chosen varies widely—from 25% in the Malaysia Australia Trade Agreement and the Malaysia New Zealand Trade Agreement and upward, with 50% being the modal percentage. Many preferential agreements (e.g., the North American Free Trade Area (NAFTA) and the Common Market of the Southern Cone [MERCOSUR]) use a combination of rules.

In cases where a country is a member of more than one RTA it may use different rules or combinations of rules in different regional agreements; for example, the United States uses CTH as the principal rule and the percentage rule or specific process rule for some commodity groups in NAFTA, whereas it uses the percentage rule in the free trade area (FTA) with Israel and in its General System of Preferences and other non-reciprocal preference schemes. Many other countries do likewise. (For the Asia Pacific Economic Cooperation (APEC) members the preferential and non-preferential rules are set out in APEC 1997). Similarly, where non-preferential rules are used in a country for more than one purpose, different rules are applied for different purposes.

These diverse rules impose a number of costs on economies that trade. Compliance costs are high for both private traders and government officials administering the schemes. On the basis of a median estimate of 3% of the value of the goods traded, Stephenson (1997) estimates that, for APEC economies, these costs could have amounted to US$70 billion in 1995.

Rules of origin, which are designed to prevent trade deflection in FTAs, are protective by nature (Shibata 1967). Thus they reduce the extent of the regional liberalization of trade that would otherwise occur. They matter most for those potential imports for which the substantive tariff rates or tariff equivalent rates are high. In an era when other border instruments are coming under increasing discipline, they may be used as a substitute for tariffs or non-tariff barriers. There is evidence that they have been used deliberately to increase the protection of certain industries in the EU (e.g., semiconductors, automobiles) and in NAFTA (e.g., telecommunications, color televisions, automobiles) (see Jensen-Moran 1996).

For preferential rules the costs may be specific to one rule for assessing origin. These differences are important when choosing between different rules.

As the extent of the transformation of goods in process varies greatly, value added is the economically meaningful and natural measure of transformation. The percentage rule is therefore the natural one for measuring transformation, but it leads to trade diversion as a means of increasing area content and satisfying the rules. Governments wishing to protect some industries may favor a percentage rule if the percentage can be set at a high level; for example, in NAFTA imports of automobiles and automobile parts must have an area content of at least 62.5% of the net cost if they are to satisfy the rules of origin. Thus the compliance costs with the percentage rule may be higher than they would be with the CTH rule.

The CTH rule also leads to trade diversion and has particular disadvantages. It cannot handle changes in technology that make the tariff classification and changes in tariff headings obsolete. It also introduces the possibility of political-economy manipulations of the rules. The level of disaggregation for tariff headings can give more protection to the domestic industry in the country of sale. In the Canada-U.S. Free Trade Agreement and in NAFTA the tariff headings for rules of origin purposes are defined from the two-digit to the nine-digit level of the classification. According to Palmeter (1993, p. 329), "no ascertainable rule of principle was employed in determining what level of change would be required for any particular product."

The specific process rule is extremely impractical in that it requires the specification of processes for every traded good if it is used as a general rule.

2. FRAGMENTATION AND RULES OF ORIGIN

Fragmentation of international trade has affected the application of rules of origin in a number of ways. Rules of origin based on substantial transformation are necessary only because of fragmentation of international trade. If all goods traded internationally were produced by processes that were fully integrated in one country, applying rules of origin would be straightforward.

Increasing fragmentation is part of the explanation for the rising importance of rules of origin. (For a review of increasing fragmentation in international goods trade see Feenstra 1998.) The usual explanation for the increasing importance of preferential rules of origin is the increase in the number of countries that are members of RTAs or nonreciprocal preference schemes. A wave of new arrangements of this kind sprang up in the 1990s. Of the 135 current WTO members, only Japan, Korea, and Hong Kong do not belong to at least one RTA. The usual explanation for the increasing importance of nonpreferential rules of origin is the increase in the use of border instruments (particularly antidumping action, countervailing duties, and safeguard actions, all of which are forms of contingent protection) that require rules of origin. The increasing fragmentation of international trade has meant that a larger proportion of goods traded have value added in two or more countries and that on average the proportion of value added that does not occur in the country of last substantial transformation has increased. It is the conjunction of increasing preferential trade and increasing fragmentation of trade that has made the problems of rules of origin more important in world trade.

Increasing fragmentation has made the determination of origin more difficult and arbitrary. Moreover, global sourcing is constantly changing as the costs of production of intermediate inputs in different countries change. Consider one country that is supplying some intermediate good to a second country with which it has formed an FTA. If the costs of goods sourced in the former country increase relative to other potential sources, the second country may source more of its intermediate inputs from a third country or countries. This may then mean that its exports to the partner country no longer satisfy the area content requirement. Similarly, the area content may fail to be satisfied if the bilateral exchange rate between the partner countries change, even if the source of materials does not change. It is a frequent complaint in FTAs with percentage rules that small changes in unit costs of production and/or in exchange rates mean that an export no longer satisfies the area content. A marginal change in costs leads to a large discrete change in the tariff burden. This makes the effect of the rules uncertain.

It is possible that a final good may be subject to border taxation in more than one country. This can occur if a product is manufactured by a chain of processing operations that are carried out in two or more countries before the product is delivered to the last country in which it sold and if the cross-border transactions are taxed at least twice. Such multiple taxation is a form of international cascading of commodity taxes. It is probably not common, because intermediate inputs in most countries are admitted duty free or at low duty rates.

Increasing transformation has increased the costs of trade diversion resulting from rules of origin. Producers may increase area content in order to satisfy a rule of origin either by switching the country of source of intermediate inputs to another country within the area or by increasing the extent of processing within the area. The effect is to distort and raise the costs of production in the exporting country. Krueger (1993) provides an example of a change in sourcing. She employs a three-stage production process such as that used in the manufacture of final-product clothing going through two earlier stages of cotton-fiber and then cotton-textile production. In the case of NAFTA, for example, clothing producers in Mexico exporting to the United States could be encouraged by a rule of origin to source their cloth imports from new or expanded production in Mexico or the United States, whereas in the pre-FTA situation Mexico imported the cloth from the least-cost source outside the area. This can happen even if imports into the member that exports to the other member are duty free. The percentage of cloth sourced from within the area is an increasing function of the percentage of value added specified by the percentage rule. As Krueger puts it, "Rules of origin agreed upon in forming an FTA in fact extend the protection accorded by each country to producers in other FTA members countries." Other examples of such trade diversion are given in Lloyd (1993) and Falvey and Reed (1998).

The growth of foreign direct investment (FDI) that has accompanied fragmentation has introduced a new margin of choice, namely, the choice of mode of delivery to a foreign market, either by home production and direct exports or by establishing a foreign affiliate.[2] This in turn has introduced new possibilities of distortion in the world allocation of production. FDI outside an RTA may be diverted to a country in the arrangement in order to take advantage of preferential market access. The choice of mode for the corporation supplying a foreign market is also clearly affected by rules of origin. Strict rules of origin may by themselves lead a corporation to switch from the mode of direct export to that of international production. Jensen-Moran (1995) provides examples from the rules in force in the EU and NAFTA, mainly in high-tech industries.

3. A PROPOSAL FOR REFORM OF RULES OF ORIGIN

Plainly rules of origin in the multilateral trading system need to be reformed. The aim of negotiations should be to increase the predictability of duties and to reduce the degree of protection implicit in strict rules.

Reforming the rules will involve two elements: choice of the rule used to assess origin and choice of standards within each rule, that is, the level of disaggregation of tariff headings, or of the percentage for the percentage rule. So far, most discussion of this topic has focused on the former element, but both are important and difficult to deal with.

With regard to the choice of the method of assessment, each has its advantages and disadvantages, as noted above. According to Palmeter (1993, p. 328), "Different methods have been used to determine origin, but none has proved totally satisfactory, and none is likely to do so." The root of the problem is that a good is regarded as being produced in one country, and the problem is to determine which one. That is, the tariff rate for good i, which is assessed as originating from country j, is

$$\begin{aligned} t_{ij} &= t_i \text{ if country } j \text{ is not a member of the area, and} \\ &= 0 \text{ if country } j \text{ is a member of the area} \end{aligned} \tag{1}$$

where t_i is the scheduled most favored nation (MFN) rate of duty expressed in *ad valorem* terms.

However, with value adding in two or more countries, origin should not be conceived as an all-or-nothing concept. Lloyd (1993, p. 708) notes that "the essential problem is that, with international trade in intermediate inputs and the splitting of the value added in traded goods between two or more independent countries, countries produce the value added components but trade the product in which the value added is embodied. The proportions of value added in a single member and in the area collectively can vary all the way from just over zero to 100 per cent." The natural way of addressing this problem is to allow the duty payable to vary in proportion to the value added outside the area. This led Lloyd to recommend a tariff whose base is not the price of the imported article if the good is assessed as originating outside the area but a tariff based on the proportion of the value added outside the area. This can vary over the interval [0,1]. For good i, let v_i and v_i^* denote the proportions of the value added that are added by factors inside and outside the area, respectively. Then the tariff rate that is actually levied on a dollar's worth of imports of the commodity by a member country with a value added tariff will be simply

$$\bar{t}_i = v_i^* t_i \qquad v_i + v_i^* = 1. \tag{2}$$

$\bar{t}_i$ might be called the value-added tariff rate. Given t_i, the duty payable varies continuously with the proportion of non-area value added.

This form of duty will eliminate the large changes in duty under the present rules when an import switches from being assessed as originating within the area to originating outside the area, and it will therefore make the rules less uncertain. This form of duty is not arbitrary, unlike both the CTH and percentage rules, because it does not involve the choice of a parameter, as with the percentage rule or tariff headings. It is therefore uniform across goods and cannot be manipulated by governments to give more protection to selected goods.

The value-added tariff approach to rules of origin was endorsed by Ho (1998) and by the Pacific Economic Cooperation Council (PECC) Task Force on Rules of Origin (PECC 1998). Ho (1998) declared that "the idea of one single country of origin is becoming increasingly out of date and out of touch with reality." The task force concluded that "the time has probably not yet come for a system of multi-country ROO (rules of origin, my own) to replace the current system of single country ROO." They saw the main disadvantages of the method as the difficulty of accepting an entirely new concept, and the administrative difficulties associated with it. Ho (1998) extended this form of adjustment to quantitative restrictions requiring a rule of origin and argued that the administrative difficulties were not as great as supposed.

4. ASSESSING THE VALUE-ADDED TARIFF

The central question in assessing the proposal for a value-added tariff is producers' response to a change in the tariff base, that is, the amount of trade diversion and creation that takes place under a value-added tariff. This section compares producer response under a value-added rule to that under a percentage rule.

A comparison of the costs of trade diversion under this tariff with those under the percentage (or CTH) rules is complex. There are a large number of distinct cases, depending upon whether production takes place in one member, two, or none before the RTA is formed (see Falvey and Reed, 1998, and Rodriguez, 2000) and the possibilities of substituting sources or processes between countries.

The key difference between these two alternative rules of origin is the incentive provided by each to producers within the area. Suppose that an FTA has just been created between two or more countries. There will be a reduction in duty on imports into a member country for those imports that satisfy the rules of origin. In the case of the percentage rule, this reduction is, from equation (1),

$$\Delta_{t_{ij}} = t_i - 0 = t_i \tag{3}$$

if and only if the percentage of value added in the area is at least equal to the minimum specified in the percentage rule, that is, $v_i \geq v_{\min}$ t_i is the (*ad valorem*) tariff rate for the good that applies to non-area imports. It is assumed that the same percentage applies to all goods. In the case of the value-added tariff, this rule will also result in a reduction in duty. From equation (2), this reduction is

$$\Delta_{t_{ij}} = t_i - v_i^* \ t_i = v_i \cdot t_i . \tag{4}$$

Note that in both cases the reduction is proportional to the scheduled tariff rate, t_i. The rule of origin is more important for goods that otherwise are subject to high rates of duty.

The reductions in the tariff paid by the exporter (or an agent) in equations (3) and (4) will be the increase in the export price if the importing country is small. (Otherwise, the reduction in tariff will be divided between the exporter and the importer.) The rules under equations (3) and (4) give different incentives.

For the percentage rule, the duty reduction only operates in the neighborhood of transformation that will affect the determination of the country in which the good is considered to have originated. Indeed, for any good produced in the area, if the percentage rule ($v_{\min}$) is set low enough, the rule will not be binding.[3] But, in the neighborhood where it is binding, the rule provides a very strong incentive to substitute area content because of the all-or-nothing nature of the resulting tariff reduction. (The CTH rule has a similar effect, but it is difficult to analyze because of the arbitrariness of the classifications and the consequent changes in tariff headings.)

The value-added tariff will not eliminate the incentive to change inputs or domestic processing. In fact, because the tariff burden declines continuously as the proportion of value added in the area increases, the value-added tariff rule provides a continuous marginal incentive to substitute area content for non-area content. Hence the value-added tariff might induce producers to increase value added beyond that specified in a percentage rule as $v_{\min} < 1$.

The precise effects within an area of each rule of origin will depend on the costs of substituting sources or processes between countries for each good. One model is particularly useful for this purpose. This is the continuum of stages model used by Dixit and Grossman (1982) and employed in the context of rules of origin by Lloyd (1993), Falvey and Reed (1998), and Rodriguez (2000). Assume that the production of some good takes place in a continuum of stages that can be arrayed on the unit line, with 0 representing the first stage and 1 the production of the final good. For each stage, z, there is a unit cost function $c(z)$. These unit costs of producing each stage may vary among countries. It is not necessary to retain the Dixit and Norman assumption that technologies of producing stages are identical across countries, nor their assumption that there are only two factors and that stages can be arrayed in order of capital intensity. However, the assumption of constant returns to scale is retained.

Suppose there are three countries, countries 1, 2, and 3. If each country produces the good by a fully integrated process within the country, the unit costs of producing each stage in each country are $c_1(z)$, $c_2(z)$, and $c_3(z)$. respectively. The unit costs of producing the final good in each country are then

$$C = \int_0^1 c_i(z)\,dz \quad i = 1, 2, 3. \tag{5}$$

These functions are strictly increasing in z, of course. For country i, the unit cost of that stage is given by the gradient of this function. Comparing these gradients gives the comparative costs of production for each stage.

Competition ensures that the global costs of production are minimized. Under a regime of free trade in all stages and all countries, the stages are allocated among countries from the solution to

$$\text{Min}\left\{\int_{R_1} c_1(z)\,dz + \int_{R_2} c_1(z)\,dz + \int_{R_3} c_1(z)\,dz\right\}, \tag{6}$$

where R_i are the range of stages produced in each country. With different gradients, the production of stages will be broken up among countries. That is, there will be fragmentation. All three countries will produce some stages under free trade if their gradients are strictly less than those of the other two for some stages. Under a regime of nonfree trade, the stages are allocated among countries from the solution to

$$\text{Min}\left\{\int_{R_1} d_1(z)\,dz + \int_{R_2} d_1(z)\,dz + \int_{R_3} d_1(z)\,dz\right\}, \tag{7}$$

where d_i are the costs including tariffs paid on imported goods in process. If there are positive tariffs on some goods, at least in each country the choice of stages will be distorted. The effects of changes in trade policy can now be modeled.

This approach has been used by Rodriguez (2000) for the case in which the $c_i(z)$ functions lie on a line. Figure 1 reproduces these functions. If we assume there is no fragmentation, country 2 will produce the good and export it to the other two countries, because the area under its function is less than that of the other two countries; in effect, the lower costs of the midstream stages in this country more than compensate for its comparatively high upstream and

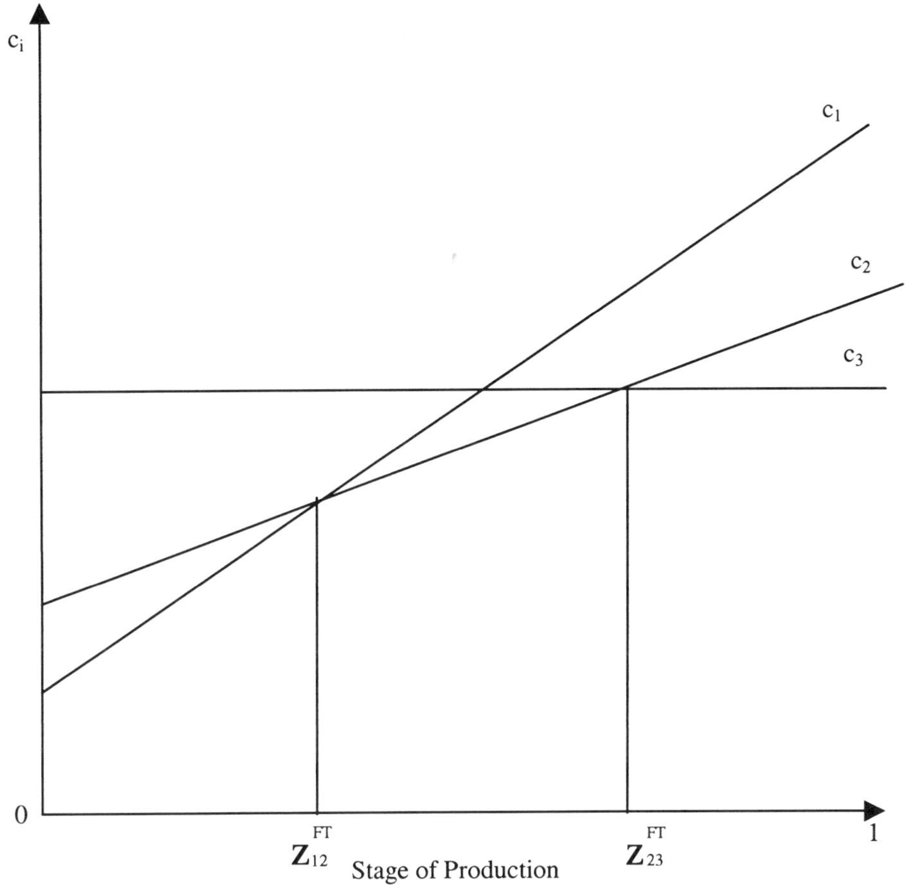

Figure 1: *The Free Trade Equilibrium.*

downstream stage costs. If, instead, we allow fragmentation and free trade, the stages will be spread over the three countries according to the comparative costs of production.

In the case of linear $c_i(z)$ functions, the limits of integration are identifiable. The stages are allocated among the countries from the solution to

$$\text{Min}\left\{\int_{0}^{z_{12}} c_1(z)dz + \int_{z_{12}}^{z_{23}} c_2(z)dz + \int_{z_{23}}^{1} c_3(z)dz\right\}. \tag{8}$$

Country 1 produces all stages from 0 to Z_{12}^{FT} and then exports the good in process at stage Z_{12}^{FT} to country 2. Country 2 produces all stages from Z_{12}^{FT} to Z_{12}^{FT} and then exports the good in process at stage Z_{12}^{FT} to country 3. Country 3 produces all stages from z_{12}^{FT} to 1. The subscripts FT denote values in the "free trade" equilibrium to distinguish them explicitly from values in other situations that are examined below. The two stages, z_{12}^{FT} and z_{12}^{FT}, are the borderline stages (goods), the first determining the division of labor between countries 1 and 2 and the second the division between countries 2 and 3. These commodities partition the continuum into three continuous segments. The borderline commodity may be produced in one or both countries, depending on demand.

Now consider a distorted trade situation in which each country imposes a uniform tariff on all goods t_1, t_2, and t_3, respectively. In this situation, the stages are allocated among the countries from the solution to

$$\text{Min}\left\{\int_0^{z_{12}}(1+t_3)(1+t_2)c_1(z)dz+\int_{z_{12}}^{z_{23}}(1+t_3)c_2(z)dz+\int_{z_{23}}^{1}c_3(z)dz\right\}. \tag{9}$$

This situation is represented in Figure 2. The solid lines represent the free-trade cost functions and the dotted lines the distorted-trade-situation costs. The two borderline stages are now z_{12}^{DT} to z_{12}^{DT}, where DT denotes values in the distorted-trade situation. The tariff escalates as imports of upstream stages are taxed when they enter country 2 and taxed again when they are embodied in the goods that enter country 3. The effect is to shift the borderline stages upstream. Country 1 produces less in terms of the stages, but countries 2 and 3 produce more. All stages continue to be traded internationally. Thus there is a reallocation of the stages per unit of output of the final good. In addition, the volume of international trade in the final good (and all of its embodied stages) is reduced from that in the free-trade situation, since the final good faces a tariff when it is imported into countries 1 and 2 (and its price in country 3 rises relative to that of other goods). This is a consumption effect that imposes a welfare loss in addition to that caused by increased unit costs of production as stages shift as a result of tariffs.

Now consider an FTA between countries 2 and 3. The tariffs are unchanged from the distorted trade situation except that imports from a fellow FTA member are exempt if they satisfy the rule of origin.

Suppose initially, following Rodriguez (forthcoming), that the rule of origin is a percentage rule, $v_{\min}$ ε (0,1). Imports from country 2 into country 3 (including stages produced in country 1 and embodied in the exports of country 2) do not pay the tariff if the rule is satisfied. In this situation the stages are allocated among countries from the solution to

$$\text{Min}\left\{\int_0^{z_{12}}(1+t_3)(1+t_2)c_1(z)dz+\int_{z_{12}}^{z_{23}}(1+t_3)c_2(z)dz+\int_{z_{23}}^{1}c_3(z)dz\right\} \tag{10}$$

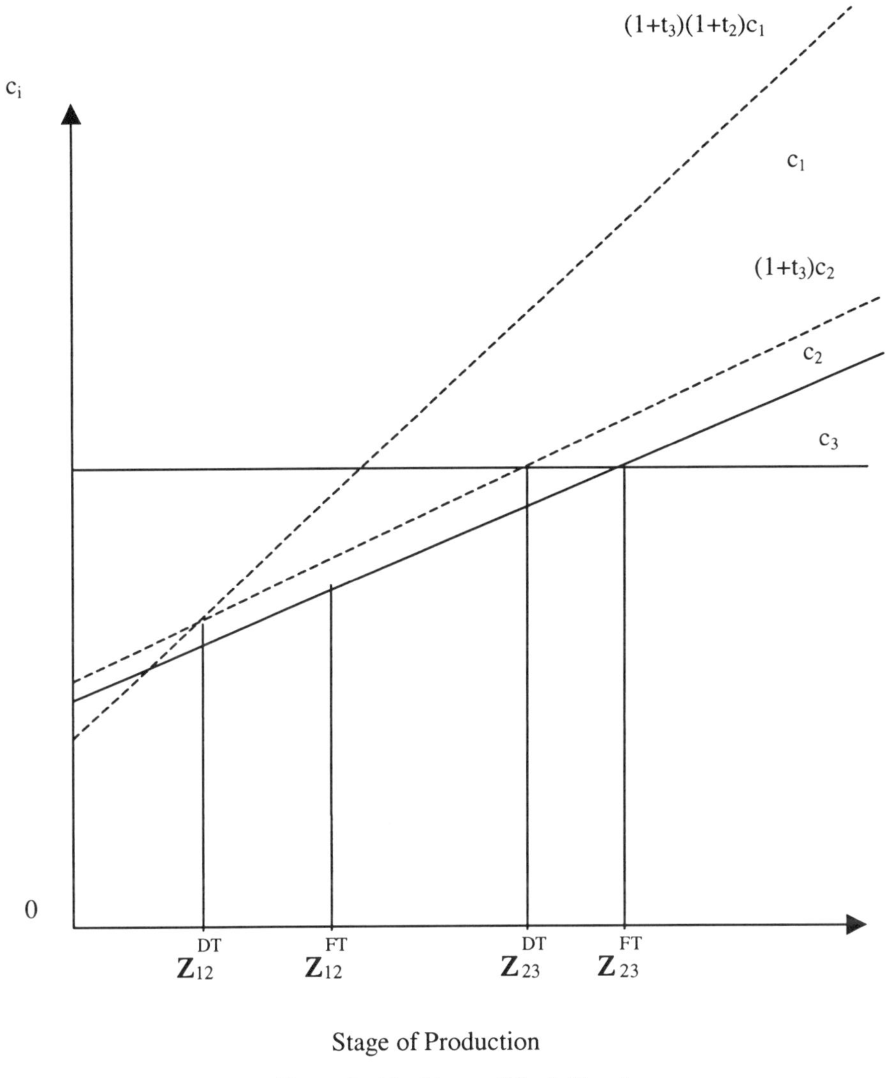

Figure 2: *The Distorted Trade Situation.*

$$\text{subject to } t_3 = o \text{ iff } \int_{z_{12}}^{z_{23}} c_2(z)dz \Big/ \left[\int_0^{z_{12}} (1+t_2)c_1(z)dz + \int_{z_{12}}^{z_{23}} c_2(z)dz \right] > v_{\min}$$

$$= \overline{t_3} \text{ otherwise,}$$

where $\overline{t_3}$ is now the MFN rate, which is distinct from the preferential rate.

There are three possible outcomes in this example.[4] In the first, the rule is satisfied by the pre-FTA value-added percentage in the FTA countries. The goods in process are now traded freely among members, and the margin of production between countries 2 and 3 returns to that which would prevail under free trade. The margin of production between countries 1 and 2 also shifts upstream in the direction of the margin that would prevail under free trade. These shifts lower costs and are a form of trade creation between members. An FTA whose rules of origin are not binding is welfare improving for its members and for the nonmember country.

The second outcome occurs if the minimum percentage can be satisfied by an increase in the area content after FTAs form. As a consequence of the intra-area fall in tariffs, the z_{12} borderline stages shift upstream, and the z_{23} borderline commodity shifts downstream. This is a form of trade diversion within the area and is another illustration of the protective effects of rules of origin, which is noted in the earlier literature. Rodriguez (2000) shows that it leads to inefficient production *within* the area, a phenomenon (which he calls trade regression) that is distinct from trade creation and trade diversion.

The third outcome is one in which the rule cannot be satisfied by profit-maximizing producers. There is no change from the pre-FTA situation. Given a configuration of cost functions, $c_i(z)$, which outcome will occur depends on the choice of $v_{\min}$.

The same method of analysis can be used for a value-added tariff. The base of the tariff is the non-area content, $\int_0^{z_{12}}(1+t_2)c_1(z)dz$. With this rule of origin, the stages are allocated among countries from the solution to

$$\text{Min}\left\{\int_0^{z_{12}}(1+t_3)(1+t_2)c_1(z)dz+\int_{z_{12}}^{z_{23}}c_2(z)dz+\int_{z_{23}}^{1}c_3(z)dz\right\}. \tag{11}$$

This rule is simpler than the percentage rule in economic terms. There is no side constraint.

The effect of substituting a value-added tariff for a percentage rule can be obtained from a comparison of the solutions to equations (10) and (11). The value-added tariff imposes a penalty on all country 1 production, whereas the percentage rule does so only if the rule is not satisfied. In this model, in which there are continuous substitution possibilities, this will reduce country 1 production of some goods under a value-added tariff compared to a percentage rule that is satisfied. On the other hand, the value-added tariff lowers the tariff on all goods compared to the pre-FTA situation and does not provide an incentive to distort production within the area.

The case of linear $c_I(z)$ functions is special. Linearity also restricts the possibilities; for example, there is no two-way trade in intermediates between FTA members. In general there is no reason why these functions should be linear or even increasing or continuous. There are many other possibilities. However, we can compare the results of the percentage rule with those of a value-added tax.

One cannot state that in general the value-added tariff is preferable to the percentage rule in terms of its effects on members' welfare, or vice versa. There will be goods for which a percentage rule is less distorting than a value-added tariff. There will be other goods for which a percentage rule increases area content substantially, whereas a value-added tariff does not overcome the cost disadvantage of producers within the area.

The comparison will depend largely on the choice of the percentage rule, that is, the choice of v_{min}. If a minimum percentage is set low enough it will not be binding for any goods traded within the area. This is the best choice of rule for member countries. However, rules of origin are contentious precisely because all FTAs wish to discriminate against non-area producers. In practice, the chosen percentage will be a binding constraint on producers of some goods but not on producers of others.

For those cases where a percentage rule has been satisfied by increasing area content, one can be more definite. The total reduction in duty under the value-added tariff rule of origin must be less than (or at worst equal to) that under a percentage rule. The remission of duty under a percentage rule is 100%, whereas it can be a 100% remission of duty under the value-added rule only if the area content is 100%. In this environment the value-added tariff is less distorting than the percentage rule or other binding rules that determine origin in one country only.

In addition to being non-arbitrary and non-manipulable, the value added-tariff has other advantages over traditional rules of origin. Furthermore, the same value-added tariff can replace both preferential and non-preferential rules of origin. The one possible disadvantage of the value-added tariff is its greater complexity as compared to the CTH rule and, to a lesser extent, the percentage rule.

5. CONCLUSION

Rules of origin distort production within FTAs and where they apply to non-preferential trade. They need to be reformed. The WTO should pursue reform under the Uruguay Round mandate or under a new mandate if and when a new round is agreed upon. With respect to preferential rules, Article XXIV might be amended to include rules of origin and to require that the rules of origin be a part of the WTO review of new arrangements notified to the WTO. However, harmonization is not the real goal. The goal is to reduce distortions of production that result from protective rules of origin.

The value-added tariff has a number of attractions. It is uniform among commodities, non-arbitrary, and non-manipulable. Less radical proposals than the value-added tariff include a narrowing of the choice of method of assessment. One proposal is that an RTA be required to opt for a single method of assessment (Stephenson 1997). This would rule out industry-specific rules that provide protection to particular industries (e.g., the NAFTA rules for automobiles, clothing, and textiles).

Even if rules of origin are included in the next round, reforms are likely to proceed slowly. GATT countries' inability to agree on preferential rules of origin in the Uruguay Round is an indication of the difficulty. Since then, more rules have been set out in new RTAs and in non-preferential trade, often to provide protection for particular producers. It is likely that the problems of rules of origin will intensify as fragmentation and preferential trade increases.

References

Asia Pacific Economic Cooperation (APEC), (1997), *Compendium on Rules of Origin 1997* (Singapore: APEC Secretariat).

Dixit, A. K and Grossman, G. M. (1982), "Trade and Protection with Multistage Production," *Review of Economic Studies*, 43, 583–94.

Falvey, R. and Reed, G. (1998), "The Economic Effects of Rules of Origin," *Weltwirtschafliches Archiv*, 134, 209–229.

Feenstra, R. C. (1998), "Integration of Trade and Disintegration of Production in the Global Economy," *Journal of Economic Perspectives*, 12, Fall, 31–50.

Ho, Lok Sang (1998), "Country of Origin Rules: Its Origin, Nature and Directions for reform," *Pacific Economic Review*, 3, June, 161–66.

Jensen-Moran, J. (1996), "Trade Barriers as Investment Wars: The Coming Rules of Origin Debate," *The Washington Quarterly*, 19, 239–253.

Krueger, A. O. (1993), "Free Trade Agreements as Protective Devices: Rules of Origin," NBER Working Paper No. 4352.

Lloyd, P. J. (1993), "A Tariff Substitute for Rules of Origin in Free Trade Areas," *The World Economy*, 16, November, 699–712.

Lloyd, P. J. (1994), "Offshore Production and the Base of Taxation," *The World Economy*, 17, September, 719–35.

Pacific Economic Cooperation Council (PECC), (1998), "Report by the Task Force on 'Rules of Origin'," (Santiago, Chile: PECC XII).

Palmeter, D. (1993), "Rules of Origin in Customs Unions and Free Trade Areas" in K. Anderson and R. Blackhurst, eds., *Regional Integration and the Global Trading System* (New York: Harvester Wheatsheaf).

Rodrigez, P. (2000), "Rules of Origin under Multistage Production," *The World Economy*, forthcoming.

Shibata, H. (1967), "The Theory of Economic Unions: A Comparative Analysis of Customs Unions, Free Trade Areas and Fiscal Unions" in C. S. Shoup, eds., *Fiscal Harmonization in Common Markets* (New York: Columbia University Press).

Stephenson, S. (1997), "The Economic Impact of Rules of Origin in the Asia-Pacific Region," *Asia-Pacific Economic Review*, 3, August, 14–27.

Vermulst, E., Waer, P. and Bourgeois, J. (1994), eds., *Rules of Origin in International Trade: A Comparative Study* (Ann Arbor: University of Michigan Press).

Endnotes

[1]Endnotes The International Convention on the Simplification and Harmonization of Customs Procedures, known as the Kyoto Convention, was agreed upon in 1973. This convention was designed under the auspices of the Customs Cooperation Council, an intergovernmental organization. It established that the origin of goods should be determined by the last/final country in which "substantial transformation" of the goods took place. But it did not impose rules to determine the choice of origin rules, and it was unenforceable.

[2]Another related problem arises from fragmentation: MFN imports into a country may contain indirect domestic inputs because domestic factors have been used in a foreign affiliate or plant. Some countries have recognized this problem by introducing special tariff provisions to exempt such goods from duty. For example, the United States has had Offshore Assembly Provisions dating back to the 1930 U.S. tariff, and, more recently, other countries have introduced provisions as offshore assembly operations have increased. Indeed, this problem can be coupled with the problem of handling indirect inputs in the context of imports from an FTA member. Both problems could be resolved by a single border tax that taxed only the value added by residents of foreign countries, with value added in a partner country of an FTA being regarded as non-foreign area content in the case of the trade area.

[3]If the value percentage were lowered toward zero, an FTA would converge to an area in which members traded freely with all countries, not just with other members. This could be a modality for multilateral trade liberalization, which is an alternative to progressive uniform reductions in the scheduled tariff rates, which some countries have used.

[4]In the examples analyzed here all three countries produced some stages in all trade situations. There are other possibilities in which some countries produce no stages or in which countries begin to produce stages after an FTA is formed.

Chapter 15

INTRAFIRM FRAGMENTATION: FUJITSU, LTD.'S PRODUCTION OF HARD DISK DRIVES

Fukunari Kimura[1]

1. FUJITSU, LTD., AND HARD DISK DRIVES

Japan's Fujitsu, Ltd., is one of the world's leading information technology companies producing semiconductors and other electronics devices. Among its other manufacturing activities, it produces hard disk drives (HDDs). The company has been very successful in this area, and it currently commands about 15% of the world HDD market. Its final assembly lines for HDDs are located in the company's affiliates in the Philippines (high end) and Thailand (low end). With upstream production in Japan, Vietnam, and the Philippines, Fujitsu has developed a sophisticated system of intrafirm fragmentation.

Since the mid-1990s the HDD market has been one of the most active markets in the computer-hardware-manufacturing industry. Although prices have been falling rapidly, as a result of the explosive increase in demand, the market has been expanding by 20% per year. Rapid technological progress in the field exerts pressure on firms interested in engaging in research and development (R&D). At the same time, parts of the HDD production process, particularly the final assembly process, are highly capital intensive, and capital must depreciate quickly if manufacturers are to catch up with product innovation. Although the cost of employing university-educated workers is a small part of total production costs, a great many such workers are needed to maintain equipment and ensure quality control. In addition, some areas of the HDD production process use a great deal of lower-skilled labor for manual assembly, testing, and other tasks. In short, R&D firms need a massive amount of capital as well as strong applied research capability to survive in the market.

In the context of the fragmentation literature, Fujitsu's HDD production is interesting for three reasons. First, fragmentation must always be backed up by differences in location advantages across production sites. We have traditionally assumed that less-developed countries (LDCs) have a comparative advantage in providing cheap unskilled labor and that labor-intensive production activities are thus suitable for LDCs in general. This is true in many cases, but Fujitsu utilizes more sophisticated location advantages. Fujitsu's plants are actually very capital- and human-capital intensive. Second, the cost reduction of service links is key in fragmentation, but intrafirm service links have different characteristics from arm's-length service links. Fujitsu takes advantage of cost reduction in both kinds of service links. Third, vertical integration provides opportunities to capture externalities, but at the same time it can make a firm less competitive. Fujitsu keeps its intrafirm links loose and endeavors to promote a competitive environment.

The Philippines' Location Advantages

Fujitsu's Philippines HDD assembly plant was established in February 1995. Fujitsu holds 100% of the plant, which is located in the special export processing zone, Carmelray Industrial Park, Canlubang Calamba, Laguna, 40 kilometers south of Manila. The plant started production soon after it opened. Within two years was producing 500,000 HDD per month, and it employed 5,000 workers, of whom 40 were Japanese. The plant now has about 9,000 workers, more than twice the number of the company's Thailand plant, where 4,000 workers engage in HDD assembly.

Fujitsu's Philippines plant makes sophisticated use of the country's advantages. First, in terms of cost share, the assembly process is highly capital intensive; it uses many processing and testing machines. But, as previously mentioned, rapid product innovation requires quick depreciation of capital. This means that it is preferable to use machines intensively and accelerate depreciation. It is relatively easy to do this in the Philippines, where workers can be found to work on a three-shift, 24-hour basis; in developed countries (DCs) running this type of round-the-clock operation is becoming increasingly difficult and expensive. Although it may sound paradoxical in light of the traditional Heckscher-Ohlin-type argument, the Fujitsu case suggests that LDCs could be suitable for extremely capital-intensive activities.

Second, of all Association of Southeast Asian Nations (ASEAN) countries, the Philippines has the largest number of engineers with higher education, and the salaries they can command are much lower than those of their counterparts in DCs. This is convenient for high-tech companies', which need to employ a large number of engineers in order to maintain quality control. The Fujitsu plant had about 400 engineers on its payroll in early 1998 and has added roughly 100 new ones every year since. Other advantages of the Philippines are that English-speaking unskilled workers are abundant, and that there is no serious conflict between capital and labor.

Third, HDD assembly requires various kinds of customized parts and components, and thus a plant's proximity to parts and components suppliers is an advantage. Although indigenous manufacturers are still immature, a large number of upstream suppliers in Japan and other DCs have already established affiliates in the Philippines and have started forming a large agglomeration.

Fourth, Fujitsu has aggressively taken advantage of Philippine government policies such as a preferential taxation scheme for new investment and a zero-tariff scheme in the export-processing zone. And the industrial park was constructed and is managed by a Japanese general trading company (GTC).

Although the plant's electricity comes from its own power generator (buying electricity from outside is a back-up), otherwise, the physical infrastructure, including air transportation, works quite smoothly. This creates a favorable environment not only for Fujitsu but also for upstream suppliers, and that has aided in the formation of an industrial agglomeration.

In early 1998 Fujitsu publicized a plan to establish a new plant in the First Philippine Industrial Park Ecozone in Tanauan, Batangas, to produce magneto-resistive (MR) heads. Such MR heads formerly came from Korea, but the new plant has been gradually taking over their manufacture. The plant now employs about 300 workers. MR heads contain firm-specific high-level technology and thus generate a great deal value added. The fact that Fujitsu has relocated its production activities indicates that it is well aware of the location advantages of the Philippines.

Vietnam's Location Advantages

Fujitsu's Vietnam plant is one of the few export-oriented subsidiaries in Vietnam. It produces printed-circuit-board assemblies (PCBAs) and multilayer printed wiring boards (PWBs), which are in the upstream of the value chain. The plant was also established in 1995 and currently employs 2,700 persons, 19 of whom are Japanese. It is located in the Bien Hoa Industrial Zone II, Dong Nai Province, which is about a one-hour drive from Ho Chi Minh City.

The Vietnam plant is a 100% import, 100% export plant. Because Vietnam is in the early stage of industrialization and has a less-than-favorable policy environment, there are no parts and components providers in the vicinity. Hence, the plant imports all parts and components from Japan and other foreign countries. We can assume that most of the parts and components used in Vietnam are less customized than those used in the Philippines and thus that the Vietnam plant's proximity to suppliers is less important than the Philippine plant's. Print boards produced here are exported mainly to Fujitsu's own assembly plants in the Philippines and Thailand.

Despite the lack of agglomeration benefits, Fujitsu's Vietnam plant has taken advantage of labor-market condition in Vietnam. There is an excess of engineers in the country, and, therefore, by virtue of the company's image as an information technology leader, it can employ high-quality engineers at a very low cost. The starting salary for an engineer in Vietnam is less than US$200 per month. The company currently employs about 200 engineers in printed-board manufacturing and is trying to push the scope of work toward the upstream, including manufacturing design.

The supply of unskilled labor in Vietnam is also abundant. Job candidates come through the local labor bureau, but strict internal selection process will follow. The starting salary for unskilled laborers is less than $60 per month (or about $90 after adding various benefits), and the quality of labor is high.

Workers mount components, test products, and repair defective products. Virtually no labor-relation problem exists, and the feasibility of implementing a 24-hour operation again helps accelerate capital depreciation.

Although the policy environment for incoming foreign direct investment is not very favorable in Vietnam compared with other ASEAN countries, Fujitsu is considered a high-caliber company in Vietnam. As a 100%-owned subsidiary, the managerial grip is solid, and Ho Chi Minh City's policies are probably more flexible than those in the northern part of the country.

Service Links

The cost reduction of service links is an essential condition for fragmentation. In the case of HDD production, one important technological condition is that most parts and components are small and have high values per weight, so they can be carried by air transportation. Hence, much as it is in semiconductor manufacturing, the fragmentation of production blocks for HDD production is relatively easy.

Fujitsu has also taken advantage of the reduction of telecommunication costs, both for arm's-length and intrafirm links. In the case of arm's-length links, the company makes extensive use of Internet communications to seek the highest quality parts and components at the lowest possible cost. The markets of noncustomized parts have recently become better integrated.

Global communications among the Fujitsu units located in different parts of the world is also very important to the company's operation. The company has developed its own electrified network connecting the R&D headquarters in Kawasaki, production plants in Nagano and Yamagata, and production plants in Thailand, the Philippines, Vietnam, and elsewhere. E-mail communication takes place at all levels, and teleconferences are conducted daily. Although communicating in this way is expensive, Fujitsu seems to have captured economies of scale in establishing its communications infrastructure, which is a key component of intrafirm fragmentation.

Maintaining a Competitive Environment

Vertical integration provides an opportunity to internalize externalities, but at the same time it carries risks of slackness due to insufficient competitive pressure. Fujitsu has tried to introduce a number of arrangements to keep the intrafirm environment competitive. One such arrangement has been the introduction of competition between intrafirm procurement and outside suppliers. In the case of printing boards for HDD, roughly 70% of the printing boards used in the company's assembly plants in the Philippines and Thailand are provided by the Vietnam plant, while the rest are procured via intentional outsourcing. Obviously the company needs to watch possible collusion between insiders and outsiders, but the arrangement seems to be working well so far.

Future Prospects

The future of manufacturing activities related to information technology is never certain. Although the market has recently grown explosively, no one knows what will happen in five or ten years. The HDD technology has successfully increased the density of memory in a limited space. However, scanning technology depends on the physical speed of disk rotation, which potentially limits the memory storage function. Some other technologies such as flash memory may overcome such bottlenecks in the future.

If the HDD market stops growing or even shrinks, what will that mean? The fear that this will occur is actually accelerating the speed of capital depreciation, which generates a business environment in which only big companies abundant in capital can survive. Quick depreciation means that producers can switch production sites without much cost. HDD producers must always be aware of the advantages of various locations as well as of market conditions, and they must consider the possibility of moving production sites or switching products. Only those manufacturers whose managers can make bold, quick decisions can survive in today's HDD market.

Endnotes

---

[1]Most of the information on Fujitsu was obtained from the company's Web site and the author's visits to the Philippines and Vietnam plants in March 1998 and August 2000, respectively. The author would like to thank company presidents Nagamine and Kawashima and their staff. Any errors in the study are, of course, on the author's responsibility.

Chapter 16

THE DEVELOPMENT PATTERN OF TAIWAN'S BICYCLE INDUSTRY

Wan-wen Chu

1. INTRODUCTION

Over the past 30 years Taiwan's bicycle industry has thrived. It accumulated production experience following Word War II, during the import-substitution period (1951-68). When export opportunities emerged in Taiwan in the early 1970s the industry was able to react quickly in terms of both output expansion and entrepreneurial responses. With government help, it consolidated itself after the U.S. demand surge subsided in the mid-1970s. It became the world's top exporter in 1980, replacing Japan. By 1986, under mounting pressure to upgrade, the industry had begun producing high-end products by using more imported components. Subsequent cooperative efforts among assemblers, parts producers, and the government helped to restore the parts self-sufficiency ratio to its previous level. Meanwhile firms moved most low-end production to the People's Republic of China (PRC).

External relations have been important in the development of this industry. Export orders and related assistance provided by foreign buyers initiated the growth process. Reliance upon imported components has been critical. However, to really understand the industry it is necessary to examine first the determinants of its structure, as we do in Section 2. Section 3 examines the industry's expansion path. A more detailed discussion as to why and how the growth occurred appears in Section 4. Section 5 contains some concluding remarks.

2. STRUCTURE OF THE BICYCLE INDUSTRY: VERTICAL DISINTEGRATION

Taiwan's bicycle industry consists of two sectors: bicycle assembly and parts production. A bicycle contains over 100 different parts. Total vertical integration entails a firm that produces and assembles all parts in house. Nonintegrated assemblers purchase all their parts from various independent parts producers. Nonetheless, each case is related to how the parts industry is organized. For example, nonintegrated assemblers can exist only if they can always find necessary parts on the market.

Economies of scale and scope are certainly involved. The various parts required differ, and hence each has its own distinct production technology and scale economy. The size of the scale economy depends on the part. A totally integrated assembler has to produce a quantity large enough to fulfill all these scale economies at once, so as to achieve production efficiency. This scale requirement also implies that the assembler must be able to acquire enough of the sales opportunities, capital, and managerial and skilled human resources that the numerous production processes involved demand. This would no doubt be a daunting task for an assembler at the early stage of development.

If the structure is not integrated, the scale requirement takes a different form. The benefits of aggregate economy manifest themselves. The fulfilment of the scale economy of the production of various parts now depends upon the total output of the industry, not on that of a single firm.[1] This is also called Marshallian externality; that is, the average cost of an assembled bicycle declines as larger industry output allows for a finer division of labor and scale economies of parts production to be realized. It also implies that, on the other hand, if industry output is too low, the advantage of using the market may be limited.

Williamson's (1985, pp. 90–96) notion of asset specificity is relevant here. Bicycle structure has remained basically unchanged for a long time, though frames and style vary. As a result, bicycle parts have been quite standard as well. With standardized parts, when the number of suppliers and the amount of such parts available on the market increases, the risk and uncertainty associated with relying on outside sourcing decreases. Actually, for zero-asset specificity, Williamson assumes that the cost of using the market will be unambiguously less than that of using internal organization, for incentives provided by the market are always stronger than those of bureaucracy, other things being equal. So far, this trade-off refers to the production of just one bicycle part, and, according to Williamson's theory, the balance is already in favor of using the market. And, as previously mentioned, a bicycle comprises over 100 parts.

The governance costs of internal organization rise as the number of parts produced in house increases. The burden on the management's capacity is related to the size of the major assembling part of the operation. Bicycle production, however, involves rather simple technology and very labor-intensive assembly work, and has always been categorized as a light industry and dominated by smaller firms.[2] Even the largest bicycle assembler in Taiwan is a small to medium-sized enterprise (SME).

Thus, in the trade-off between the cost of using the market and that of using internal organization, both asset specificity and governance capacity factors tilt the balance toward the use of the market.[3] The benefits of aggregate economy and powerful incentive provided by the market dominate.[4]

Therefore, if Williamson's assumption holds, the bicycle industry will not be vertically integrated. The actual extent of integration may vary among countries, depending upon the particular conditions of the country involved and the resulting trade-off between the costs of the two options. The conjecture here is that a competitive bicycle industry will not be a vertically integrated one, and this is confirmed by evidence from Japan and Taiwan.

The structure of the bicycle industry in the leading country in the world market should provide us with a good indication as to how this trade-off works. The leading bicycle exporting country in the postwar period was Japan before 1980 and Taiwan afterwards. The industrial structure of the bicycle industry is similar in the two countries. Almost all assemblers produce only the bicycle frame, in which changes in bicycles' style are mainly embodied, and procure other parts from the network of numerous parts suppliers.

Parts producers are mostly specialized and independent, but they form long-term supply relationships with assemblers. The parts industry as a whole is quite prosperous, and it exports on its own to maintain scale economies. It is said that no country's bicycle industry could prosper without a healthy parts sector. Actually, in Japan, when the volume of bicycle exports began to decline as a result of rising wages after 1980, the bicycle-parts sector was able to sustain its competitive position in the international market through intensified research and development (R&D) efforts. Both countries are known to have a significant number of SMEs, which either form networks of their own or build relationships with larger firms. That is, the presence of the network not only complements the working of the market but also reduces the cost of using it in the two countries.

3. EXPANSION PATH

This section briefly describes the development history of the bicycle industry in Taiwan.[5] We discuss the characteristics of this growth process in the next section.

Import substitution, 1951–68. Simple bicycle assembly began in Taiwan around 1946. When the country was first permitted to trade with Japan in 1949, bicycle importing began to interrupt this local production. The government then banned the import of bicycles and allowed the imports of only 12 key parts. The next few years, 1952 to 1954, saw the emergence of four major bicycle assemblers and of manufacturers of various bicycle parts. However, competition from "underground" bicycle assemblers and stagnant local demand undermined the four major producers. Between 1958 and 1965 all four ceased production, and bicycle output remained below 30,000 until the export drive began in 1969.

The export boom: 1969–77. Taiwan began to export bicycles to the U.S. market in 1969. Exports rose sharply from 107,000 in 1970 to more than 1 million sets in 1972. However, the surge in export to the North America provoked response from local producers (i.e., charges of dumping and unsafe products). Luckily for Taiwanese exporters, the antidumping case did not stand in the United States. The U.S. government passed a new set of safety guideline for bicycles in 1974, which helped the industry to improve its products.

After the energy crisis. The Taiwanese government helped the bicycle industry establish a set of product standards and undertook the responsibility of inspecting exports. These and other measures helped restore market order and stabilize Taiwan's bicycle exports to the United States. Exports increased continuously from 1978 until 1986. In 1980 Taiwan overtook Japan as the world's top exporter, and Taiwan reached its exporting peak in 1986, when it exported 10 million bicycles. The growth in quantity was, however, accompanied by a slight decline in unit price.

Transformation since 1986. Sharp currency appreciation and other major macroeconomic changes occurred in Taiwan around 1986, and they caused a subsequent decline in bicycle export volume. These changes, however, prompted the industry to upgrade its product mix, and as a result the export unit price rose rapidly, by 62%. Taiwan's bicycle exports now rank first in the world, and its output ranks fourth. Nonetheless, after 1990, most bicycle producers moved their low-end production to the PRC while keeping the high-end production in Taiwan, further reinforcing the declining trend in the quantity of Taiwan's bicycle exports.

4. CIRCUMSTANCES OF GROWTH

A Surge in U.S. Demand

Before 1969 Taiwan's bicycle industry sold mainly to the domestic market; afterwards its main market was overseas. The ratio of export to output increased from a negligible amount to over 90%, and has remained at that level up to the present time. The switch in market orientation occurred around 1971 and was due to the sudden surge in demand in the U.S. market. In 1972 several American bicycle importers came to Taiwan and placed original equipment manufacturer (OEM) orders for almost 1 million bicycles. The U.S. domestic market demanded around 4 to 5 million bicycles per year in the early and mid-1960s and around 7 million in the late 1960s. The energy crisis in the early 1970s brought about a big surge in demand; in 1972–74 demand jumped to around 13 to 15 million bicycles per year. The figure, however, dropped quickly back to 7 million in 1975, after the energy crisis ended. The surge in demand was partly met by imports, which amounted to around 4 to 5 million bicycles a year in 1972–74 and then dropped back to 2 million after that. Because demand exceeded supply by about 2 to 3 million bicycles in the 1972–74 period, some importers came to Taiwan to place large orders during those years.

The effect of the influx of export orders was evident in Taiwan. The sectoral growth rate in terms of quantity averaged about 13% per year from

1977 to 1990, after the initial great leap forward. That rate, of course, is much higher than that of the 1953–68 period, during which the average was about 4%. By becoming the world's major bicycle producer/exporter, the industry was able to achieve the kind of fast growth it would not have been able to attain if it had concentrated only on the domestic market.

Not Direct Foreign Investment, but Learning by Exporting

During the early stage of export-led growth, U.S. buyers provided Taiwanese manufacturers with various kinds of help. Besides placing orders, the buyers usually supplied advice, product specifications, production blueprints, and even parts. They also monitored progress, checked products, offered financing via letters of credit, and marketed the products in developed countries. The local producers were responsible for only the production stage of the operation in the early learning-by-exporting period. Bicycles produced for the U.S. market differed from those produced for the local market in terms of quality, price, and style. More than a technology transfer was involved, however. The United States actually initiated the growth process by making the decision to source from Taiwan.[6]

This is a case study of a relatively small industry. Bicycle and bicycle parts exports occupied about 2.2% of Taiwan's overall exports in 1990. However, this pattern of learning by exporting is representative of Taiwan's labor-intensive industries during this early stage. Overseas buyers played an important role in the export growth of Taiwan's other light manufactures, such as textiles and apparel. It is this group that moved its production base from Taiwan and other Asian Newly Industrializing Economies (NIEs) to the PRC and Southeast Asia in the 1990s.

Why Did Taiwan Receive the Export Order?

The surge in U.S. demand and the appreciation of the Japanese yen helped push U.S. buyers to look past Japan for sourcing. Why did they choose Taiwan? Besides factors advantageous to development in general, which cannot possibly be covered here, there must have been factors specific to this industry that made Taiwan a particularly favorable choice.

One such factor was the local industry's ability to respond quickly to opportunities. The fact that the volume of its bicycle exports increased by 10 times in two years attests to this ability, which consists of two aspects, one concerning production capability and one concerning entrepreneurial resourcefulness. The former probably resulted from bicycle and parts manufacturers' accumulated learning during the import substitution years. The latter was clearly manifested as numerous SMEs quickly emerged to undertake production when the opportunity arose, but its cause relates to historical, institutional and policy factors and cannot be fully discussed here.

Between 1971 and 1976 the number of bicycle production firms in Taiwan increased by 60%, rising from 279 to 447. In other words, hundreds of new enterprises sprang up to seize the opportunities available.[7] This influx of entries reflects widespread entrepreneurship in the overall economy, which was probably due to Taiwan's accumulated commercial experience.[8]

In Taiwan these new and old enterprises together were able to quadruple the export level of bicycles and triple the output level of parts within one year's time in 1971–72. The development of the two sectors, bicycles and parts, has been quite even. Though we do not know the number firms for each year between 1971 and 1976, judging from the way production multiplied starting in 1972, many firms must have entered right at the beginning of the export boom. The response was immediate.

Moreover, most of these enterprises were specialized. Each scrambled to grab a piece of the pie, and the economy of agglomeration made it work. That is, the sudden arrival of opportunity spurred enough small businesspeople to set up shops in a short period of time, which in turn fulfilled the large orders and the critical scale requirement and allowed for a sufficiently fine division of labor at that time. Unspecialized manufacturers probably lost out in the competitive process. Over time the Taiwanese bicycle industry became sufficiently competitive and its prospects looked sufficiently promising to allow manufacturers who were already there to concentrate on this line of business and thus remain specialized. Thus the size of the export orders and the immediate entrepreneurial response made it possible for Taiwan to establish the bicycle-assembly and parts sectors successfully at the same time.

At present over 90% of Taiwan's bicycle assemblers produce no parts themselves, except for the frame. The parts producers are also highly specialized, with over 90% of them producing only one type of bicycle part.[9] The division of labor within a bicycle assembly plant is rather Taylorist; that is, each worker performs only one simple task on an assembly line. Besides the main assembly line, which assembles the bicycle, there are two separate lines that assemble the wheel and the frame.

On average, each bicycle producer has about 60 parts suppliers, and each parts producer supplies up to 20 bicycle assemblers. Thus the degree of dependence on a particular supplier or buyer is not particularly high. In the past 20 years parts producers have also managed to export about 50% of their products, which helps them maintain their independence from assemblers as well as their economies of scale.

The fact that the sector's output could increase 10 times in just a couple of years, with only a slight decrease in the parts self-sufficiency rate, indicates that the economy can respond quickly to expansion opportunities. This is true in terms of entrepreneurship, technology, labor supply, and supporting infrastructure in both hardware and software. The bicycle industry was also the right kind of export industry for Taiwan at its stage of development. The technology involved was simple enough for Taiwan to master but would have been beyond many other less developed countries, and the required inputs could also be readily supplied by supporting industries.

Transformation and Self-sufficiency in Parts

Between 1966 and 1970 the parts self-sufficiency rate averaged 86.3%, the parts sector's export ratio 53%, and the growth rate of its export 32%. Thus the parts industry was not only able to satisfy most local bicycle-production needs but was also able to sell abroad in that early period before the bicycle export boom.

When bicycle exports began to take off in 1971 the parts self-sufficiency rate showed a slight decline. The decline occurred partly because the local parts sector could not expand fast enough but also because the bicycles to be exported were higher quality and hence required higher-quality parts, some of which had to be imported. The government relaxed import restriction on certain bicycle parts in 1970, obviously in order to satisfy the bicycle sector's export needs.

The parts self-sufficiency rate averaged 67% between 1972 and 1974, a significant drop from 86% between 1966 and 1971; but it regained ground and averaged 79% from 1975 to 1985. This rate remained stable in 1975–85 because the industry was dependent upon imports of key components. Japan still holds an edge in some technologically sophisticated components and high-end products, such as the derailleur, brakes, chain wheels, and hubs.

Another dramatic decline in this rate occurred after 1986. The New Taiwan (NT) dollar appreciated against the U.S. dollar by about 40% from 1986 to 1988. This occurred at a time when 78% of the industry's exports went to the U.S. market. Other important changes also occurred in the overall economy, such as rising wages and economic liberalization. The negative impact was evident in that exports dropped from over 10 million bicycles in 1986 to 7 million in 1988.

A turnaround, however, occurred in 1990, when the export volume recovered to over 9.4 million, and the unit export price rose from NT$1,630 per set in 1986 to NT$2,644 in 1990. The average growth rates for bicycle export volumes and values were about the same in the early 1980s, but that of export volumes turned negative after 1987, while that of values was 13% in 1987–90.

At the same time the parts self-sufficiency rate dropped significantly. It averaged around 78% in 1981–85 but decreased to 52.9% in 1990. The obvious reason for this was that upgrading required using more expensive parts imported from Japan. Japanese parts manufacturers, especially Shimano, have become quite competitive in recent years. Shimano often bundles its products in sales, but it also provides assembling instructions.

Thus buying parts from Shimano also makes upgrading easier for some Taiwanese bicycle assemblers.

The lack of vertical integration made the industry quite flexible in responding to external changes. Institutional flexibility made the switch to using Japanese parts, and hence upgrading, easier, but it also weakened the ties between the assembly and parts sectors in Taiwan. With trade protection no longer possible, the government promoted cooperative R&D efforts to develop new parts technology and to replace Japanese imports. These efforts raised the self-sufficiency rate in terms of parts production to 72.7% by 1996.[10]

Another development has contributed to the further fragmentation of the industry. Most of Taiwan's major bicycle assemblers built production facilities in the PRC during the 1990s. For example, the leading firm, Giant Co., began assembling bicycles near Shanghai in 1994, with a planned annual capacity of 1.5 million bicycles. Many parts manufacturers followed suit. At the moment most of them divide their operations in such a way that they produce low-end products in the PRC and high-end ones in Taiwan. In 1996 bicycle imports from the PRC and Taiwan occupied 51.5% and 41.2% of total U.S. bicycle imports in quantity, and 28.6% and 61.9% in value, respectively.[11] A large portion of the exports from the PRC come from Taiwanese subsidiaries. The ratios clearly indicate that bicycles exported from the PRC have a lower unit price than those exported from Taiwan. It is likely that some of these Taiwanese subsidiaries will turn to the potentially large local PRC market after concentrating on export for a period of time. This testifies to the strength of the trend toward globalization.

5. CONCLUSION

Is the case of Taiwan's bicycle industry one of increasing fragmentation of production on a global scale? In some respects it is. Taiwan's bicycle industry prospered by seizing the opportunity to satisfy demand in the United States and other developed countries. It has always had to import a certain percentage of components from Japan to accomplish this task. After the local wage level rose and currency appreciated, Taiwanese firms moved most low-end production to the PRC.

On the other hand, the industry was able to prosper because assemblers could rely upon a well-developed local parts supplier network, which can provide around 70% of the components required most of the time. When low-end production moved to the PRC, the assemblers and the parts suppliers moved together. As a result, the core capabilities of this industry have not been overly fragmented.

References

In English

Amsden, Alice H. (1985), "The Division of Labour is Limited by the Rate of Growth of the Market," *Cambridge Journal of Economics*, 9, 271–84.

Coase, R. H. (1937), "The Nature of the Firm," *Economica*, 4, 386–405.

Chu, W. (1997), "Causes of Growth: A Study of Taiwan's Bicycle Industry," *Cambridge Journal of Economics*, 21/1, 55–72.

----- and J-J. Li (1996), "Growth and Industrial Organization: A Comparative Study of the Bicycle Industry in Taiwan and South Korea," *Journal of Industry Studies*, 3/1, 35–52.

Cycle Press Inc., *Cycle Press*. Various issues, Tokyo.

Levy, B. (1991), "Transaction Costs, the Size of Firms and Industrial Policy," *Journal of Development Economics*, 34, 151–178.

Stiglitz, J. E. (1994), *Whither Socialism?* (Cambridge: MIT Press).

Wade, R. (1990), *Governing the Market: Economic Theory and the Role of Government in East Asian Industrialization* (Princeton: Princeton University Press).

Williamson, O. (1985), *The Economic Institutions of Capitalism* (New York: the Free Press).

In Chinese

China Credit Information Corporation, *Taiwan's Industry Yearbook*, various years, Taipei.

Chu, W., and A. C. Tung (1990), "The Impact of Currency Appreciation on Taiwan's Industries" (Taipei: Council for Economic Planning and Development).

Dai, C. Y., and H. J. Lee (1989), "Studies on the Development Strategy of Taiwan's Bicycle and Parts Industries," commissioned by the Industrial Development Bureau (IDB), Taiwan Institute for Economic Research (TIER), Taipei.

Feng, C. H., and W. W. Yang (1992), "A Study of the International Bicycle Market," Mechanical Industry Research Laboratories, Industrial Technology Research Institute (ITRI), Hsin-chu.

Lee, Y. C. et al. (1988), "Studies on How to Assist the Upgrading of Taiwan's Traditional Industries: the Bicycle Sector," commissioned by IDB, ITRI, Hsin-chu.

Lee, H. J. et al. (1989), "Studies on the Development Strategy of Taiwan's Bicycle and Parts Industries," commissioned by the IDB, TIER, Taipei.

Liao, S. L. (1994), "Further Studies on the Development Strategy of Taiwan's Bicycle and Parts Industries," commissioned by the IDB, TIER, Taipei.

Machinery Industry Research Laboratories (MIRL) (1997), *Development Trend of the Transport Equipment Industry,* commissioned by IDB, ITRI, Hsin-chu.

Endnotes

[1]If there is a monopoly this difference between the industry and the firm will dissolve, but a monopoly's adverse effect on competition is not desirable.

[2]Compared to the bicycle industry, the automobile and motorcycle industries are much more capital intensive and technologically complicated, and they use nonstandardized parts and brand-name designs. Thus both governance capacity and asset specificity factors allow more room for in-house parts production in these two heavier industries.

[3]The cost of using the market and the cost of using internal organization depend upon the economic and social conditions of the country concerned, and hence will vary from country to country, and even from time to time.

[4]In the discussion so far it has been assumed that the standardization of parts implies zero-asset specificity. More important, the conclusion is also based on the assumption that, with zero-asset specificity, the cost of using the market is less than that of using internal organization. The case for the use of the market is made stronger by the fact that there are more than 100 parts involved in making a bicycle.

[5]See Chu (1997) and Chu and Li (1996).

[6]The role of foreign buyers here closely resembles that of the multinational corporation (MNC) making direct investment in LDCs. Except for the fact that the foreign buyers did not commit their capital directly, the roles foreign buyers and the MNC played are similar.

[7]It is likely that many new firms also replaced old ones, and hence the total number of entrants could be much larger. Most of the leading assemblers of today emerged at that time. In contrast, at the same time, the number of firms in the Korean industry declined.

[8]See Levy (1991) for a discussion of the possible effects of earlier commercial experiences in Taiwan.

[9]See Chu and Tung (1990).

[10]See MIRL (1997, 3–29).

[11]See MIRL (1997).

Chapter 17

TAIWAN'S INTEGRATED CIRCUIT INDUSTRY

An-Chi Tung

1. GROWTH OUT OF FRAGMENTATION

Until the 1980s Taiwan's integrated circuit (IC) industry was small and relatively undeveloped, but in the past 20 years it has expanded impressively. Its growth has been key to the development of Taiwan's electronics industry, which has been the most dynamic manufacturing industry in East Asia in the past two decades.[1] ICs' share in Taiwan's total manufacturing production rose from 0.37% in 1981 to 3.44% in 1999. Since 1995 the island has been the world's fourth-largest semiconductor producer, after the United States, Japan, and South Korea; in 1999 Taiwan's shares of dynamic random access memory (DRAM) and foundry services in world marketplace rose to 16% and 67%, respectively.

The remarkable growth of the IC industry has much to do with the vertical disintegration of IC production, which now comprises four distinct stages. Design is in the upstream, fabrication is in the midstream, and packaging and testing are in the downstream, and design can be further split into circuit designs and intellectual property (IP) services. In the second half of the 1960s the separation of packaging from manufacturing brought the lower end of the IC industry to Taiwan. In the late 1980s a Taiwanese firm pioneered the foundry concept by disintegrating manufacturing from design. The worldwide impact was immense, as this action boosted the "fabless" design industry and also initiated the foundry business. The late 1990s saw a finer degree of labor division within the design sector. This new development prompted the foundry sector to upgrade and expand and spurred the growth of the entire industry in Taiwan.

Because of the possibility of vertical disintegration, the value of IC production or exports varies according to the breadth of coverage. Taiwan's total IC products were valued at $U.S. 3.65 billion in 1998. (This value included the production value of design and manufacturing but did not include foundry services). Yet the production value of the industry was US$8.45 billion (or US$9.82 billion, if the value of foreign-owned packaging firms' dices are included). The industry total was 2.69 times the product total, since the value of foundry services (33.10% of industry total), packaging (19.05%), and testing (4.62%) were added on top of design (16.55%) and nonfoundry manufacturing (26.67%). Similarly, the export of IC products manufactured by domestic firms amounted to only US$1.84 billion in 1998, whereas total IC exports amounted to US$6.87 billion. The difference between the two figures consisted of packaging exports (US$3.6 billion) and reexports (US$1.45 billion). But the value of foundry exports (about US$1.82 billion), that is, the value of foundry services provided to foreign design firms, did not show in these figures at all (ITIS project, Ministry of Economic Affairs 1999). Therefore the growth of the IC industry must be understood in the context of the fragmented production structure.

2. Ascent of the Industry

The commencement of IC production in Taiwan dates back to 1966, when the U.S. firm General Instruments set up the first IC plant on the island.[2] Philips, Texas Instruments, RCA, and others followed suit later on. These foreign companies were attracted by Taiwan's low wages, which yielded up to a 50% reduction of total manufacturing costs (Tyson 1992, p. 91). Taiwan welcomed foreign direct investment (FDI) because it allowed for a more efficient use of the abundant domestic labor supply. At this point, however, the overseas firms kept their design and manufacturing technologies in the home countries. It follows, then, that backward linkages and the Schumpeterian process of imitation and swarming on the part of local firms were observed only sporadically.

When Taiwan's inexpensive labor supply was exhausted in the early 1970s, the state started to nurture domestic production capacity in the IC industry, with a view toward upgrading the economy's overall technological level. In 1975 Taiwan acquired complementary metal oxide semiconductor (CMOS) fabrication technology from RCA via the semi-public Electronics Research and Service Organization (ERSO). In 1980 the technology, along with the necessary human capital and funding supports, was transferred to a local private firm United Microelectronics Corporation (UMC). For the first time a local firm was able to produce for the fledgling domestic consumer electronics markets. UMC's immediate success resulted in a wave of new investments in the design sector. Some new design houses were set up as subordinates to Silicon Valley start-ups that had been established in Northern California by overseas Chinese.

ERSO moved on to develop very large-scale integrated circuit (VLSI) fabrication technology and transfer it to yet another spin-off from itself, Taiwan Semiconductor Manufacturing Corporation (TSMC) in 1987.[3] Two new industries arose. TSMC, a pure foundry, became the first fabrication plant in the world that did none of its own designing. The company thereby lowered the entry barrier for fabless design houses on the island and elsewhere. Small chip design companies that could not afford to build their own fabrication plants no longer needed to compete to buy unused capacity from rival integrated device manufacturers (IDMs), which conduct both design and fabrication activities. The number of design houses in Taiwan soared, rising from 4 in 1986 to 40 in 1987.

Taiwan's IC industry diversified into DRAM production in the 1990s. Almost a decade earlier the Korean *chaebols* had moved into the memory chip markets and become internationally respected suppliers. By contrast, Taiwanese firms, which were constrained by their small size and lack of financial strength, specialized in either niche markets (such as application-specific ICs) or low-end products. However, as incumbent IC firms proved their profitability, large domestic capital started to rush into new corners of the IC industry,[4] especially into DRAM production, which requires massive initial capital input as well as continuing R&D investments. The value of DRAM manufacturing in Taiwan rose 30.3 times during 1991–99.

A number of interrelated developments in the late 1990s had wide-ranging and long-term effects on the industry. First, the disintegration of IP services from traditional chip design operations prompted the rise and proliferation of dedicated IP design companies (i.e., the "chipless" companies). As IC manufacturing technologies had far outpaced design technologies, the provision of reusable, portable, modularized, and reprogrammable IP became necessary. In response, the leading foundries sought to upgrade. TSMC, for instance, which was originally a straight contract manufacturer, has become a provider of specialized services in order to be a "virtual fab" to customers. In turn, some IDMs started to outsource to dedicated foundries rather than constantly expanding and upgrading, basing their decision to do so on cost and efficiency concerns.[5] In the 1990s the foundries were already accounting for 5% to 10% of global semiconductor production; the figure could hit 35% by 2010 (Dataquest 1999). Much of the expansion will be on the part of TSMC and UMC, the two largest foundry plants in the world, which were already responsible for 52% of the world market in 1999. In the meantime, domestic DRAM fabrication grew steadily, despite the oligopolization of the world DRAM market and the fact that two local DRAM manufacturers dropped out.[6] The combined shares of DRAM fabrication and foundry services in IC manufacturing thus increased from 37.7% in 1991 to 94.58% in 1999.

3. TSMC, THE INNOVATOR

TSMC revolutionized the entire IC industry in the 1980s by identifying the need for independent chip makers. A decade later the same company participated in the third stage of production fragmentation, and it may reshape the chip industry once again.

By the early 1980s there was a strong need for an advanced chip manufacturing plan, as UMC, the only manufacturer at that time, was unable to meet the technical requirement of the new design houses that worked on VLSI technologies. Yet it was not economical to build a new plant small enough to take care of just the needs of these design houses, and it was financially unfeasible to build an IDM large enough to reap scale economies. Morris Chang, a 25-year veteran of Texas Instruments, was chairperson of UMC and president of Industrial Technology & Research Institute (ITRI) at that time.[7] He persuaded the state of Taiwan to help him build the world's first chip manufacturing plant devoted mainly to contract work for foreign designers. Skeptics, domestic and foreign, questioned whether the investment would ever pay off. TSMC managed to win the manufacturing contract from Intel before long, and the former built a reputation of quality and punctuality. By eliminating chip designers' need to build their own fabrication plants, Chang also paved the way for an explosion of fabless chip startups in Taiwan as well as in Silicon Valley. The proliferation of these fabless startups drastically speeded up the development of new designs.

TSMC derives its competitive advantage mainly from specialization. By concentrating on fabrication, it gains in Smithian production efficiency. Furthermore, because it does not compete with clients, the firm's customers always trust it with trade and technical secrets. Huge customers (Intel and Motorola, for example) often transfer technological capabilities to TSMC out of their own interests. Therefore, TSMC is able to upgrade continuously and stay at the technological front line.

TSMC registered a 29.82% to 55.95% gross profit rates in 1991–98 and has been the world leader in contract manufacturing since 1987. That success attracted copycats such as UMC, Singapore's Chartered Semiconductor. UMC has always been a formidable competitor; it became the world's second-largest foundry after it was transformed from an IDM into a pure foundry in 1995. The perennial competition between the two firms helps minimize X-inefficiency. But neither TSMC nor UMC is easy for foreign foundries to compete with. There is a learning curve for every type of chip, and the first mover enjoys certain advantage over followers (Gruber 1992). More important, all sectors within the disintegrated industry are well coordinated in Taiwan. Transaction costs are thus lowered. We will elaborate on this unique type of virtual integration in the next section.

The trend of embedded ICs, which cram the jobs of many chips onto a single chip (system-on-a-chip, or SOC), presents a new challenge to the industry. TSMC is now going a step further by transforming itself from a pure manufacturer into a provider of semiconductor design services. The firm wants to help customers design complex SOCs, which save space, power, and cost. TSMC's trump card is its access to circuit designs from many sources; it has created libraries of chip design building blocks that can be quickly assembled and produced into one single, integrated product on the firm's state-of-the-art production lines. The plan could accelerate the growth of TSMC, which was already the fourth-largest semiconductor firm in the world in terms of market value by 2000, behind only Intel, Texas Instruments, and Applied Materials of the United States (*China Times*, January 28, 2000). More crucial, the plan poses a challenge to the IDMs, which are now beginning to factor outside foundry capacity into their expansion plans so as to be able to concentrate more on design and marketing. In turn, the foundries need to upgrade further to meet IDMs' technical requirements. In this sense, the entire industry is currently being revamped.

4. OVERCOMING COORDINATION FAILURES

Although Taiwan's IC industry is vertically disintegrated, individual sectors are well coordinated and virtually integrated. This balance and coordination is unique. IDMs dominate in Korea and Japan, and although the United States has more fabless and chipless firms than other countries, its IC industry is not as strong as Japan's and Korea's in fabrication and packaging. According to Jones and Kierzkowski (1990) there is a trade-off between higher transaction costs under production division and higher marginal cost under integrated production. The coordination among individual sector in Taiwan's IC industry lowers transaction costs and makes fragmented production feasible and desirable.

In 1998 the shares of design, manufacturing, packaging, and testing in the production value of Taiwan's IC industry were 15.77%, 63.61%, 15.95%, and 4.67%, respectively. Intrasector trade becomes heavier over time as the supply chain gradually builds up. In 1990, 34.0% of all designs by independent design houses were fabricated by domestic manufacturers; the shares rose from 72.3% to 96.9% in 1994–98. In 1990–98, 89.9% to 99.6% of domestic manufactured chips were contracted to local packaging firms. In other words, a major part of the value added stayed at home.

Coordination takes shape through formal and informal alliances, which result in virtual integration at the industry level despite disintegration at the firm level. First, agglomeration externalities are generated in the industrial cluster in the Hsinchu Science-based Industrial Park (HSIP).[8] The park has a regional network-based industrial system that promotes collective learning. The system also benefits from a network of individuals who either attended the same universities or worked together at ERSO or in Silicon Valley. Companies compete intensely while at the same time learning from one another and transferring technologies to upstream or downstream business partners through both informal communication and formal collaborative practices.

Second, firms at different fragments of the industry have federated with one another since the late 1980s. For example, TSMC has teamed up with Taiwan's major packaging firm, Advanced Semiconductor Engineering (ASE). Winbond, an IDM, has formed an alliance with the packaging company Walsin Advanced Electronics, Ltd., through a common majority shareholder, the Walsin Lihwa business group. The closely linked team structure serves to substantially reduce transaction costs.

Third, domestic firms entered into various strategic alliances with international players in 1990. Mosel, a DRAM manufacturer, established

Promos Technologies, Inc., with Siemens in the form of technology alliances in the early 1990s. In 1996 TSMC, jointly with three American design houses (ADI, ISSI, and Altera), set up a foundry plant, WaferTech, in the United States. More recently ASE, the packaging giant, procured Motorola's plants in Taiwan and in Korea. VIA, Taiwan's leading design company, acquired the CPU divisions of the United States' National Semiconductor (Cyrix) and IDT. UMC consolidated four of its former federate companies and forged technological links with IBM and Germany's Infineon. Statistics on world strategic alliances indicate that in the period 1996–98 Taiwanese firms initiated 52 out of 259 cases of technological cooperation or joint production agreements in the world. Following Taiwan's lead was the United States, with 79 cases, and Japan, with 70 cases. Korea initiated 12 cases (ITIS project, Ministry of Economic Affairs, 1999).

Among all cases of international coordination, of particular interest is the symbiotic relationship between Taiwan and Silicon Valley. The division of production between American design houses and Taiwanese manufacturers is well known, and two-way interactions have frequently been observed as well. About 15% of the engineers in Silicon Valley are of Chinese heritage; many came from Taiwan and did their graduate work in the United States. These immigrants have formed a network that has helped facilitate both trade and investment links and the flow of knowledge across the Pacific Ocean (Saxenian 1999; Zhu 1997).

5. CONCLUDING REMARKS

To sum up, the emergence of competitive advantages in Taiwan's IC industry is based upon the following characteristics: vertical disintegration, virtual integration, and innovation. Taiwan has the most and the largest dedicated foundries in the world and more fabless design companies than all other economies apart from the United States. By specializing in one production function, these firms gain in production efficiency. The transaction costs that normally arise from such fragmented production structure are much reduced, owing to the unique production network and human network that exist in Taiwan. Most important of all, innovation makes vertical disintegration possible and virtual integration profitable.

References

Gruber, Harald (1992), "The Learning Curve in the Production of Semiconductor Memory Chip," *Applied Economics*, 24, 885–94.

Jones, Ronald, and Henryk Kierzkowski (1990), "The Role of Services in Production and International Trade," in R. Jones and A. Krueger eds., *The Political Economy of International Trade* (Cambridge, Mass.: Blackwell), 31–48.

Saxenian, Annalee (1999), *Silicon Valley's New Immigrant Entrepreneurs,* (California: Public Policy Institute of California).

SIA, *World Semiconductor Trade Statistics (1999),* (California: Semiconductor Industry Association).

Su, Li-Ying (1994), *With Storm and Sun* (Hsinchu: ERSO). (In Chinese.)

Tung, An-Chi (2001), "Taiwan's Semiconductor Industry: What the State Did and Did Not Do," Review of Development Economics, 5(2), in press.

Tyson, Laura (1992), *Who's Bashing Whom? Trade Conflict in High-Technology industries* (Washington, D.C.: Institute for International Economics).

Zhu, Li-Zhi (1997), *Gathering of Wind and Cloud, Thundering in The Silicon Valley* (Taipei: The Third Wave). (In Chinese.)

Endnotes

[1]IC accounted for 87.1% of total semiconductor production in the world in 1999, while optoelectronics and discrete devices made up another 3.9% and 9.0%, respectively (SIA 1999).

[2]For a more detailed description of the development process of the semiconductor industry and how the state helped overcome market failures, see Tung (2000).

[3]Philips also had a 27.3% stake in TSMC in 1987.

[4]By contrast, both UMC and TSMC had difficulty raising capital from domestic capitalists at the time of their establishment.

[5]For example, Motorola plans to outsource at least 50% of its semiconductor production by 2002, up from 16% in 1999 (*Far Eastern Economic Review* 1999, October 14).

[6]Vanguard, which used to pride itself on technology autonomy, found it increasingly difficult to compete with its own technology. The former TI-Acer joint venture met with similar problem after TI withdrew from DRAM production entirely. Both firms became pure foundries in 2000.

[7]ERSO is a division of ITRI.

[8]The state established HSIP in 1980. Two of the country's top universities are located nearby. Most park members are semiconductor firms (118 out of 292 in 1999), and most of Taiwan's semiconductor firms are located in the park.

Chapter 18

A CASE STUDY OF TECHTRONIC INDUSTRIES CO., LTD.

Henryk Kierzkowski

Hong Kong is a near-perfect example of an open economy in which the survival of existing firms and the success of new entrants depend on their ability to identify a niche in the international market and to respond quickly to changes in the external environment as well as to constantly shifting comparative advantages. There is a very narrow margin for error or misjudgment of the market. Since the domestic market is so small, firms must engage in sustained and vigorous export to do well. Thousands of Hong Kong firms have risen to this challenge over the last several decades, and, as a result, Hong Kong's economy has become stronger, its citizens have grown more affluent, and wages have increased dramatically.

The very success of the Hong Kong economy has undermined the competitiveness of the first generation of its export leaders. Sunrise industries have become sunset industries. A new generation of industrial and export leaders is emerging. They offer much more sophisticated products, engage in research and development (R&D), and no longer regard cheap local labor as a key to success. TechTronic Industries Co., Ltd. (TTI), is an example of this new generation of companies.

TTI was established in 1980. Like many Hong Kong companies, in the first stage of its existence it specialized in general trading, mainly in power packs and unsophisticated household appliances. Doing this gave the company a chance to learn before moving into manufacturing. Interestingly, the professional backgrounds of the company's founders, Roy Chung and Horst Pudwill, lay in a much more advanced field of electronics. Perhaps positioning their firm as they did reflected their view that although Hong Kong was no longer competitive in traditional labor-intensive products, it was also not yet competitive in sophisticated high-tech consumer goods.

By 1985 TTI had moved into the production of rechargeable products.[1] It soon started supplying Bissell Inc., USA, with floor-care products. It repeated this strategy of teaming up with a major producer and distributor of rechargeable consumer products time and again. In 1986, for example, it entered a long-term business agreement with Singer to supply cordless power tools to Sears.

TTI initially developed a competitive advantage in rechargeable power tools such as fastening tools, screwdrivers, drills, and reciprocating and circular saws. However, it soon expanded its product mix to lawn and garden tools such as grass shears and hedge trimmers. It also initiated the production of carpet cleaners and steam cleaners, hand-held vacuum cleaners, and broom vacuums. Most recently it has begun producing health care and dental care products and solar-powered and electronic products such as electronic measuring devices and solar lightning devices.

The figures below show the company's rapid expansion of sales since 1985 and its turnover by product and geographic location. With turnover now exceeding US$300 million, TTI has established itself as a major manufacturer of cordless power tools and other electric products. The Asian financial

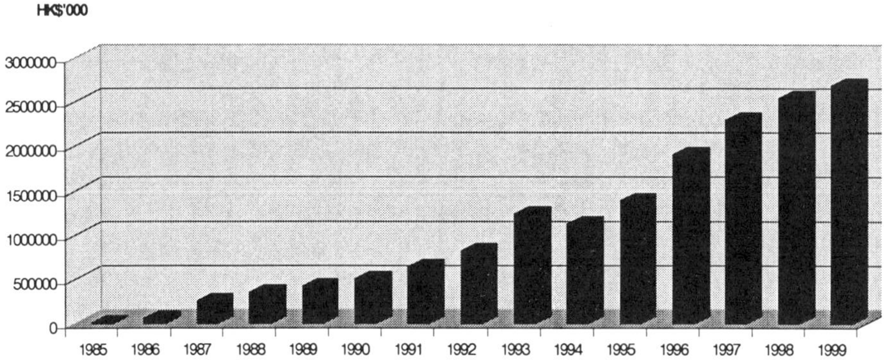

Figure 1

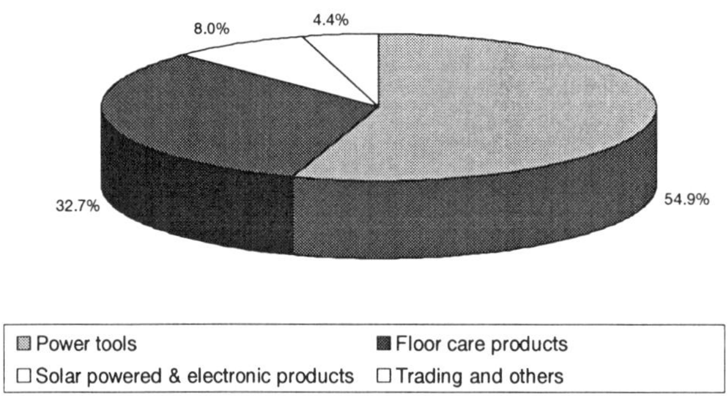

Figure 2

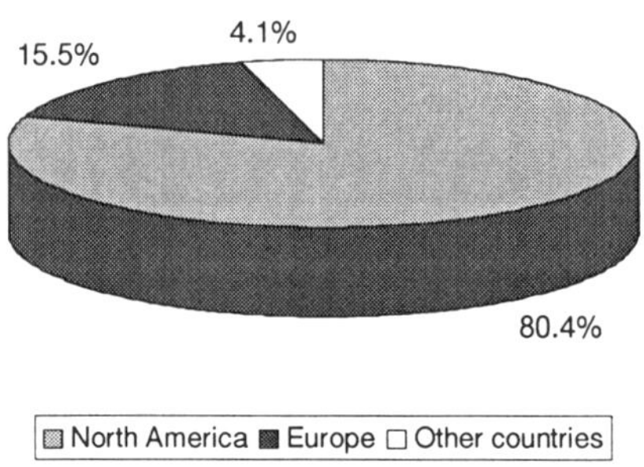

Figure 3

crisis of 1997 hardly affected it. The United States was the main market for its products when the company was formed, and it remains so today. However, the company is now making an effort to increase its degree of geographic diversification. Western Europe, Australia, and Japan offer new sales possibilities.

According to its original equipment manufacturing (OEM) strategy, TTI produced and assembled products or their parts to very strict specifications. Sears or Phillips provided the brand name and design, and TTI's task was to fulfill orders on time, meet quality standards, and keep costs down. The company has gradually begun moving toward an original design manufacturing (ODM) framework. Under this framework it designs a unique product for a specific Western company (e.g., a toothbrush for Phillips) and produces and assembles it outside Hong Kong. Designing a product is of course much more challenging than producing standardized parts, and it requires highly skilled R&D and design personnel. In addition, close contact with consumers must be maintained to monitor changes in taste and anticipate shifts in demand. For the purpose of keeping in close touch with customers TTI has set up R&D facilities in the United States, Switzerland, Singapore, Hong Kong and Japan.

Products such as rechargeable screwdrivers, vacuum cleaners, and electric toothbrushes undergo constant technical improvements and design modifications. In a market where several global producers offer very similar products, minute design changes can have a major impact on sales. One of TTI's most popular products is a simple rechargeable screwdriver. Introduced in 1988 and sold in the United States under the brand name Craftsman, it is still popular today thanks to more than a dozen major improvements that it has undergone since.

TTI's move from OEM to ODM reflects its success, but the shift has posed new problems. Under the former arrangement the company could supply a large number of Western firms with basically the same product or component. However, under the ODM scheme it must develop an exclusive production relationship with a particular company. Few companies are willing to outsource R&D, knowing that doing so could mean that their competitors could gain access to their research. But choosing a single company, no matter how large and well established, to market its products could put TTI in a vulnerable position.

It would appear that TTI's best choice is to enter original brand manufacturing (OBM). Once it has accumulated enough production and R&D experience, it could become the Sears of East Asia. It seems that the company is gradually adopting this strategy. Its best OBM opportunities are in new markets, particularly in the People's Republic of China (PRC), where well-established North American and European incumbents do not enjoy the name recognition and other advantages available at home, and where geographic proximity gives TTI an edge. The emerging markets of Eastern and Central Europe are another area where TTI might be able to establish itself. An added attraction of that region is that it could be a stepping stone to Western Europe.

Even in the well-established markets of Western Europe and North America, TTI can gradually improve its position through mergers and acquisitions. TTI recently purchased Ryobi Limited's North American power tool businesses.[2] This purchase will allow TTI, a long-term Ryobi supplier, to maintain far more control of its power tool business in North America.

Finding the right market niche and strategy constitutes only one element of TTI's success. The others are an optimal allocation of productive tasks, and, in particular, the use of outsourcing. As was alluded to earlier, Hong Kong has lost its competitive advantage in many manufacturing activities. Instead it has become a hub for numerous production networks.

When TTI employed 30 people its labor force was located in Hong Kong. Today the company employs over 6,000. Its manufacturing base has shifted to various locations in the PRC (Dongguan and Shenzhen) and the island of Batam in Indonesia. Even R&D and product development activities are largely conducted outside Hong Kong: in the PRC, the United States, Switzerland, and Singapore. Needles to say, modern technology permits the company to coordinate and closely monitor these activities. As an example, 22 R&D computer stations with 3D CAD (3-Dimentional Computer Aided Design) facilities in the PRC are fully integrated with 9 similar stations in Hong Kong.

While TTI supplies companies in North America, Europe, and Japan with products and subsystems, it also relies on others, mainly based in East Asia, to supply it with parts and components when cost savings can be achieved by doing so. The key components supplied from outside Hong Kong include batteries (from Panasonic, Sanyo, and Energizer), motors (from Johnson and Mabuchi), chucks (from Jacobs and Rohm) and switches (from Capax and Satori). TTI is likely to continue in this vein for the foreseeable future, because high development costs and possible economies of scale in the production of batteries, motors, and chucks encourage concentration of output in the hands of a limited number of global producers supplying multiple users. It is also worth pointing out that batteries, motors, and chucks are examples of horizontal fragmentation; they can be used across a number of industries, which further encourages specialization and concentration.

Endnotes

[1]The company was listed on the Hong Kong Stock Exchange in 1990.

[2]Note that Ryobi Limited is a publicly held company with headquarters in Tokyo.

Chapter 19

LI & FUNG, LTD.: AN AGENT OF GLOBAL PRODUCTION

Leonard K. Cheng[1]

"We, the traders, are the interface between the users and the producers..., and when the producers and users are geographically, linguistically, and culturally separated, we add much of our value by getting around the problems inherent in the relationship between far apart business partners.... To excel in this business, you have to master the four Cs: Connections, Communications, Control of quality, and Consolidation of shipments." Victor and William Fung, quoted in Long and Seet (1996).

1. INTRODUCTION

The trade theory that deals with international production fragmentation assumes that producers in different parts of the world specialize in segments of various value chains such that the final price of each product is as low as possible. The slicing of a value chain into segments to be allocated to producers in different countries depends on technical feasibility and on the costs of linking the successive production processes. In practice, how do producers in different countries determine their respective positions along the value chain?

In theory only efficient producers survive, but in reality existing producers in the same country have different production and management capabilities. How are their capabilities discovered and verified? How can far-away business partners count on one another's good performance? If it takes time and effort to build and maintain trust, who should be included in a network that links trustworthy buyers and producers?

Also, who provides the services necessary for linking the successive stages of production? Do anonymous service providers supply their services to a competitive market? Or, at the other extreme, are the services provided in a hierarchy, such as a multinational firm? Or is the problem solved with a combination of hierarchy and market, including market transactions that are based on long-term relationships and trust? If the total costs of production and transaction are to be minimized, how should a trader or sourcing agent structure contractual arrangements for transactions between producers responsible for different stages of production?

Finally, what kind of internal organization and incentive system encourages global rather than local cost minimization?

A study of Li & Fung, Ltd., a modern trading company based in Hong Kong, will help answer the questions raised above.

2. LI & FUNG, LTD.'S RISE TO PROMINENCE

The history of Li & Fung, Ltd., has been documented in Long and Seet (1996) and O'Connell (1996), so it will not be repeated here. For our purposes it suffices to recap the evolution of the company's business scope. When the company was first founded in 1906 in Canton, it traded in porcelains, antiques, and handicrafts. In the 1950s and 1960s the company diversified into garments, toys, electronics, and plastic flowers. Today it continues to specialize in low-cost, labor-intensive consumer products, with garments, its most important product, comprising 80% of all the goods it sourced in 1997. It also sources plastics, handicrafts, fashion accessories, sporting goods, and household items.

3. THE ROLE OF A TRADING COMPANY OR SOURCING AGENT

A trading company or sourcing agent needs a thorough knowledge of what its customers want and how to best mobilize its suppliers to meet customer demand. When Li & Fung was founded, one of its essential functions was to serve as an interpreter for buyers and sellers. Most of the goods it sold had already been produced before buyers came to inspect them. Later the trader began to manage the supply chain to satisfy the demands of buyers who specified the product they wanted. Today buyers may provide only specific ideas and concepts, and the sourcing agent comes up with products for buyers to choose from or modify.

Regardless of whether products are clearly or vaguely specified by buyers, the agent's role as a manager of the supply chain remains the same. The agent makes itself available to order and inspect the required raw materials, identify and contract with appropriate manufacturers, inspect the manufacturers' production processes and the final products, and, finally, pack and ship products to buyers. In some cases, the sourcing agent may manage the buyers' own suppliers for a fee.

By 2000 Li & Fung had set up 48 offices in 32 countries. With about 3,600 employees, the company serves 600 customers and 6,000 suppliers throughout the world. In 1999 its total sales amounted to US$2.1 billion.

4. KNOWING WHAT BUYERS WANT

In the old days, buyers had to travel to producing countries to find out what was available. More recently, they specified the products they wanted, and the sourcing agents decided how to fill orders. Now buyers give sourcing agents ideas only, and the agents, in turn, ask their designers to specify products based on these ideas and produce samples. The agents then finds out how the buyers like the products.

For seasonal goods and fashion items, buyers like to wait as long as possible before placing an exact order to ensure that they will not misjudge what the final consumers want. This desire to wait for consumer preferences to be revealed implies that the time left for production can be very short indeed. As a result, on-time delivery of products is crucial. Multinational firms' or sourcing agents' ability to shorten the production cycle has made it possible to have six or seven seasons instead of two or three seasons per year. That in turn has stimulated demand for more fashion-oriented products.

The negotiation of a contract with customers is complicated. According to Long and Seet (1996, p. 9), "Product group managers, in conjunction with buyers, deliberated raw materials sourcing, manufacturing site, product quality standards, reliability of manufacturers, factory skill level, lead times, quota availability, packaging, and costing."

5. UNDERSTANDING PRODUCERS' CAPABILITIES

Before the sourcing agent is in a position to decide how to meet its buyers' needs, it must understand the capabilities of the manufacturers in its network and how they can work together to produce the best product at the lowest cost. Detailed knowledge of producers' capabilities, together with information on trade restrictions imposed by the importing country on the various potentially producing countries, enable the agent to dissect the value chain for the purpose of minimizing the total costs of production and transaction. Allocating the successive production stages to producers in different parts of the world must be done in accordance with these producers' comparative advantages and is subject to trade restrictions, product quality, and time of delivery. This type of "dispersed manufacturing," an innovation of Li & Fung, is an important example of international production fragmentation.

6. BUILDING TRUST AND ENSURING QUALITY AND TIMELY DELIVERY

Knowing the capabilities of producers in different parts of the world is only the first step in the dispersed manufacturing process. Equally important is building trust between the sourcing agent and its geographically scattered suppliers. Smooth coordination among all the producers involved is essential to both product quality and on-time delivery. For dispersed manufacturing or fragmented production to work, the producers responsible for the successive stages of production must be reliable enough not to turn an efficiency gain into a disastrous loss.

As far as coordination goes, the sourcing agent serves as a central node of trust in the network that links suppliers and buyers. Suppliers and buyers do not have to worry about the trustworthiness of the other parties because they are only buying from or working for Li & Fung. To ensure that things go wrong as infrequently as possible, the company has devoted many resources to quality assurance. According to O'Connell (1996, p.6), "almost one fifth of Li & Fung's staff specialized in quality assurance."

7. BUILDING AND RESHAPING CUSTOMER AND SUPPLIER NETWORKS

Like any other company, Li & Fung seeks out potential customers that appear attractive. In addition, it invests in certain companies as a minority shareholder in the hope that these companies will broaden its customer base. As for the network of suppliers, Li & Fung may contact new suppliers in specific countries at the request of customers. It also expands its network in response to changes in the global competitive landscape. The risks associated with untried new suppliers are usually higher than those associated with suppliers already in the network. Nevertheless, the risks to the company of not extending its network to new competitive sources of supply could be fatal.

Shifts in the comparative advantages of different economies resulting from changes in technology and economic policy call for the development of new networks and the realignment of new and existing networks. Given Li & Fung's focus on low-cost, labor-intensive consumer products, it has been in "constant search for new sources of supply" that offer cheap labor with acceptable quality and delivery time. It has expanded its network of producers in the following ways:

a) Li & Fung sends engineers to evaluate the production capabilities of new economies with the most potential. If technical capabilities are confirmed, the next step is to build trust; this starts when the company places the first order and closely supervises the production process to ensure that product quality and delivery time meet with expectations.
b) With its 1995 acquisition of Inchcape Buying Services, an established British trading company, Li & Fung expanded its customer network in Europe and its producer network in South Asia, including India, Pakistan, Bangladesh, and Sri Lanka.
c) By investing in new companies in advanced economies, Li & Fung has sought to overcome the weaknesses of its networks and to create new networks. It chose to be a minority investor because it felt it had no comparative advantage in managing firms in the advanced economies.
d) As the number of Li & Fung's customers and suppliers grows, information technology has become an increasingly important tool in the management of the network network.

8. WHY SOURCING AGENTS EXIST

A buyer can place an order directly with a supplier, but a sourcing agent like Li & Fung can engineer a better deal for that buyer for the following reasons:

a) By virtue of its overall business volume, a sourcing agent enjoys an advantage in the procurement of raw materials. It can get better prices and faster delivery than an ordinary buyer.
b) A sourcing agent can reward good producers with repeat orders (regardless of whether the orders originate from the same buyer). Thus, producers give priority to orders from sourcing agents. Compared with a single buyer, the sourcing agent is in a better position to ensure product quality and delivery time.
c) A sourcing agent's large network of producers enables it to minimize costs over a much larger feasibility set. For a single buyer to invest in the building of trust with so many producers to serve only itself would be too costly. Even large multinational companies that are not disadvantaged by a lack of economies of scale may find it necessary to leverage a sourcing agent's supplier network in order to build their own networks.

Other things being equal, a sourcing agent like Li & Fung is more appealing to companies with many product lines, especially if the efficient production of those products requires a large supplier network.

Even though ensuring product quality and delivery time is crucial, Li & Fung does not want to own manufacturing facilities. As the company's chairman, Victor Fung, has emphasized, "We do design. We buy and inspect the raw materials. We have factory managers, people who set up and plan production and balance the lines. We inspect production. But we don't manage the workers, and we don't own the factories" (Magretta 1998, p. 109). The company believes it has no comparative advantage in managing production workers around the world. That job is best left to local entrepreneurs and managers.

Indeed, the company does not even want its suppliers to be overly dependent on it, although sometimes it advances credit to them to shorten the production cycle. As a general policy, it uses only 30% to 70% of factories' production capacity, but that is enough to make Li & Fung its manufacturers' largest customer. More important, owning factories would obliterate the flexibility Li & Fung enjoys in responding to changes in demand and supply conditions. Flexibility is crucial to the survival of sourcing agents, who after all are in the business of finding the most competitive suppliers to meet their customers' demand.

It should be emphasized that dealing directly at the factory level is only one link in the entire supply chain. Nowadays buyers focus on the total costs to their final customers, not just on the ex-factory cost. William Fung, Li & Fung's managing director, has summed it up as follows: "Our role is to come up with a total solution for the customer, including flexibility and just-in-time solution. We are involved from raw materials to production planning, to delivering the product to the retailers' stores, all with the purpose of minimizing the product's total costs. By managing the whole supply chain, we have added tremendous value." Moreover, the division of labor is based on the participants' respective comparative advantages, as a trade theorist would predict. "In our way," explained Fung, "the raw materials suppliers supply the raw materials, the manufacturers manufacture the goods, and we handle all the logistics. No one in our supply chain is asked to do more than what he is capable of doing" (Long and Seet, 1996, p. 10).

9. INTERNAL ORGANIZATION AND INCENTIVE SYSTEM

Obviously, a trading company's internal organization and incentive system are crucial to the company's efficient operation. At Li & Fung, the following three criteria are important in the design of both:

a) The company should minimize costs by using the company's entire network, not just part of the network in a particular country or region.
b) The company should be flexible in response to changes in customer demand and to supply conditions.
c) Operations that are not essential to trading should not become a burden to the trading units.

To meet the first criterion, Li & Fung makes its profit center the division that caters to the needs of major customers or to groups of small but similar customers. There are about 60 divisions in the company. By steering away from an organizational structure with a country or regional emphasis, Li & Fung strives to minimize costs globally. Above the divisions are product groups that are supervised by the company's executive directors. This organizational structure is different from that of many other trading companies that possess extensive networks. Those companies are often organized according to geographical locations, with country units serving as profit centers.

The trading staff at Li & Fung receive annual bonuses based on their division's performance. Division managers' bonuses constitute a large portion of their total compensation and may even exceed their salaries. Product-group managers and executive directors are also compensated according to the performance of the divisions under their management. Nevertheless, to make product-group executives more sensitive to problems faced by a country director, "every product group executive also has responsibility for one country" (Magretta 1998, p. 110).

The second and third criteria are reflected in the division of labor between the headquarters and the divisions. The former is responsible for overhead services, while the latter takes care of trading and sourcing. In addition, headquarters maintains strict financial control over divisions. According to the company's management, Li & Fung's divisions have both the flexibility of small trading companies and the general support of large companies.

In addition to its headquarters located in Hong Kong, Li & Fung has over 40 branch offices in 32 countries. Its first branch office opened in Taiwan in 1968. These offices serve as an extension of the headquarters to areas with a major clustering of suppliers and allow the company to better serve the divisions. The company does not calculate the profits and losses of its branch offices, and branch office managers are not evaluated on their offices' profitability. This managerial arrangement is intended to free branch offices from biases in favor of producers in the country in which they are located when division managers seek advice from them about the country's ability and efficiency in handling orders.

The branch offices hire mostly native-born staff because they are good at identifying the best factories in their own areas and in establishing an effective local network that becomes part of the company's larger global network. In addition, this practice allows these offices be run according to the best local practice (Long and Seet 1996, p. 12).

References

Long, Diane and Richard Seet (1996) "Li & Fung (A)," *Harvard Business Case Studies,* N1-396-107, 15, April.

Magretta, Joan (1999) "Fast, Global, and Entrepreneurial: Supply Chain Management, Hong Kong Style: An interview with Victor Fung," *Harvard Business Review,* September-October, 102–14.

O'Connell, Jamie (1996) "Li & Fung (Trading) Ltd.," *Harvard Business Case Studies,* 9-396-075, 28, June.

Endnotes

[1]I would like to thank Dr. William K. Fung for his assistance in the preparation of this case study, in particular for his insights about supply chain management.

Chapter 20

THE CHANGING PATTERN OF PRODUCTION FRAGMENTATION IN SINGAPORE AND ITS ECONOMIC CONSEQUENCES

Hian Teck Hoon and Kong Weng Ho

1. INTRODUCTION

The slicing up of the value chain, to use the term coined by Krugman (1995), or the phenomenon of production fragmentation, the term preferred by Jones and Kierzkowski (1990, 2001a, 2001b), is a major factor behind the growth of world trade in the postwar era. Because of the importance of intermediates or "middle products" in the production of final goods whose production characteristics can be differentiated by varying factor intensities, it is economically attractive for firms to base different production blocks in different countries based upon their relative factor endowments.[1] Consequently, in the absence of significant trade barriers, relatively labor-abundant countries produce labor-intensive parts and components, while relatively capital-abundant countries concentrate on producing capital-intensive items. Such vertical fragmentation of the production process based upon countries' comparative advantage, however, requires the crucial function of service links (see Jones and Kierzkowski 1990). Determining whether firms choose to outsource part of the value chain or to engage in the whole vertically integrated process within national borders requires a comparison of the total cost involved in the two modes of organization. Crucially, the lower *marginal cost* of production made possible by taking advantage of international specialization must be balanced off the higher *fixed cost* of providing service links to coordinate activities across nations.

Heightened professional interest in the international location of economic activities in the past decade or so has been stimulated in part by the coincidence of increased international economic integration—brought about both by substantial cost reductions in transport and communications and by the lowering of trade barriers in many countries—and in part by widening wage-earnings gaps between skilled and unskilled workers in many Organization for Economic Cooperation and Development countries. (In Europe, where skill premiums have not risen by as much, the unemployment rate among workers at the lower end of the skills range has risen more rapidly than in the United States.) Debate has raged, mainly at an empirical level, about the blame that should be apportioned to the role played by international trade as opposed to skill-biased technology. Our own review of the Singapore economy, which we think about from the perspective of current-day concerns about trade integration and international production disintegration, has led us to recognize the importance of this phenomenon to this nation as much as 35 years ago. Indeed, Singapore's whole economic transformation since independence has relied upon engaging in the process of international production fragmentation. As we hope to document, however, Singapore's role in this process has changed over the years. It began as a base for traditional entrepôt activity, went on to become a modern manufacturing base, and then moved from a phase in which the emphasis was on lowering marginal costs to attract multinational corporations (MNCs) to base part of their value chain there to one in which the emphasis was on lowering the fixed cost of providing a service link to MNCs. We believe that studying the change and its attendant economic consequences in the Singapore case will help us understand the general principles at work in today's global economy.

The rest of this chapter is organized as follows. In Section 2 we document the importance of international production fragmentation for Singapore from as early as the 1960s. While Singapore has historically been an entrepôt economy, bringing in raw materials from the other parts of Southeast Asia and reexporting them to the rest of the world after minimal processing, that role waned partly because neighboring countries developed their own transport facilities and engaged in direct trading with buying countries. It was not necessary that the industrialization program the country launched after it gained full independence in 1965, having left the Federation of Malaysia, should be based upon housing production blocks along the value chain, which would have involved massive imports of upstream intermediates and subsequent exports of downstream intermediates. However, as the work of Sandilands (a British economist who spent a few years at the National University of Singapore) and his collaborators demonstrates, the move away from being a traditional entrepôt economy took precisely this form, and the country became in effect a reexport economy. We show that Singapore's engagement as a production base largely took the form of playing host to MNCs from the United States, Japan, and Europe. Arms-length transactions with contract manufacturers became a more important phenomenon in Singapore only in the 1990s. In Section 3 we seek to explain how Singapore made a transition from being simply a production base for some components of MNC subsidiaries to serving more as a service link in an integrated approach—with the Singapore headquarters facilitating production block activities in the relatively labor-abundant regional economies. In Section 4 we discuss the economic consequences of international production fragmentation for Singapore's labor market. Section 5 briefly discusses how local contract manufacturers have emerged in Singapore over the past few years, drawing upon two interviews we conducted with a major local contract manufacturer of electronic components and a major MNC subsidiary based in Singapore. Our case study is discussed in greater detail in a later chapter. We provide some concluding remarks in Section 6.

2. EXTENT OF INTERNATIONAL PRODUCTION FRAGMENTATION

Singapore's natural harbor and its strategic geographical location, together with its free port policy under British rule, made entrepôt trade a significant activity there by the time it achieved self-government in 1959. In 1960 entrepôt exports made up almost 94% of Singapore's total exports (see Lim et al. 1988, p. 276). Though lacking in natural resources, Singapore's ability to import raw materials from other resource-rich Southeast Asian countries and to reexport them after simple processing allowed for the growth of industries such as tin smelting, rubber processing, saw milling, and petroleum refining.[2] Although the importance of traditional entrepôt activities gradually declined, ancillary activities such as storage, banking, insurance, transportation, and communications were developed along with entrepôt trade, laying the groundwork for Singapore's later development as a crucial service link for coordinating production activities in the region.

According to official published data, domestic exports, which in 1960 accounted for only 6% of total exports, rose to account for nearly 65% in 1985 (see Lim et al. 1988, p. 276) before receding a little to nearly 58% in 1998 (see Table 1). Sandilands and Tan (1986) and Lloyd and Sandilands (1986) have, however, done some careful statistical work based upon the import coefficients derived from the input-output tables for 1973 and 1978 published by the Department of Statistics. They show that, in fact, published domestic exports have an extremely high import content that is not netted out to obtain their true value added. Deducting the import content of all industries from the Singapore input-output tables shows that in the early 1980s around three-quarters of the value of official "domestic exports" was actually the value of imported components, materials, and other inputs (see Lloyd 1996, p. 509). A more accurate measure of the contribution of exports to total value added is termed "net domestic exports" by these authors and is obtained after deducting the import content of domestically produced goods. Lloyd and Sandilands (1986, pp. 190–1) show that their measure of the share of net domestic exports of goods in GDP rose from 9% in 1964 to 22.7% in 1982. In contrast, the official measure of domestic good exports, which fails to net out the import content of intermediates, rose from 36.3% of GDP in 1964 to 101.2% of GDP in 1982. Later work by Sandiland's students using input-output tables for 1983 and 1988 shows that Singapore was still very much a reexport economy in the early 1990s in the sense that it imported large amounts of intermediates, which it then reexported after contributing value added (see Woon 1991 and Tan 1994). Our own calculations reported in Figure 1 show that import requirements made up nearly 70% of the total value of exports of goods and services in 1983 but that the percentage receded to nearly 60% in 1990. The percentage of intermediate inputs imported from foreign sources was also very high in manufacturing—close to 80%. The share of intermediate inputs that are imported is reported for

Table 1: *GDP and Selected Trade Statistics of Singapore*

Year	GDP (S$m)	Total Exports (S$m)	Domestic Exports (S$m)	Entrepôt Exports (S$m)	Total Imports (S$m)	Share of Domestic Exports in Total Exports (%)	Share of Entrepôt Exports in Total Exports (%)
60	2150	2965	190[a]	2775[a]	3498	6.4[a]	93.6[a]
61	2329	2750	—	—	3358	—	—
62	2514	2860	—	—	3431	—	—
63	2790	3291	—	—	3995	—	—
64	2715	2772	—	—	3479	—	—
65	2956	3004	766[a]	2238[a]	3807	25.5[a]	74.5[a]
66	3331	3374	—	—	4066	—	—
67	3746	3491	—	—	4407	—	—
68	4315	3891	—	—	5084	—	—
69	5020	4741	1676	3065	6244	35.34	64.66
70	5805	4756	1832	2924	7534	38.53	61.47
71	6823	5371	2373	2998	8664	44.19	55.81
72	8156	6149	3101	3048	9538	50.43	49.57
73	10205	8907	4412	4495	12513	49.54	50.46
74	12543	14155	8414	5741	20405	59.44	40.56
75	13373	12758	7540	5218	19270	59.10	40.90
76	14651	16266	9362	6904	22405	57.56	42.44
77	16039	20090	11652	8439	25522	58.00	42.00
78	17830	22986	13226	9759	29601	57.54	42.46
79	20523	30940	18200	12740	38334	58.82	41.18
80	25091	41452	25805	15647	51345	62.25	37.75
81	29339	44291	29452	14839	58248	66.50	33.50
82	32670	44473	29158	15315	60245	65.56	34.44
83	36733	46155	29206	16949	59504	63.28	36.72
84	40048	51340	33051	18289	61134	64.38	35.62
85	38924	50179	32576	17603	57818	64.92	35.08
86	38962	48986	32062	16924	55545	65.45	34.55
87	43145	60266	39071	21195	68415	64.83	35.17
88	50714	79051	49555	29496	88227	62.69	37.31
89	58943	87117	55252	31865	96864	63.42	36.58
90	67705	95206	62754	32452	109806	65.91	34.09
91	75280	101880	66031	35849	114195	64.81	35.19
92	80637	103351	66337	37014	117530	64.19	35.81
93	92905	119473	75394	44079	137603	63.11	36.89
94	106490	147327	88533	58794	156396	60.09	39.91
95	118491	167515	98473	69042	176314	58.78	41.22

Table 1: *(Continued)*

Year	GDP (S$m)	Total Exports (S$m)	Domestic Exports (S$m)	Entrepôt Exports (S$m)	Total Imports (S$m)	Share of Domestic Exports in Total Exports (%)	Share of Entrepôt Exports in Total Exports (%)
96	128974	176272	103589	72683	185183	58.77	41.23
97	141262	185613	107535	78078	196605	57.94	42.06
98	141216	183763	105918	77845	169864	57.64	42.36

Sources: *Singapore Department of Statistics, various issues,* Yearbook of Statistics, Singapore; *Singapore Department of Statistics, 1983,* Economic and Social Statistics Singapore 1960–1982; *Lim et al., 1988,* Policy Options for the Singapore Economy, *page 276.*

Notes: *Entries in cells marked with a superscript "a" are taken and computed from Lim et al., 1988. Data for cells marked "—" are not available. Exports and imports here refer to exports and imports of goods alone, with services excluded.*

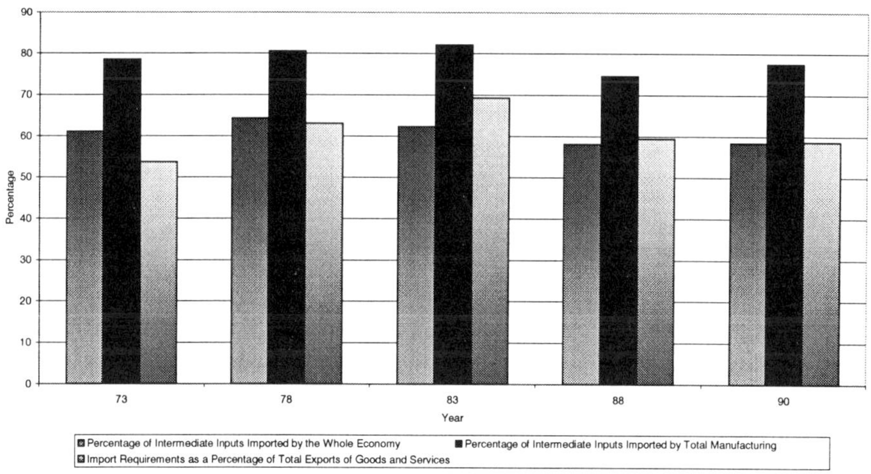

Source: *Computed from Singapore Department of Statistics, various issues, Singapore Input-Output Tables.*

Figure 1: *Measures of Fragmentation from Input-Output Tables.*

key industries within the manufacturing sector in Table 2. In the petroleum industry, this share was almost 95%. The share was also extremely high in the electronics industry, which makes up about half of the manufacturing sector.

Another set of statistics reported in Figure 2 makes it amply clear that the Singapore economy has been a prominent player in the phenomenon of

Table 2: *Share of Intermediate Inputs Imported*

	Share of Intermediate Inputs Imported (%)				
Year	**1973**	**1978**	**1983**	**1988**	**1990**
Selected Industries in Manufacturing					
1. Food, beverages & tobacco	55.62	58.35	55.80	56.84	49.18
2. Textiles, wearing apparel & leather	77.28	69.73	63.70	72.87	65.70
3. Wood & wood products including furniture	76.60	60.55	58.25	65.10	55.37
4. Paper & paper products, printing & publishing	60.30	63.67	52.34	66.12	57.93
5. Chemical & chemical products	59.74	56.18	54.53	57.19	52.71
6. Petroleum & petroleum products	94.25	97.08	97.05	94.09	93.24
7. Rubber & plastic products	82.08	72.63	66.40	56.47	51.02
8. Basic metals	46.20	76.36	67.83	77.74	71.86
9. Fabricated metal products	56.35	67.81	59.43	62.67	55.47
10. Machinery	62.48	56.06	56.57	63.26	59.57
11. Electronic products & components	84.09	85.22	78.57	76.42	72.19
12. Transport equipment	52.12	54.56	63.93	54.68	58.63
13. Instrumentation, photographic & optical good	75.78	60.53	65.66	71.19	67.06
Sum of the above 13 industries	74.75	80.44	81.56	74.79	70.96

Source: *Computed from Singapore Department of Statistics, various issues, Singapore Input-Output Tables.*

Notes: *We have consolidated the relevant industries in the I-O tables into the major industrial categories in manufacturing as those in Singapore Department of Statistics, various issues, Foreign Equity Investment in Singapore.*

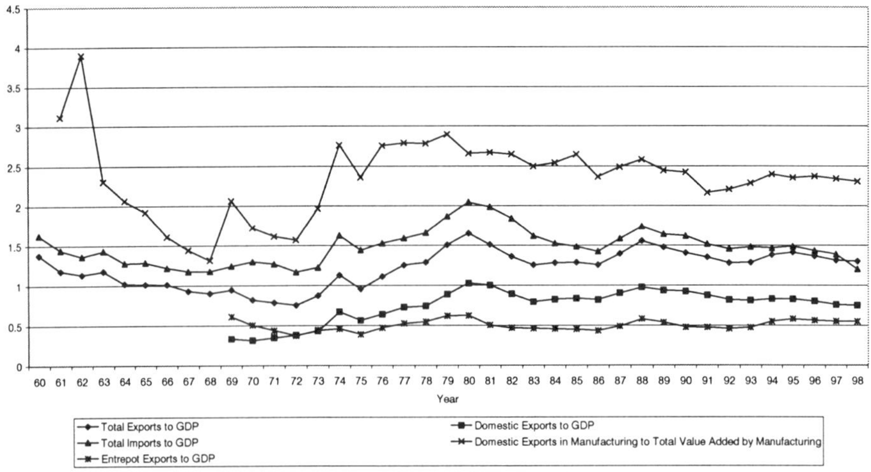

Sources: *Singapore Department of Statistics, various issues,* Yearbook of Statistics; *Singapore Department of Statistics, 1983,* Economic and Social Statistics Singapore *1960–1982.*

Figure 2: *Ratios of Exports and Imports of Goods to GDP.*

international production fragmentation from as early as the 1960s. In a world where vertically integrated production processes occur only within national borders, and international trade takes place only in final goods, the maximum ratio of total exports to GDP is one—achieved when a country exports all of its output—and the maximum ratio of total imports to GDP is also one—achieved when all domestic aggregate demand is satisfied from buying abroad. A value that is greater than one must mean that the country is engaged in a great deal of international trade in intermediates. As data in Figure 2 show, the share of total exports has exceeded one in the past 25 years while the ratio of imports to GDP has exceeded one in the past 40 years. The importance of production fragmentation in manufacturing is especially evident in the data.

As we argue in the next section, the pattern of international production fragmentation has changed over the course of Singapore's recent economic history. In the first phase after independence, Singapore attracted MNCs to base part of their value chain there by virtue of lowering the marginal cost of producing there. The early years saw the development of Singapore-based factories that produced mainly labor-intensive goods. As the capital endowment increased, more and more capital-intensive lines of production became profitable. Overall, the first phase coincided with the rapid growth of the manufacturing base, a fact confirmed by an examination of Figure 3. As we can see, the manufacturing sector grew from over 10% of GDP to nearly

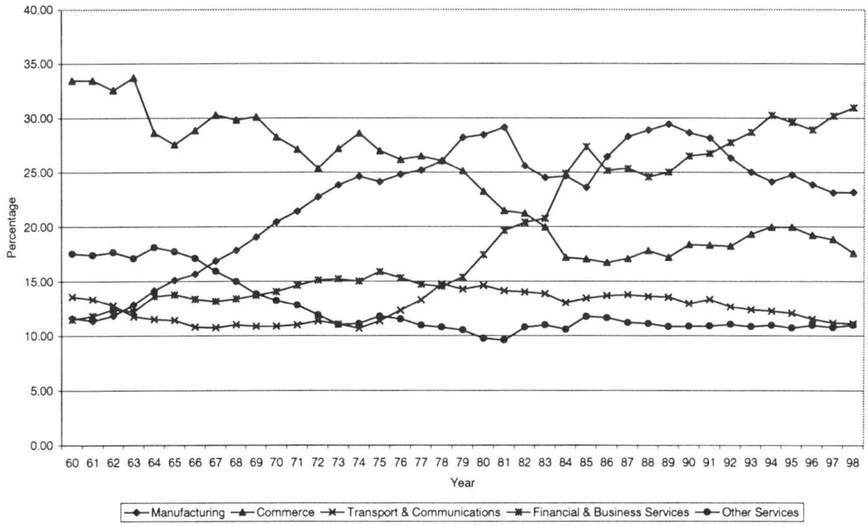

Sources: *Singapore Department of Statistics, various issues,* Yearbook of Statistics; *Singapore Department of Statistics, 1983,* Economic and Social Statistics Singapore *1960–1982.*

Figure 3: *Contributions to GDP by Major Sectors.*

30% by the end of the 1970s before stabilizing at about 25%. The fact that production disintegration took the form of MNC subsidiaries being set up in Singapore to produce parts and components for parent companies, rather than of foreign companies purchasing from local contract manufacturers, can be gleaned from Figures 4 and 5. As we can see, the ratio of domestic exports to manufacturing value added is highest for wholly foreign-owned and more-than-half foreign-owned companies. The same is true for the other measure; domestic exports as a percentage of total sales is highest for wholly foreign-owned and more-than-half foreign-owned firms.[3]

We present results of regressing the extent of international production fragmentation on foreign direct investment (FDI) in Table 3. There, DExpSale is domestic exports as a percentage of total sales in manufacturing and is used as a measure of the degree of fragmentation. FDEShare is foreign direct equity investment as a percentage of the sum of local equity investment and foreign direct equity investment in manufacturing. FDEShare is used as a measure of the amount of FDI. The regression result clearly shows that the amount of FDI has a positive impact on the degree of production fragmentation in manufacturing. A1% increase in FDI increases fragmentation by 0.58% and this positive impact is statistically significant at the 5% level of significance. We have corrected for the serial correlation problem, and the adjusted R^2 is 0.74.

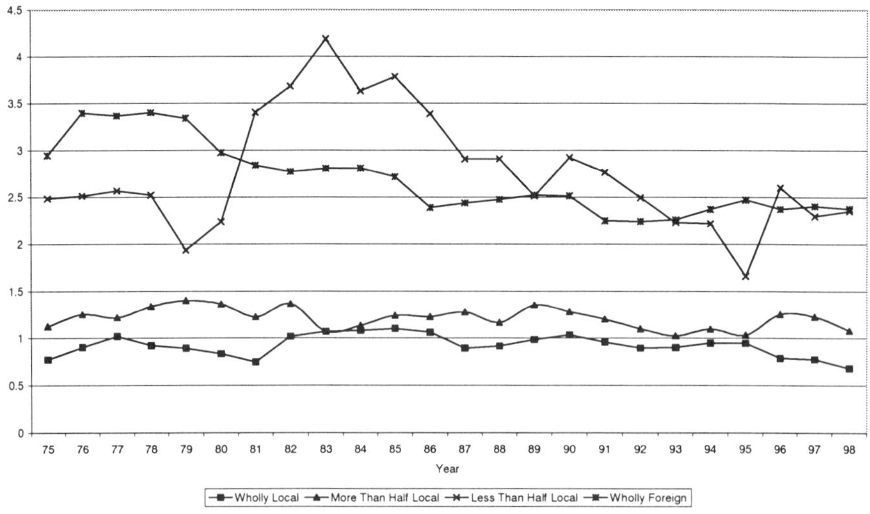

Sources: *Economic Development Board, various issues,* Report on the Census of Industrial Production.
Note*: Data by ownership for years earlier than 1975 are not available.*

Figure 4: *Ratio of Domestic Exports to Vaule Added in Manufacturing by Ownership.*

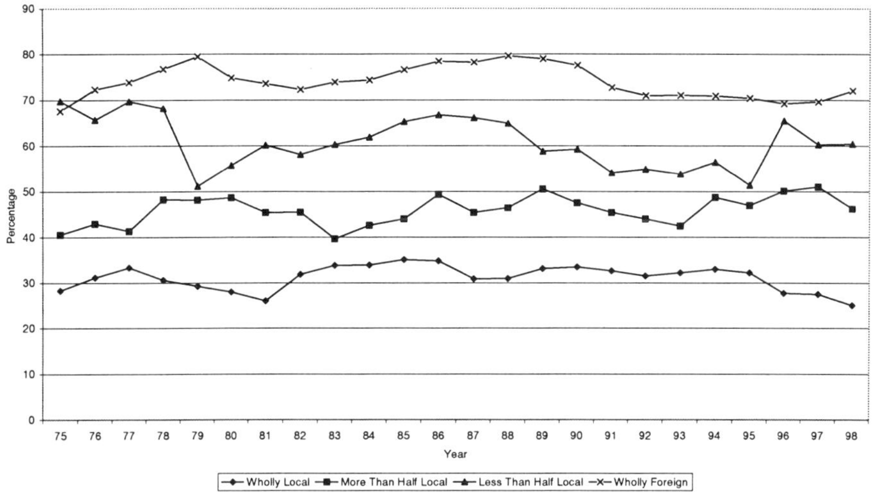

Source: *Economic Development Board, various issues,* Report on the Census of Industrial Production.
Note: Data by ownership for years earlier than 1975 are not available.

Figure 5: *Domestic Exports as a Percentage of Total Sales in Manufacturing by Ownership.*

Table 3: *Fragmentation and FDI*

	Dependent Variable: DExpSale **Sample(adjusted): 1981 to 1996** **Convergence Achieved After 4 Iterations**			
Variable	**Coefficient**	**Std. Error**	**t-Statistic**	**Prob.**
Constant	44.48345	7.898327	5.632009	0.0001
FDEShare	0.575433	0.245614	2.342838	0.0357
AR(1)	0.673518	0.208495	3.230375	0.0066
R-squared	0.773930	F-statistic		22.25222
Adjusted R-squared	0.739151	Prob(F-statistic)		0.000063

Notes: *DExpSale is domestic exports as a percentage of total sales in manufacturing and is used as a measure of the degree of fragmentation. FDEShare is foreign direct equity investment as a percentage of the sum of local equity investment and foreign direct equity investment in manufacturing. FDEShare is used as a measure of FDI. AR(1) is the correction for serial correlation in the regression.*

Sources of data: *DExpSale is computed from Singapore Department of Statistics, various issues,* Report on the Census of Industrial Production. *FDEShare is computed from Singapore Department of Statistics, various issues,* Foreign Equity Investment in Singapore.

In the second phase, particularly after the serious recession of 1985, the economy aimed to develop itself as an important service link for foreign companies to coordinate their production activities in regional economies. The sea change came in the form of a key recommendation by the Economic by the then Minister of State Brigadier General (NS) Lee Hsien Loong. The

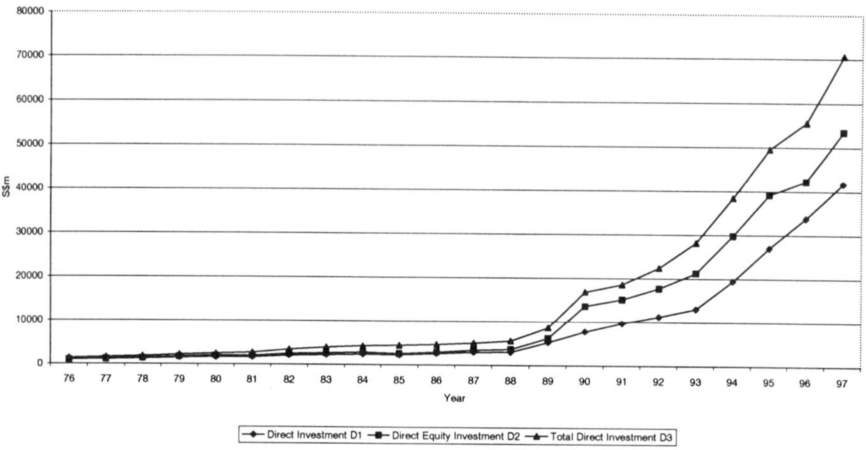

Source: *Singapore Department of Statistics, various issues,* Singapore's Investment Abroad.

Figure 6: *Stock of Singapore's Direct Investment Abroad 1976 to 1997.*

committee recommended that equal emphasis be given to the development of manufacturing and services, which were the twin engines for future growth (see Report of the Economic Committee 1986, p. 139). A series of proposals were made to deregulate services, privatize state-owned service companies, attract foreign professionals, and develop local enterprises. The 1990s saw major moves toward deregulating the financial and banking sector and the telecommunications sector. Coordinated efforts were made to promote Singapore as having the comparative advantage—by virtue of its strategic location, excellent communications and transportation infrastructure, and a long tradition in commerce—in providing service links to MNCs. The period coincided with labor-abundant regional economies opening up to foreign capital, so efforts were made to attract MNCs to base their headquarter services in Singapore to coordinate production activities in the region. One reflection of the transition from the first phase of engagement in international production disintegration (in which Singapore served as a production base to foreign capital) to the second phase (in which the country became an outsourcing country itself) is the rise of outward investment by Singapore firms shown in Figure 6. As we can observe, a steady rise in Singapore's outward investments began to take shape only in the very late 1980s.

3. INTERPRETATION OF SINGAPORE'S EXPERIENCE IN TERMS OF THE JONES-KIERZKOWSKI FRAMEWORK

In the preceding section, we attempt to document how engaged the Singapore economy has been in the phenomenon of international production fragmentation since early in its history. The transition from being a traditional entrepôt economy to serving as a manufacturing production base for MNCs taking advantage of international specialization began to take place in the mid-1960s. The growth of the latter activity is reflected in the steady increase in the size of the manufacturing sector throughout the 1960s and 1970s. Toward the second half of the 1980s an increasing emphasis began to be placed on developing Singapore as a crucial service link to MNCs seeking to locate their production blocks in the regional economies. In thinking about the changing patterns of international production fragmentation in the Singapore economy, we have found it useful to apply the framework introduced by Jones and Kierzkowski (1990). It is helpful to conceptualize the driving force behind the increased production fragmentation in the early phase—during which Singapore was transitioning from being a traditional entrepôt economy to being a production base—as arising primarily from the lowering of the marginal cost of production for MNCs, that is, attracting MNCs to base part of their value chain in Singapore by taking advantage of relatively cheap labor. But even during this phase the fixed cost of developing service links was important. One must ask why MNCs would find it attractive to base more production in Singapore rather than in other regional economies that offered cheap labor. Singapore's concerted effort to lower the price of services, however, played a more significant role in the later phase when the country was transitioning from being just a manufacturing production base to being based upon manufacturing and services.

Jones and Kierzkowski's framework is illustrated in Figure 7. There, we let the total cost schedule, TC_1, represent the total cost of production within the domestic economy, say, the United States, in a vertically integrated process. All activity in this organizational mode takes place within a single production block. For simplicity, we assume that no fixed cost is incurred when the production process is fully integrated vertically. The total cost schedule, TC_2, represents the total cost at different output levels associated with a service link coordinating activities across two production blocks based within the *domestic* economy. The fixed cost is associated with maintaining a service link, and the geographic specialization of production activity within national borders allows the marginal cost to fall. The third schedule, TC_3, is associated with maintaining a service link to coordinate activities of two production blocks based in *two countries,* say, one in the home base—the United States—and another in Singapore. Coordinating these production activities across international boundaries costs more than doing so within national boundaries. Consequently, the fixed cost associated with TC_3 is higher than that associated with TC_2. This increase is, however, compensated for by a lower marginal cost associated with TC_3 compared to TC_2, reflecting the benefits of international specialization based on comparative advantage.

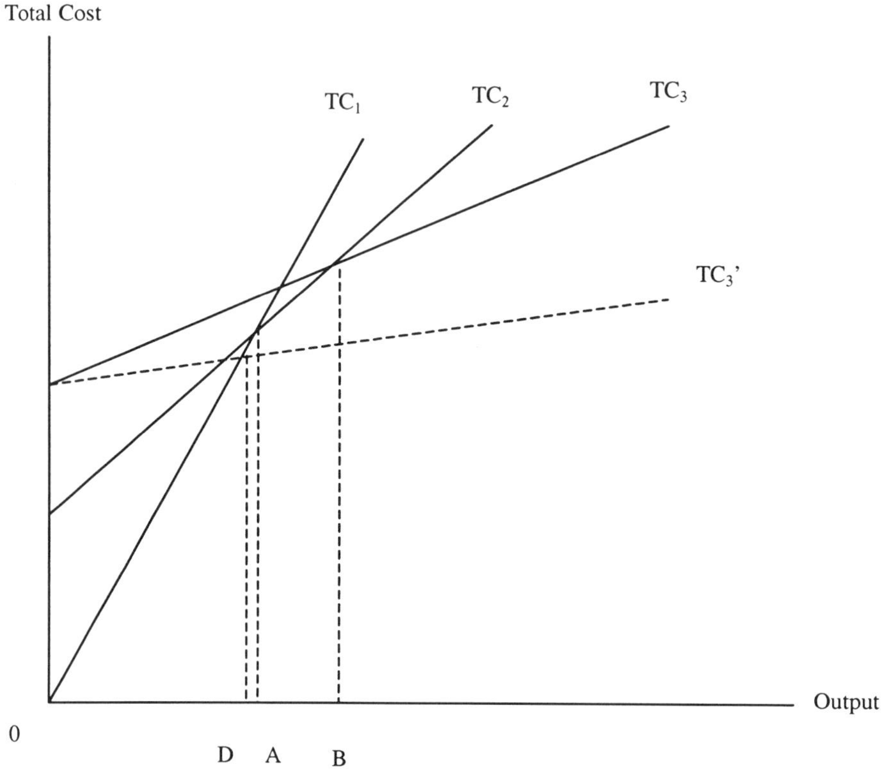

Figure 7: *Lowering of Marginal Cost.*

As drawn, we see that international production fragmentation occurs only when demand is high enough to justify the firm's producing an output level greater than *OB*. Given the fixed cost implied by the total cost schedule TC_3, suppose that the marginal cost of producing in Singapore is reduced so that the new schedule is given by TC_3. In that case international production fragmentation can take place at an output level above *OD*, which lies below *OA*. Consequently, the U.S. economy never goes through a stage of fragmenting the production process domestically but moves directly from a vertically integrated production process to fragmenting internationally. In our view the early industrialization efforts were essentially directed at lowering the marginal cost of production for MNCs. While labor was cheap by virtue of its relative abundance in the 1960s, the legislations introduced that brought about a sharp decline in labor strikes and fostered a healthy industrial relations climate also acted to lower the marginal cost of production by raising the marginal productivity of labor.

Why did MNCs come to Singapore in the 1960s rather than go to neighboring countries? We believe that the cost of coordinating production activities between the United States and Singapore was lower than that of coordinating the same activities in the United Sates and a domestic production base. This is so for two reasons. First, Singapore was unusual in adopting a liberal foreign-investment policy early on. During the 1960s inward-looking policies were still very popular among developing economies. Second, Singapore had the advantage of being traditionally strong in international commerce, so by the standards of that time it was relatively efficient at providing services to support manufacturing, compared to its neighbors. Consequently, Singapore had a head start in serving as a production base for MNCs looking for a place for the labor-intensive parts of their value chain.

Figure 8 shows another way of facilitating the disintegration of production activity globally. There, we show that lowering the fixed cost of providing services, while keeping marginal cost unchanged, helps to bring about international production activity. Hence the schedule shifts in a parallel manner from TC_3 to TC_3". With a lower fixed cost of coordinating manufacturing activity internationally, a U.S. firm finds it attractive to base part of its production activity in Singapore after its scale reaches *OE*. We argue that the country became attractive as a base for providing service links to coordinate

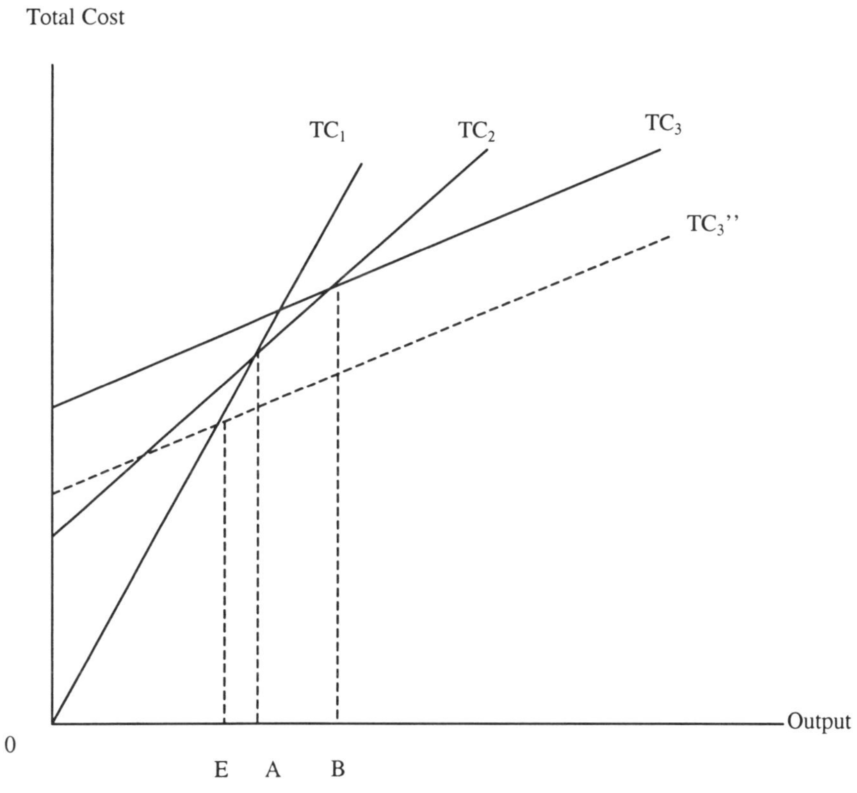

Figure 8: *Lowering of Fixed Cost.*

activity across countries partly because of recent advances in telecommunications and deregulation of financial services in Singapore but also by virtue of Singapore's historical development as a center for international commerce. In this new phase, compared to other regional economies, Singapore's marginal cost rose for labor-intensive activities. Singapore then became attractive as a place in which to locate headquarter activities, with production blocks being based in the region. Of course, the rise in the stock of human capital in Singapore also made the country a desirable production base for human capital-intensive parts of the value chain.

4. ECONOMIC CONSEQUENCES OF FRAGMENTATION

One of the main consequences of production fragmentation during Singapore's transition from a traditional entrepôt economy to a manufacturing base has been pressure to raise workers' real wages. Here, we find it convenient to abstract from the fixed cost of providing a service link and to focus on the shift in the range of intermediates that became profitable for Singapore to produce. We build upon the Dixit-Grossman 1982 model to analyze the effects of foreign investment on wages as well as the pattern of production specialization by stages in a small open economy. We then discuss some regression results from examining the influence of some measure of fragmentation on wages.

We suppose that there are two sectors: manufacturing, in which a range of intermediates is produced using labor and capital, and a final-goods sector, which produces the numeraire good. The numeraire good is produced with labor and land. The stages of manufacturing are indexed by z, ranging over the interval [0,1]. The upstream end of the range is $z = 0$, and the downstream end is $z = 1$. We assume that all goods indexed by $z < 1$ are pure intermediates. We assume zero transport costs. We choose to rank goods by increasing labor intensity, so that

$$\partial[a_{Lx}(z)/a_{Kx}(z)]/\partial z > 0,$$

where $a_{ix}(z)$ is the amount of input i required to produce one unit of manufacturing intermediate z.

We let $\bar{w}$ and $\bar{r}$ be the exogenously given world wage and rental rate, respectively, and w_1, r_1 denote Singapore's wage and rental rate, respectively. The relative price of manufacturing in terms of the numeraire good, p, is also given by the world market and treated exogenously by economic agents in Singapore. Manufacturing-sector output is given by x, while output of the numeraire good is y. We denote z^* as the marginal intermediate good produced by both Singapore and the rest of the world. We think of Singapore as being relatively labor abundant, so it will produce goods ranging between z^* and 1. Our main hypothesis is that an increase in capital K will lower z^* so that with inflows of foreign direct investment (FDI), Singapore ends up producing an increased range of intermediate goods that are becoming more capital intensive. We denote labor force by L and land by T. Return to land is denoted q_1.

The following system of six equations enables us to determine the following variables: w_1, r_1, q_1, z^*, x, and y.

$$\int_0^{z^*} C(\bar{w},\bar{r},z)\,dz + \int_{z^*}^{1} [a_{Lx}(z)w_1 + a_{Kx}(z)r_1]\,dz = p; \tag{1}$$

$$a_{Ly}w_1 + a_{Ty}q_1 = 1; \tag{2}$$

$$a_{Ty}\,y = T; \tag{3}$$

$$x\int_{z^*}^{1} a_{Kx}(z)\,dz = K; \tag{4}$$

$$x\int_{z^*}^{1} a_{Lx}(z)\,dz + a_{Ly}\,y = L; \tag{5}$$

$$C(\bar{w},\bar{r},z) = a_{Lx}(z^*)w_1 + a_{Kx}(z^*)r_1; \tag{6}$$

where $C(\bar{w},\bar{r},z)$ denotes the unit cost of producing input z in the rest of the world. The two key comparative-static results we are interested in concern the signs of dz^*/dK and dw_1/dK, that is, what effect does an increase in capital stock, resulting, say, from FDI, have on the range of intermediates produced and the wage paid to Singapore workers? We show that an increase in K lowers z^*, so that with a greater inflow of FDI, more capital-intensive intermediate components are produced in Singapore and that an increase in K raises Singapore's wages. After some manipulation we obtain the following two derivatives:

$$\frac{dz^*}{dK} = \frac{\int_{z^*}^{1} a_{Lx}(z)\,dz}{[L-(a_{Ly}/a_{Ty})T]\{[K/(L-(a_{Ly}/a_{Ty})T)](a_{Lx}(z^*)/a_{Kx}(z^*))-1\}} < 0;$$

$$\frac{dw_1}{dz^*} = \frac{\int_{z^*}^{1}\left\{\begin{aligned}&\frac{a_{Kx}(z)}{a_{Kx}(z^*)}\left[\frac{dC(\bar{w},\bar{r},z^*)}{dz^*} - w_1\frac{da_{Lx}(z^*)}{dz^*}\right] - \\ &\frac{a_{Kx}(z)}{a_{Kx}^2(z^*)}\left[C(\bar{w},\bar{r},z^*) - a_{Lx}(z^*)w_1\right]\frac{da_{Kx}(z^*)}{dz^*}\end{aligned}\right\}dz}{\int_{z^*}^{1} a_{Lx}(z)\left[\frac{a_{Kx}(z)a_{Lx}(z^*)}{a_{Lx}(z)a_{Kx}(z^*)} - 1\right]dz} > 0.$$

We refer the reader now to Tables 4A and 4B, which illustrate our attempt to empirically test the influence of the degree of international production fragmentation on the real wage. REALWAGE is the average annual real wage in manufacturing, deflated by Consumer Price Index (CPI). DexpSale is the domestic exports as a percentage of total sales in manufacturing and is used as a measure of the degree of fragmentation. DExpVA is domestic

Table 4A: *Real Wage and Fragmentation (DExpSale) Adjusted for Capital Intensity*

Dependent Variable: REALWAGE Sample(adjusted): 1961 to 1998 Convergence Achieved After 9 Iterations				
Variable	**Coefficient**	**Std. Error**	**t-Statistic**	**Prob.**
Constant	8.547055	1.883764	4.537223	0.0001
DExpSale	−0.102278	0.040841	−2.504261	0.0172
DexpSale*RvWorker	0.004465	0.000363	12.29281	0.0000
AR(1)	0.718060	0.138398	5.188362	0.0000
R-squared	0.987488	F-statistic		894.4985
Adjusted R-squared	0.986385	Prob(F-statistic)		0.000000

Table 4B: *Real Wage and Fragmentation (DExpVA) Adjusted for Capital Intensity*

Dependent Variable: REALWAGE Sample(adjusted): 1961 to 1998 Convergence Achieved After 8 Iterations				
Variable	**Coefficient**	**Std. Error**	**t-Statistic**	**Prob.**
Constant	8.712907	1.299153	6.706604	0.0000
DexpVA	−3.743049	0.796043	−4.702071	0.0000
DexpVA*RvWorker	0.146255	0.008599	17.00826	0.0000
AR(1)	0.673842	0.137024	4.917675	0.0000
R-squared	0.988681	F-statistic		989.9489
Adjusted R-squared	0.987682	Prob(F-statistic)		0.000000

Notes: *REALWAGE is average annual real wage in manufacturing, deflated by CPI (Consumer Price Index). DExpSale is domestic exports as a percentage of total sales in manufacturing and is used as a measure of the degree of fragmentation. DExpVA is domestic exports as a percentage of value added in manufacturing, which is used as another measure of the degree of fragmentation. RvWorker is real value added per worker in manufacturing. The interactive terms, DExpSale*RvWorker and DExpVA*RvWorker, denote measures of fragmentation adjusted for capital intensity. AR(1) is the correction for serial correlation in the regressions.*

Sources of data: *CPI is computed from Singapore Department of Statistics, various issues,* Yearbook of Statistics, Singapore. *REALWAGE, DExpSale, DExpVA and RvWorker are computed from Singapore Department of Statistics, various issues, Report on the Census of Industrial Production.*

exports as a percentage of value added in manufacturing, which is used as another measure of the degree of fragmentation. RvWorker is real value added per worker in manufacturing. The interactive terms DExpSale*RvWorker and DExpVA*RvWorker denote measures of fragmentation adjusted for capital intensity. Both regressions in Tables 4A and 4B show that an appropriate measure of production fragmentation (i.e., one that is adjusted for capital intensity) has a statistically significant positive impact on workers' real wage. From Tables 4A and 4B, we can see that the impacts of DExpSale*RvWorker and DExpVA*RvWorker on REALWAGE register highly significant t-ratios of 12.3 and 17.0, respectively. The adjusted R^2s are both 0.99 with serial correlation corrected in the regressions. Without correction in serial correlation, the R^2s remain high at 0.98, and the *t*-ratios are as significant as those in Tables 4A and 4B. Interestingly, the original measures of fragmentation without adjustment for capital intensity have negative impacts on real wage when the adjusted measures are introduced in the regressions. However, if we simply regress the real wage on the original measures of fragmentation without the capital-intensity adjustment, their impacts are positive and significant statistically.

In our model above, a lowering of the threshold z^*, implying a greater degree of fragmentation, is associated with a higher degree of capital intensity in production. Hence, empirically, it is necessary to adjust our measure of fragmentation for capital intensity. We are constrained by the availability of data in finding alternative proxies for the measure of capital intensities. We have only managed to get from the same data source containing the other variables in the regressions the values of real value added per worker. This is what we have used as the proxy for adjustment for capital intensity in production.

We have also carried out similar regressions by ownership structure: wholly foreign firms, more-than-half foreign-owned firms, more-than-half local-owned firms, and wholly local firms. The results are similar to those of Tables 4A and 4B and are presented in Tables 5A (wholly local firms), 5B (more-than-half local-owned), 5C (more-than-half foreign-owned), and 5D (wholly foreign). The variables are prefixed with "Ll," "L," "F," and "Ff" to denote the different types of ownership corresponding to Tables 5A, 5B, 5C, and 5D, respectively.

5. MNCS VS ARMS-LENGTH TRANSACTIONS

An interesting question is that of why international production fragmentation should take the form of setting up subsidiaries to produce a part of the value chain as opposed to buying from contract manufacturers in arms-length transactions. The Singapore experience is noteworthy in this regard because

Table 5A: *Real Wage and Fragmentation of Wholly Local Firms*

Dependent Variable: LIREALWAG Sample(adjusted): 1976 to 1998				
Variable	**Coefficient**	**Std. Error**	**t-Statistic**	**Prob.**
Constant	1281.945	103134.3	0.012430	0.9902
LIExpVA	−4.975609	3.048635	−1.632078	0.1191
LIExpVA*LIRvWorker	0.172507	0.085814	2.010238	0.0588
AR(1)	0.999545	0.037490	26.66171	0.0000
R-squared	0.987265	F-statistic		490.9972
Adjusted R-squared	0.985255	Prob(F-statistic)		0.000000

Table 5B: *Real Wage and Fragmentation of More-Than-Half Local-Owned Firms*

Dependent Variable: LREALWAG Sample(adjusted): 1976 to 1998				
Variable	**Coefficient**	**Std. Error**	**t-Statistic**	**Prob.**
Constant	29.45864	6.489724	4.539275	0.0002
LexpVA	−22.41347	4.902526	−4.571820	0.0002
LExpVA*LRvWorker	0.287813	0.022268	12.92478	0.0000
AR(1)	0.048637	0.264347	0.183988	0.8560
R-squared	0.936393	F-statistic		93.23686
Adjusted R-squared	0.926350	Prob(F-statistic)		0.000000

Table 5C: *Real Wage and Fragmentation of More-Than-Half Foreign-Owned Firms*

Dependent Variable: FREALWAG Sample(adjusted): 1975 to 1998				
Variable	**Coefficient**	**Std. Error**	**t-Statistic**	**Prob.**
Constant	14.24046	2.220136	6.414227	0.0000
FexpVA	−4.971197	0.712926	−6.972954	0.0000
FExpVA*FRvWorker	0.134281	0.008112	16.55340	0.0000
R-squared	0.935217	F-statistic		151.5791
Adjusted R-squared	0.929047	Prob(F-statistic)		0.000000

Table 5D: *Real Wage and Fragmentation of Wholly Foreign Firms*

Dependent Variable: FfREALWAG Sample(adjusted): 1975 to 1998				
Variable	**Coefficient**	**Std. Error**	**t-Statistic**	**Prob.**
Constant	23.71273	3.334971	7.110327	0.0000
FfExpVA	−7.437867	0.923680	−8.052429	0.0000
FfExpVA*FfRvWorker	0.078320	0.005652	13.85662	0.0000
R-squared	0.967915	F-statistic		316.7557
Adjusted R-squared	0.964859	Prob(F-statistic)		0.000000

Notes: *Refer to the sources and notes for Tables 4A and 4B. The regressions here are run for different ownership structure.*

Singapore's early development as a production manufacturing base took the form of many MNCs setting up subsidiaries in Singapore to produce their parts and components. In the 1990s, however, there has been an increase in the number of local contract manufacturers who produce parts and components that they sell to MNCs.

The early reliance on MNC subsidiaries rather than on local subcontractors can be explained as follows. In the early phase of Singapore's industrial development, there was a dearth of local entrepreneurs with the necessary capital and skills to produce the types of intermediates required by foreign firms. Owing to this lack, developing a pool of local subcontractors would have required MNCs to transfer technology to local firms. MNCs' willingness to transfer technology to unrelated local firms would, however, depend upon their ability to make long-term business commitments to local suppliers. Given the lack of local entrepreneurial talent at that time, few local suppliers had built a reputation for supplying parts and components of reliable quality. There were also few local middle-management employees of MNCs who could leave to start their own business.

Things have changed over the years as the government has invested massively in raising the stock of Singapore's human capital. Additionally, through the Local Industry Upgrading Programme, the government has provided more subsidies to create a pool of local entrepreneurs. The major local contract manufacturer we interviewed pointed out that since many of the top- and middle-level managers had been employees of Singapore-based MNCs, a degree of trust had developed between local subcontractors and MNC subsidiaries that had turned from producing components and parts to buying from local subcontractors. The MNC subsidiary we interviewed said that it now purchased a certain component from a subcontractor owned by a former company employee. We present other findings of our interviews with the two firms—one a major local contract manufacturer and the other a MNC subsidiary—in a separate case study that aims at understanding the considerations that enter into the reliance on arms-length transactions as opposed to affiliates for the supply of parts and components in the final chapter of this book.

6. CONCLUDING REMARKS

In this paper, we attempt to show that the phenomenon of international production fragmentation in manufactures began to be important to the Singapore economy at least 35 years ago. The country's engagement in this phenomenon has, however, evolved over the years. The Singapore economy changed from being a traditional entrepôt economy to a production base for MNCs. This move had enormous consequences for the stages of intermediate production based in Singapore as well as for Singapore wages. Our focus in this paper has been the influence of international production fragmentation

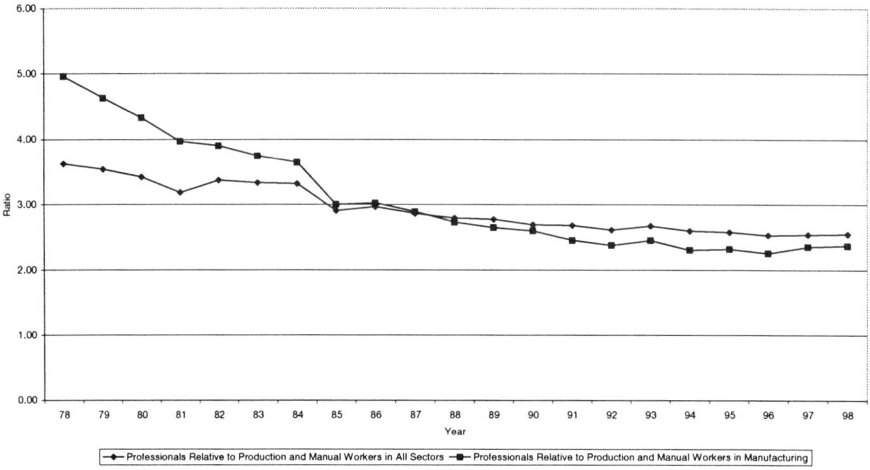

Source: *Computed from Singapore Department of Statistics, various issues,* Report on Wages in Singapore.

Figure 9: *Relative Occupational Wages 1978 to 1998.*

on average wages, as the skilled-unskilled wage gap issue has not been a problem so far (see Figure 9). However, as the nature of production fragmentation has changed to emphasize Singapore, which serves now as a production base for the human-capital-intensive parts of the value chain, as a crucial service link to coordinate production activity in the region, forces may arise that will widen the wage gap in future. This is a potentially interesting issue for future research.

References

Cheng, L., L. Qiu and G. Tan (2001), "Foreign Direct Investment and International Fragmentation of Production," in Ardnt, Sven and Henryk Kierzkowski (eds.), *Fragmentation: New Production Patterns in the World Trade* (Oxford: Oxford University Press).

Dixit, A. and G. Grossman (1982), "Trade and Protection with Multistage Production," *Review of Economic Studies* 49/4, 583–94.

Jones, R. and H. Kierzkowski (1990), "The Role of Services in Production and International Trade: A Theoretical Framework," in Jones, R. and A. Krueger (eds.), *The Political Economy of International Trade* (Oxford: Basil Blackwell).

Jones, R. and H. Kierzkowski (2001a), "Globalization and the Consequences of International Fragmentation," in Dornbusch, R., G. Calvo and M. Obstfeld (eds.), *Festschrift in Honor of Robert A. Mundell* (Cambridge, MA: MIT Press).

Jones, R. and H. Kierzkowski (2001b), "A Framework for Fragmentation," in Ardnt, Sven and Henryk Kierzkowski (eds.), *Fragmentation: New Production Patterns in the World Trade* (Oxford: Oxford University Press).

Jones, R. and H. Kierzkowski (2000), "Horizontal Aspects of Vertical Fragmentation," this volume.

Krugman, P. (1995) "Growing World Trade: Causes and Consequences," *Brookings Papers on Economic Activity*, no. 1, 327–77.

Lim, C. Y. et al. (1988), *Policy Options for the Singapore Economy* (Singapore: McGraw-Hill).

Lloyd, P. (1996), "Specialisation and Value Adding in the Global Economy," in Kapur, B., E. Quah and H. T. Hoon (eds.), *Development, Trade and the Asia Pacific: Essays in Honour of Professor Lim Chong Yah* (Singapore: Prentice-Hall).

Lloyd, P. and R. Sandilands (1986), "The Trade Sector in a Very Open Re-export Economy," in Lim C. Y. and P. Lloyd (eds.), *Singapore: Resources and Growth* (Singapore: Oxford University Press).

Report of the Economic Committee (1986), *The Singapore Economy: New Directions* (Singapore: Ministry of Trade and Industry).

Sandilands, R. and L. H. Tan (1986), "Comparative Advantage in a Re-export Economy: The Case of Singapore," *Singapore Economic Review* 31, 34–56.

Tan, T. Y. (1994), *An Analysis of Singapore's External Trade in the Light of the New (1988) Input-Output Tables,* unpublished master's dissertation, National University of Singapore.

Woon, K. C. (1991), *Net Exports and Economic Growth: Assessment of Applicability of Export-led Growth Hypothesis to Singapore,* unpublished master's dissertation, National University of Singapore.

Endnotes

[1]This perspective uses a Heckscher-Ohlin framework to predict where each production block along the value chain will be based. An alternative perspective predicts the international allocation of activities using the Ricardian comparative advantage framework.

[2]Most of these activities have diminished in importance, with the exception of petroleum refining. Indeed, in the country's economic blueprint, *Industry 21*, the petroleum and the petrochemicals industry is projected to be a key-growth industry. To highlight the government's commitment to attracting petroleum- and chemical-related MNCs to base the highly capital-intensive parts of their value chain in Singapore, a S$7 billion land reclamation project linking seven islands was undertaken. The resulting amalgamated Jurong Island will allow the petroleum and petrochemicals companies based there to reap the benefits of integration and cluster linkages.

[3]Cheng, Qiu, and Tan (2000) have argued that when technology level between home and host countries differs, the presence of FDIs encourages international production fragmentation.

INDEX